Handbook of Mechanical Works Inspection

Handbook of Mechanical Works Inspection

A Guide to Effective Practice

by

Clifford Matthews

Professional Engineering Publishing Limited
London and Bury St Edmunds, UK

First Published 1997
Reprinted 2001
Reprinted 2002

This publication is copyright under the Berne Convention and the International Copyright Convention. All rights reserved. Apart from any fair dealing for the purpose of private study, research, criticism or review, as permitted under the Copyright, Designs and Patents Act, 1988, no part may be reproduced, stored in a retrieval system, or transmitted in any form or by any means, electronic, electrical, chemical, mechanical, photocopying, recording or otherwise, without the prior permission of the copyright owners. *Unlicensed multiple copying of the contents of this publication is illegal.* Inquiries should be addressed to: The Publishing Editor, Professional Engineering Publishing Limited, Northgate Avenue, Bury St Edmunds, Suffolk, IP32 6BW, UK. Fax: +44 (0) 1284 705271.

© Clifford Matthews

ISBN 1 86058 047 5

A CIP catalogue record for this book is available from the British Library.

Printed in Great Britain by
Antony Rowe Ltd, Chippenham, Wiltshire

The publishers are not responsible for any statement made in this publication. Data, discussion, and conclusions developed by the Author are for information only and are not intended for use without independent substantiating investigation on the part of the potential users. Opinions expressed are those of the Author and are not necessarily those of the Institution of Mechanical Engineers or its publishers.

Contents

Foreword	xi
Preface	xiii
Acknowledgements	xv
Chapter 1 How to use this book	**1**
Introduction	1
Fitness for purpose (FFP)	3
Basic technical information	3
Standards	4
Test procedures and techniques	5
Non-conformances (and corrective actions)	7
Reporting	7
Key point summaries	8
Chapter 2 Objectives and tactics	**9**
The inspection business	9
Fitness for purpose (FFP)	10
Strategy	12
Develop a tactical approach	14
The end game	20
Reprise: saying 'no'	23
Key point summary	25
Chapter 3 Specifications, standards, and plans	**27**
The chain of responsibility	30
The document hierarchy	33
Specified tests	35
Using standards	38
Inspection and test plans (ITP)	40
Equipment release	43
The formula solution	45
Key point summary	48
Chapter 4 Materials of construction	**49**
Fitness-for-purpose (FFP) criteria	50
Basic technical information	51

Specifications and standards	57
Inspection and test plans	61
Test procedures and techniques	67
Key point summary	79
References	80

Chapter 5 Welding and NDT — 83
Introduction: viewpoints	83
Fitness-for-purpose (FFP) criteria	84
Basic technical information	85
Specifications and standards	96
Inspection and test plans	101
Test procedures and techniques	102
Volumetric NDT	113
Ultrasonic examination	115
Radiography	128
Destructive testing of welds	137
Key point summary	147
References	148

Chapter 6 Boilers and pressure vessels — 149
Introduction	149
Fitness-for-purpose criteria	150
'Statutory' certification	151
Working to pressure vessel codes	156
Applications of pressure vessel codes	161
Inspection and test plans (ITPs)	176
Pressure testing	181
Visual and dimensional examination	188
Vessel markings	192
Non-conformances and corrective actions	195
Key point summary	203
References	204

Chapter 7 Gas turbines — 205
Fitness-for-purpose criteria	205
Basic technical information	206
Specifications and standards	207
Inspection and test plans (ITPs)	209
Key point summary	235
References	236

Chapter 8 Steam turbines — **237**
Fitness-for-purpose criteria — 237
Basic technical information — 238
Specifications and standards — 241
Inspection and test plans — 243
Test procedures and techniques — 244
Key point summary — 256
References — 257

Chapter 9 Diesel engines — **259**
Fitness-for-purpose criteria — 259
Basic technical information — 260
Specifications and standards — 263
Inspection and test plans — 265
Test procedures and techniques — 265
Key point summary — 283
References — 284

Chapter 10 Power transmission — **285**
Gearboxes — *285*
Fitness for purpose criteria — 285
Basic technical information — 286
Specifications and standards — 287
Inspection and test plans — 289
Test procedures and techniques — 289
Couplings — *302*
Fitness-for-purpose criteria — 302
Specifications and standards — 303
Test procedures and techniques — 303
Key point summary — 308
References — 309

Chapter 11 Fluid systems — **311**
Centrifugal pumps — *311*
Fitness-for-purpose criteria — 311
Basic technical information — 312
Specification and standards — 314
Inspection and test plans (ITPs) — 316
Test procedures and techniques — 316
Compressors — *327*
Fitness-for-purpose criteria — 327
Basic technical information — 327

Specifications and standards	329
Inspection and test plans (ITPs)	330
Test procedures and techniques	331
Draught plant: dampers	*337*
Fitness-for-purpose criteria	337
Basic technical information	338
Specifications and standards	339
Test procedures and techniques	339
Piping and valves	*347*
Piping	347
Valves	350
Heat exchangers and condensers	*350*
Standards	351
Inspection and test plans (ITPs)	352
Test procedures	352
Key point summary	356
References	357

Chapter 12 Cranes — 359

Fitness-for-purpose criteria	359
Basic technical information	361
Specifications and standards	367
Inspection and test plans (ITPs)	368
Test procedure and techniques	370
Key point summary	385
References	386

Chapter 13 Linings — 387

Rubber linings	*387*
Fitness-for-purpose criteria (FFP)	387
Basic technical information	388
Specifications and standards	389
Inspection and test plans (ITPs)	392
Test procedure and techniques	393
Metallic linings	*400*
Fitness-for-purpose criteria	400
Procedures and techniques	400
Key point summary	413
References	414

Chapter 14 Painting — 415

Fitness-for-purpose (FFP) criteria	415

Basic technical information	416
Specifications and standards	422
Test procedures and techniques	425
Key point summary	431
References	432

Chapter 15 Inspection reports — **433**

Your report is your product	433
Technical presentation	436
Style	441
The report itself	442
Some logistical points	447
A specimen report	450
Specimen report: non-conformance report	455
Specimen report: pump stripdown report	456
Key point summary	457

Appendix — **459**

Index — **461**

Related Titles of Interest

Process Machinery - Safety and Reliability	Edited by W Wong ISBN 1 86058 046 7
The Economic Management of Physical Assets	N W Hodges ISBN 0 85298 958 X
Management of In-Service Inspection of Pressure Systems	IMechE Seminar 1993–1 ISBN 0 85298 852 4
The Reliability of Mechanical Systems	Edited by J Davidson ISBN 0 85298 881 8
A Practical Guide to the Machinery Directive	H P van Ekelenburg, P Hoogerkamp, D Brown, and L J Hopmans ISBN 0 85298 973 3
Journal of Mechanical Engineering Science Proceedings of the Institution of Mechanical Engineers – Part C	Edited by Professor D Dowson ISSN 0954/4062
Journal of Process Mechanical Engineering Proceedings of the Institution of Mechanical Engineers – Part E	Edited by Dr G Thompson ISSN 0954/4089

For a full range of titles published by MEP contact:

Sales Department
Mechanical Engineering Publications Limited
Northgate Avenue
Bury St Edmunds
Suffolk
IP32 6BW
England
Tel: 01284 724384
Fax: 01284 704006

FOREWORD

Over the period of 30 years during which I was involved in a range of major power projects in the UK and overseas, the importance of inspection engineers became increasingly evident. The inspection engineer must have the ability to pay attention to important detail without becoming engrossed in minutia, and yet maintain a clear understanding of the key objectives and priorities when dealing with complex equipment. This, coupled with an ability to act with authority whilst maintaining a realistic and commercially aware approach, requires a special kind of person.

The process of selection and appointment of inspection engineers was always made more difficult by the lack of common understanding of the inspector's role and his relationship with the other parties responsible for ensuring compliance with the specifications and appropriate standards. Whilst technical standards are continually changing, fitness-for-purpose issues remain largely the same. In concentrating on these issues, and providing in the technical section relating to each category of equipment criteria which assist in establishing this essential but elusive concept, Clifford Matthews develops a realistic picture of what is required of today's inspection engineer.

There is every indication that there will be a continuing need for good quality inspection engineers in the foreseeable future. In addition to the comprehensive technical content, the philosophy of this book's approach, concentrating as it does on fitness-for-purpose issues, makes this a most valuable book for the mechanical inspection engineer.

Trevor Wiltshire BSc, CEng, FlEE
Director: Ewbank Preece 1983–95

PREFACE

Ten years ago there seemed little future for the discipline of works inspection. Purchasers and end users would, it was said, soon not need to inspect engineering equipment before it left the manufacturer's works because the inspection activity would be replaced by a system of quality management. This would prevent problems occurring at source in the manufacturing process, enlightened manufacturers would produce 'zero defects' and get it 'right first time'. This all sounded fine, quite balanced in fact. It also had a seductive air of managerialism about it - making people interested in (and responsible for) the quality of their own work.

I can offer little explanation for the fact that ten years later, works inspection is still here. Purchase contracts for engineering equipment still make provision for witnessed inspection and testing in manufacturers' works, despite whatever else they say about the need for quality management systems, evaluations' and audits. The nature and extent of some works inspections may have changed, but such changes have been more the result of evolution of the inspection disciplines themselves rather than an imposed step change.

One undeniable change has been an increase in the competitive nature of the engineering inspection business. The days of large volume 'cost plus' inspection contracts have gone and will not return. The resulting downward pressure on fees has caused inspection companies to react in different ways. Some have specialized, taking care to match carefully their inspectors' skills to the main areas of demand, but others have preferred to cut their overheads and fees in the belief that they have no alternative.

I have not attempted to cover the business aspects of inspection in this book. Instead, it concentrates on the nature of the inspection work itself. The purpose of the book is to introduce the potential inspection engineer to an effective way of working – this means concentrating on the things that really matter rather than those that do not. Inspection is all about providing a good technical service to plant purchasers and end users – there is little room for the techniques of branding or marketing 'gloss' that can fit easily into some businesses.

Quality assurance is not covered in this book. I am happy to respect the views of those that place faith in its ability to provide them with the problem-free engineering plant that they want. There are numerous excellent (other) books on Quality Assurance. Please read them – but remember that we all have to live together.

Technical standards play an important part in works inspection. I have referred to them frequently throughout the book. References are given at the end of each chapter and you will find details on the sourcing of standards in the appendix.

Finally, I have tried to make this book a useful, practical handbook for the mechanical inspection engineer. If you do not find it useful, and practical, or you feel there are errors or omissions, (they can creep in) then say so. I will be happy to consider your suggested improvements. Please write to me c/o the Publishers, Mechanical Engineering Publications Limited at the address given at the beginning of the book

Clifford Matthews BSc, CEng, MBA

Acknowledgements

The author wishes to express his grateful thanks to the following people for their assistance in reviewing the technical chapters of this book.

Steven C Birks *CEng, MCGI, CGIA, FIBF, MInstNDT, MIQA, MIM, IncMWeldI*
Quality Assurance Director, Goodwin Steel Castings Ltd, Stoke-on-Trent, UK.

Peter A Morelli *HND (Mech Eng)*
Engineering Manager, David Brown Radicon Ltd, Huddersfield, UK.

John T Dallas *BSc, CEng, MIMechE*
Research and Development Manager, Weir Pumps Ltd, Glasgow, UK.

Fred P A Watson *IEng, MInstNDT*
Manager, NDT Services, International Combustion Ltd, Derby, UK.

Mark S Drew
HNC (Production Engineering). Senior Quality Test Engineer, Compair Broomwade Ltd, High Wycombe, UK.

Roberto Soloni
Senior Inspection Engineer, SGS Servici Tecnici Industriali Srl, Milan, Italy.

David P Godden
Project Manager, European Gas Turbines SA, Belfort France.

Xiao Li Fen
Director of Quality Engineering, Shanghai boiler works, Shanghai, China.

David J S Owens *BSc, MIQA*
Deputy Chief Product Quality Engineer, GEC Alsthom Ltd, Manchester, UK

Shao Xi
Inspection Manager, Vouching Technical Institute, Beijing, China.

Acknowledgement is also owing to colleagues in ABB Sae Sadelmi, Milan and General Electric, Cincinatti who have provided general comments and guidance.

Special thanks are due to Vicky Bussell for her excellent work in typing the manuscript for this book

Chapter 1

How to use this book

If you look, you can find many problems with the design and function of new mechanical equipment in large power or process plants. The purpose of works inspection is to help stop this happening. It is not at all difficult to get it wrong.

Introduction

In a large engineering construction project, a wide range of plant items have to be inspected before acceptance. As professional technicians or engineers we are likely, at some point, to be asked to inspect this equipment in the manufacturer's works. Consultants, end-users, contractors and manufacturers all have their part to play. This is fine. The problem is that it is very difficult for those of us who inspect and accept such equipment to have detailed technical knowledge relating to the wide scope of equipment we are likely to see. That having been said, what we must aim for is works inspection which is effective; that does not just 'do the job'. It is a wide field, but there is a right way and a wrong way to do it. The right way involves having a crystal-clear view of the priorities of the task and an appreciation of the tactics of the inspection process. It is then easier to supplement this with your technical knowledge and experience.

This book will not make you a technical expert on individual pieces of plant or help you to write quality assurance manuals. It is a practical 'how to do it' guide to works inspection for the non-specialist. It will help you plan and carry out successful inspections of common mechanical equipment; technical information you need to know is provided and the methodology of inspection is explained. If you follow the techniques described and start to apply them to your works inspections then this can help you do things the right way.

2 Handbook of Mechanical Works Inspection

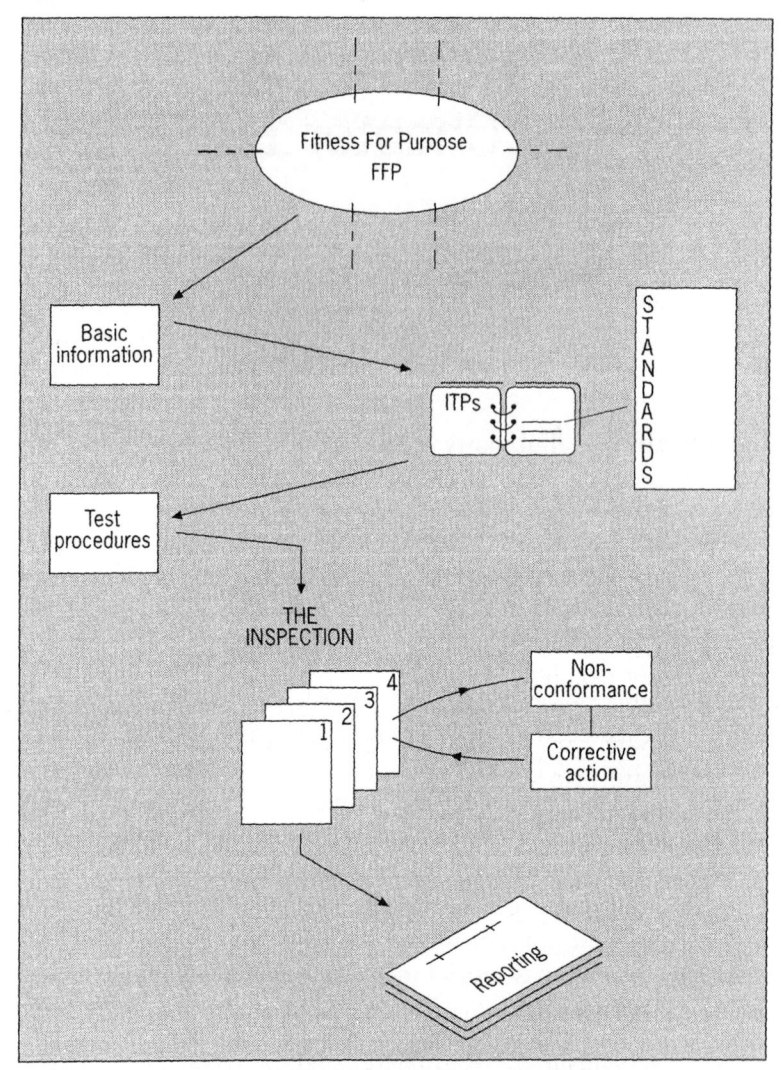

Fig 1.1 Technical chapters – structure

Chapters 1, 2 and 3 outline the objectives, tactics, and activities of works inspection. These are important principles. Chapters 4 to 14 are the technical chapters and cover the equipment items themselves. Each technical chapter is based on a common structure, as shown in Fig. 1.1. In a few cases you will see the information presented in an abbreviated two-page tabular format for clarity. Chapter 15 is about inspection reports.

To start, we must look at the most important factor in effective inspection. The prime tenet: fitness for purpose (FFP).

Fitness for purpose (FFP)

Perhaps the main objective of this book is to help the inspection engineer keep a clear focus on FFP. We will see in Chapter 2 that it is an elusive concept, carrying with it all kinds of legal and commercial implications. As engineers, we must for the moment try to see it more simply. FFP is the ability of a piece of equipment to do its job correctly – in the way that the user expects.

FFP forms, therefore, the 'skeleton' of all inspection. Each technical chapter sets out clear FFP criteria for the equipment in question. These are basically a distillation of the principles and requirements of the relevant codes, standards, and detailed engineering specifications. Most of the fine detail, however, has been omitted. What we then get is a clear view of the FFP requirements. We can work with this.

In following the methodology of the technical chapters, you should treat the FFP criteria as a practical level of requirements. Look carefully at them. Please do not treat them as if they are a full statement of the technical requirements of the equipment: they are probably not complete. They do, however, form a firm base from which to start, and then if there are any particular requirements of the relevant specification, you can apply these in the correct context and not lose sight of the main priorities.

FFP criteria will feature prominently in the type of inspection reports that I will introduce in Chapter 15. If you stick to reporting about FFP your reports will be concise and tell your clients what they need to know. This is what we are aiming for.

Basic technical information

If you ask experienced inspection engineers whether they prefer to have too much detailed technical information or too little, they will probably tell you they prefer to have too little. This makes good sense. You can nearly always obtain information that you don't have by asking – and knowing where to look.

What can we learn from this? Simply a few points about information needs during a works inspection. The range of detailed technical information relevant to mechanical equipment is quite vast. The British Library lists dozens of identifiable standards and nearly 300 textbooks and databooks relating to centrifugal pumps alone; if we discuss high pressure vessels, or special steels, the inventory gets bigger. So we must limit our information to only that which is needed to perform an effective works inspection. The range is still wide, but hopefully it will become clear as you read through the technical chapters that much of it is generic and divides relatively easily into well-defined categories.

Each technical chapter of this book provides the basic level of information and data that you need to understand works inspections performed on the equipment. Look, for example, at the technical information provided in Chapter 11 related to centrifugal pump performance testing. You will see that it is ordered specifically towards clarification of the fitness-for-purpose criteria. It will not tell you why a centrifugal pump impeller is a certain shape and size. That is design. Design is difficult to check during an inspection because it involves looking into the past. You may find it useful however to know of the practicalities of trimming the impeller diameter in order to adjust its performance, and the resulting way in which q/H and NPSH characteristics will be affected – a *rule of thumb*.

In a few chapters you will see that the technical information section presents calculation methods, where these play an important part. Air compressors and other rotating equipment are the main areas where calculation is required. Happily, no specialist knowledge is assumed, so stick with it, work slowly through the equations shown, and most should lose their mystery.

Information references are important, so each chapter provides a simplified list of references. You can use these where, as is inevitable in some cases, more detailed or expansive technical data are required. Most commonly this occurs when a non-conformance is identified (which is a common enough occurrence) so the data sources given tend to be quite well established 'destinations'. Use them as necessary, but be wary of trying to go too much further. Don't complicate the issue.

Standards

Technical standards and codes of practice are the tools of manufacturing industry. It is worth considering that without them, every programme of design and manufacture would have to start from scratch. Standards are useful to inspectors because they are objective. They help to justify decisions about FFP and also about less tangible aspects such as 'good practice'. Each technical chapter makes reference therefore to relevant codes and standards. As with the technical information discussed previously, this is intended to be a 'working sample'. Use them where they are imposed by the contract or where more detailed information is required, but remember your FFP criteria.

European and ISO (International Organization for Standardization) standards are used as the basic reference, with national equivalents shown where applicable. In many technical areas, well-established national standards used in works inspection have not yet been harmonized within the EN standards framework. Where this is the case I have referred to the accepted national standard – on the basis that it is the one which you will probably see being used during your inspection visits. The situation is changing rapidly – it is advisable to check that any technical standard you do use is the most recent issue.

Equipment *specifications* tend to be project-specific, hence it is more difficult to talk objectively about them in this book. Luckily, with good specifications (and they are not all good) there is a lot of commonality of requirements with the relevant standards and codes of practice, at least as far as FFP is concerned. Chapter 3 is devoted to discussion of equipment specifications and their nominees, the inspection and test plans (ITPs).

Looking back at Fig. 1.1 you can see that each technical chapter identifies specific actions, tests and documentation requirements that should be included in the ITP for the equipment in question, bearing in mind the fitness-for-purpose criteria. I advise that you use these requirements as a basis for what you expect to see in an ITP – but beware of encouraging further over-complication.

An important note

You will see that we have now taken the brave step into the field of terminology: namely the inspection and test plan (ITP). Discussions on nomenclature in this field are extensive, whole volumes being dedicated to defining quality assurance terms so that we can understand them better. Fine, but this is a book on the techniques of effective works inspection. I am going to suggest that we define the ITP simply, as the document used by the manufacturer and their clients to monitor the manufacture and testing activities. We can then conveniently leave other debates to one side.

Test procedures and techniques

To be effective during a works inspection you have to follow what is happening. The core element of the inspection is the test procedure or technique that is being performed and this can sometimes be difficult to follow if you are not overly familiar with the types of equipment or activities that are involved. Each chapter of the book contains, therefore, an explanation of the common tests that you are likely to be asked to witness. Some of these are well defined in existing standards; for example the non-destructive testing procedures described in Chapter 5. Other areas, such as gas turbines, are covered by standards which have a less prescriptive coverage of works test procedures, so the established practice of the equipment manufacturers themselves has increased relevance in such cases.

You can treat the test procedures shown as a distillation of the core material both of relevant standards and of good manufacturing practice. Individual steps may be omitted, or added, as the contract or specification demands, as long as the procedure retains its validity as a test of the fitness-for-purpose criteria. As a general rule to follow in using this information, you can assume that the *content* of a test procedure is, on balance, more important than the order in which things are done. There may be quite a lot of flexibility possible, indeed required, in the order in which things are done – but not so with content. Content inevitably has a greater influence on whether or not you are really testing for FFP.

There are often some peripheral activities that are an important part of works inspection. The book will show you, for each area covered, the procedural activities that surround the core inspection or test activities. I have made a conscious attempt to try and include the real issues that face the visiting inspector, such as preliminary design checks, material and component traceability, certification requirements, and documentation.

So, to summarize the book's approach, my objective is that you should view the test activities I have shown as 'typical' of a good inspection procedure. Contractual requirements do differ, but it is fortunate that most core engineering activities are reasonably well defined and don't change much. I intend the information and guidelines on inspection and test procedures to work equally well in many situations.

This may sound well ordered. It can be, but at some point we have to consider not just ourselves as inspector, but also the other parties. In large construction contracts, what is done or not done at works inspections is the result of an eventual consensus between, in reality, not less than five or six parties. There is the end user (the client), the consultant or designer (the engineer), a main contractor, a manufacturer (and subcontractors), and often a third party certification body. As inspector you will need to address the situational aspects if you want to be effective. I have tried to address this, by introducing to you a basic tactical approach, in Chapter 2. If in reading this book you feel it necessary to occasionally look forward into the technical chapters, remember that the whole question of what happens during an inspection visit is placed in context by the ideas I have tried to put forward in Chapter 2, otherwise you will only get part of the picture.

Non-conformances (and corrective actions)

As an inspector is it your job to accept equipment, or to reject equipment? If you reject it, then what happens? Before attempting an answer to this question we need to look at where non-conformances are normally found during works inspections. Every item of equipment is individual and a lot of variables operate. We can look to experience though. It is basically true that:

- Problems with the FFP of manufactured equipment repeat themselves in a more or less predictable fashion.

and

- A few common problems account for more than 80 percent of incidents where equipment is rejected.

A key objective of this book is to help you understand where these common non-conformances are likely to be. I have built this concept into the technical chapters with the hope that it will be of use to you in arriving quickly at the important points of a works inspection and not being misled by peripheral issues. The natural partner to the non-conformance is the corrective action. This is a vital part of inspection, because if all you do is identify problems without putting them right, don't expect that anyone will thank you for it. We will learn to consider the corrective action loop as an integral part of the works inspection role. This is played out within the scope of the technical chapters and we will look at ways of encouraging these things to *happen* in the works. I recommend that you set this as one of your tacit objectives in reading this book; to learn how to push these corrective actions through, not to let them ride. Once you can do this you are well on your way to becoming effective and can watch your business increase.

Reporting

The whole of Chapter 15 is dedicated to reporting. Why? Because that is what your client is actually paying for. If clients do not feel they are getting value for money they will go somewhere else – sooner or later. I have tried therefore to structure Chapter 15 around the concept of high quality, economical inspection reports. For the reports themselves I have shown you how to structure and plan a report, and how to write logical, progressive technical statements to express clearly what you want to say. There are some useful guidelines on reporting by description and reporting by exception – and how to strike a balance between the two.

Finally I have included a specimen inspection report – you can use this as a basic model on which to overlay your personal style, or that of your organization, to tailor your reports to their context. I recommend that you look forward to this chapter as often as you can when using the book. There is always room for better and more concise inspection reports.

Key point summaries

There is a summary of 'key points' provided at the end of each chapter. These provide a quick reference to the substantive content of the chapter. If you are feeling your way in a new area you should find that these key point summaries will provide a good first guide to the fitness-for-purpose criteria relevant to the equipment. Glance over them before reading each chapter, in order to help you see the structure of requirements that will be introduced. The key point summaries are formatted on separate pages for convenient reference.

Consider the suggestion that whilst it may not be particularly difficult to do works inspection, it is also not at all difficult to get it wrong. It is best to start at the beginning with the very first step: objectives and tactics.

Chapter 2

Objectives and tactics

Works inspection entails appraising equipment for conformity and fitness for purpose. To do this well you have to inspect the right things at the right time, then you must know what you are aiming for and have a tactical understanding of what is happening.

The inspection business

Ideally, there would be no need for inspection in industry. In reality, the rules of commerce tend to dictate otherwise. The trend of business in mature industrialized countries is to de-couple increasingly the activities of design and manufacture – this means that for large construction projects, those companies who design systems, and specify equipment to operate within them, rarely manufacture the systems themselves. This resulting delegation of responsibility for manufacture, if it is to be complete, needs a feedback loop – the purchaser needs to check that the equipment will do what is required of it. *Caveat emptor*. The inspection business, with its roots in commerce, is therefore an instrument of buyer confidence.

Large multi-disciplinary projects, typically power stations, process plants, and offshore installations, provide the best focus for understanding the works inspection business. Such is the complexity of design and manufacture that these projects involve many consultants, large 'package contractors', and hundreds of manufacturers and sub-manufacturers. Within the project structure at several levels are found works inspection responsibilities and activities. The client or end user will be involved in inspection, as will be the design/engineering consultant (often the client's representative), the main contractor and some of the manufacturers. In a 'control' role will be third party certification and/or statutory bodies.

The role played by each of these parties differs slightly between industries and between projects so it is not possible to produce a perfect model, prescriptively fitting all the parties into convenient 'boxes of responsibility'. What you can do, though, is look for those things which remain the same. Perhaps you would call them invariant properties, if you were a mathematician. We will now consider two of these properties.

The first property is that the fundamental business pressures that drive the activities of each of these parties *do not change*. They are located by the role of each of the parties in the commerce chain and secured by the environment of business. Starting with only a little experience, it is possible to understand progressively how each party is likely to behave, focus on the real objectives that they have, and understand what they are trying to achieve. Secondly, and on a technical level, consider this:

ALL INSPECTORS ARE LOOKING FOR THE SAME THINGS

We can expect their foci to differ, and their ways of working will be different, but they will still have technical objectives in common. The existence of common technical objectives means that solutions are normally possible to those issues that inevitably arise during works inspections. What we are interested in, though, is getting hold of the most *effective solution*. Where do we start? We need to set an early objective for ourselves:

OBJECTIVE

During a works inspection we are going to manage all the parties towards a guided consensus on FFP. We will do this without wasting our client's time or money.

BECAUSE

This will set us apart from our competitors in the inspection business.

Fitness for purpose (FFP)

Inspection responsibilities involve making decisions about FFP. To be useful, however, it is necessary to resolve this into definitions that we can work with. An inspector has a responsibility towards judgement, and also a role in enforcement. It's a rich picture so we must try and get a clear view. FFP is about *utility* and there are essentially four elements to it.

Element 1 – Function

Function is what the equipment does. Will it do the job for which it is intended? More importantly, will it do it without the implications of extra cost or wasted time (which in itself is analogous to cost)? Pieces of mechanical equipment have a basic common property in that their purpose is to *transform* an input to an output. For example, a gas turbine or other prime mover transforms a fuel input into desirable outputs, namely power, and possibly also a secondary (waste) output of heat, whilst a pressure vessel transforms an input volume of fluid into a reservoir of pressure, capable of providing useful work on demand. Similarly, a crane transforms electrical power into a capability to lift and move a load. By looking at the nature of the specific transformation required, you can investigate closely the function of the equipment and get a precise view of its capabilities.

Equipment function is never absolute of course, it is always linked to the circumstances of its integration into the process system. Further resolution can be carried out by looking at the *cost implications* of getting equipment to do what is required. This does help FFP to come into focus.

Element 2 – Mechanical performance

This is about the mechanical strength of the component; try not to confuse it with function. The main aim is that equipment and components, when in use, should be operating below their elastic yield stress with the required design factor of safety. This raises an important point which is frequently neglected. You must assess the mechanical strength of a component in relation to its likely mode of failure. As an example, it is of limited use calculating static principal stresses (δx, δy, δz) in a pressure vessel if the vessel is likely to fail by fatigue or creep-induced cracking, or because of dynamic stresses. I am not suggesting that the design process should not include principal stress calculations (they all do), but merely that you should consider the *limitations* of such an assessment in assessing the real FFP criteria.

Experience will provide you with an understanding of the ways in which things fail; there are plenty of examples. You can then start to assess mechanical strength in a more objective manner. Codes and standards providing limits on component loadings and safety factors will still be required, but this approach will help you put them into context – the difficult part.

Element 3 - Service lifetime

The length of time that a piece of equipment or a component will last is important. The engineering factors that influence lifetime are twofold: there is wear, and there is corrosion.

Nearly everything wears, but practically you can consider wear as being related mainly to moving parts: bearings, gears, pistons, seals, anything where there is relative movement between two components. Works inspectors can assess materials, tolerances, and assembly of moving parts, gaining an impression as to whether any undesirable conditions exist which will produce unacceptable wear rates. This is one area where it is necessary to rely heavily on manufacturers' drawings. The reason for this is that, as inspectors, we are often not able to assess closely the design of equipment. For even simple components, the design process is much more complicated than it appears. Unwritten rules, empirical knowledge, and accumulated experience all play a large part. It is necessary to understand, and admit, that most of this will be a closed book to you – much of the time. Try to concentrate on those areas which can be assessed, such as materials, tolerances, and assembly.

Corrosion is also universal, as man-made materials attempt to return to their natural state; paints and coatings are employed to limit its effects. The extent of protection against corrosion is therefore an important element of any FFP assessment, particularly where the materials used do not have high inherent corrosion resistance. We will look at this further in Chapters 13 and 14.

Element 4 - Quality

I have tried hard in this book not to use the term quality. It only appears a few times. There are many other excellent and authoritive books on quality assurance.

Strategy

If FFP is the objective of a buyer, then strategy is the broad plan for attaining it. Inspection strategies can take many forms with both explicit and implicit parts: as a general rule, however, most things stem from the requirements of the final client or end user of the plant. The contractual specified requirements take priority, forming the skeleton for the inspection activities that will take place during the project.

Experience shows that end-users' strategies to attain FFP nearly always exhibit two common elements. Firstly they want to dilute their exposure to technical risks by shedding some of these risks onto their consultant, main contractor, and third party bodies. Then they will try to specify a structure of inspection responsibilities and activities that will identify and *solve* technical problems, not act as problem-generators. Both these activities are guaranteed to cause some duplication of effort. Duplication is a much criticized factor of the inspection business and recent years have witnessed elaborate arguments for eliminating it, with the introduction of third party certification of manufacturers, approved vendor lists, etc. The results of all of this have been, at best, variable. The rules of commerce, which basically control who does what in the contractor/manufacturer relationship, will always remain in force, so we must come to accept the existence of duplication in inspection activities. Even better, we can *plan it in*. Make it part of the strategy.

Large projects normally mean that there are several parties sharing responsibilities for equipment supply. They have different levels of knowledge because of the varying nature of their businesses: the rules of commerce again. Figure 2.1 shows the general situation, as applicable to an equipment package such as a prime mover or process system. The total sum of knowledge necessary to achieve FFP is by definition shared, hence all parties must play a part.

Simple strategies are best. From the assumed viewpoint of managers

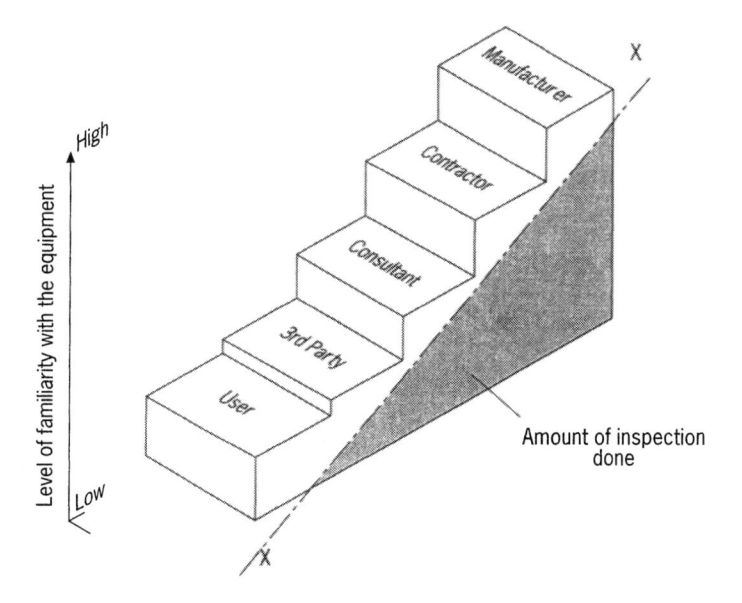

Fig 2.1 Inspection responsibilities – a guide

14 Handbook of Mechanical Works Inspection

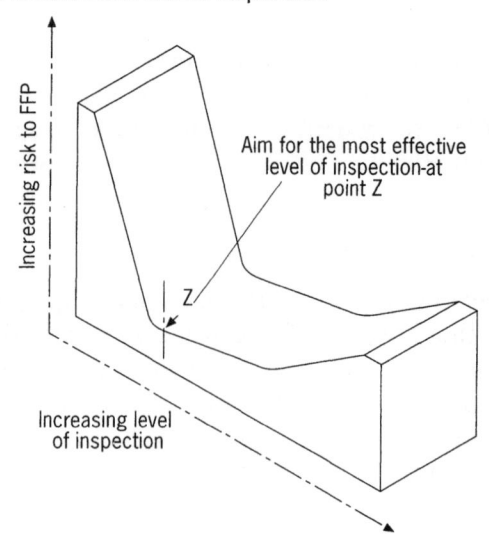

Fig 2.2 An effective level of inspection

of the inspection strategy we could propose that the line X/X represents a basic split of inspection responsibilities and the shaded area under the line depicts the total amount of inspection done. The characteristic of the curve will undoubtedly be contract-specific but now we have a model for managing the situation. Look also at the relative amounts of inspection done by the parties (the vertical scale): here we have the start of work-hour budgets, and some guidelines on relative amounts of documentation and administrative loadings.

We need Fig. 2.2 to help us keep our grip on the target of *effectiveness*. The point Z represents the most effective amount of inspection – the target. There are real risks to FFP either side of this point on the horizontal axis. Too little inspection will not find the problems. Too much inspection will cause too much discussion; issues will become clouded, politics will prevail and we will expend huge amounts of resources on generating our own problems and then having to find fine and persuasive solutions. This is very expensive. Give it only a half-chance and it will take your profit away. Remember point Z and please try to avoid too much inspection, it can be one of the major dangers to effectiveness.

Develop a tactical approach

It is important not to confuse the tactics of an inspection with strategy – the larger picture. Tactics are the strategy in action, elements of practical advice, comprising the series of rational steps that lie behind all inspection activities. It is helpful to look at four main tactical elements – they are more or less sequential – that make up an inspection activity to see how they are used.

Information: what you need to know

For works inspection of complex mechanical equipment, you can't just 'turn up on the day', not if you want to be effective. If you want to manage the situation properly and plan to obtain the best solution then you need the right information. Almost inevitably it will be impossible to have access to *all* relevant technical information relating to the equipment or contract, but basically you need the following things:

- the equipment specification

  and
- the acceptance and/or guarantee requirements (remember FFP?)

  and
- the relevant documents (standards, codes, etc.) raised by the contract specification.

Notwithstanding the need for these three cornerstones of information, you also need to think about specific background knowledge that will be needed in order to help you handle the technical discussions – and sometimes conflicts – that will inevitably arise. A better term perhaps is *qualifying information* and it is frequently this that actually makes a solution possible to the general satisfaction of the parties present. To find this qualifying information it is necessary to learn how to 'think around' the inspection activities, such as:

- Which technical arguments are likely to develop?
- What are the precedents?
- What are the real FFP criteria implicit in the specification?

The three cornerstones of information, backed up by the 'qualifiers', form the route to consensus at the inspection. It is not necessarily wrong to attend an inspection without all the information – the necessary qualification being that you know what to ask – because some of the other parties may be able and willing to help you. However, they may not, so I would be very wary of arriving at an inspection without the information I have mentioned. You could undoubtedly survive it. You would probably still be welcomed by all parties concerned, but it is undoubtedly the path of the also-rans. Keep on doing it and slowly but surely your credibility as an inspector will drain away.

Focus

Focus means keeping sight of priorities. I have made the statement earlier that the various parties present at a works inspection may each have a different focus on events, even though their technical objectives are broadly the same. I can now proceed one step further and say that it is *highly likely* that each party will have a different focus. This gives a real potential for time delays and extra costs, before a consensus agreement or solution can be found. Good inspection involves managing these different foci.

Look at how it works in practice. During a works inspection on, for example, a diesel engine, the main 'thread' of evaluating FFP is complicated by the development of many interesting, mainly technical, diversions as the parties present interact. These diversions start off slowly, but increase in number during the inspection as the situation builds. It is likely that, in the course of discussions, issues of material traceability, testing techniques, workmanship, painting and packing, weld specifications, and many others will develop as side issues of the main theme. Commercial issues and questions of interpretation also appear to further complicate the arguments. Under such conditions the side issues can become very effective at blocking your focus on FFP. Side issues are often more interesting, and easier to discuss, than FFP. They can be an easy way out. You must guard against this – fortunately the rule is simple:

- get a clear focus on FFP

then

- just keep on coming back to it, again and again.

This is a loop, and it is a good idea to make it a ten-minute loop. This means that you can play a full part in side issue discussions but every ten minutes you need to bring the subject of the discussion back to your FFP focus. Any longer than ten minutes and the side issues may well grow in priority until you are seen as unreasonable in trying to change the subject. Less than ten minutes and you may miss an important point that the side discussions reveal, so ten minutes is about right. Figure 2.3 shows a typical example of a focus loop that can occur in practice.

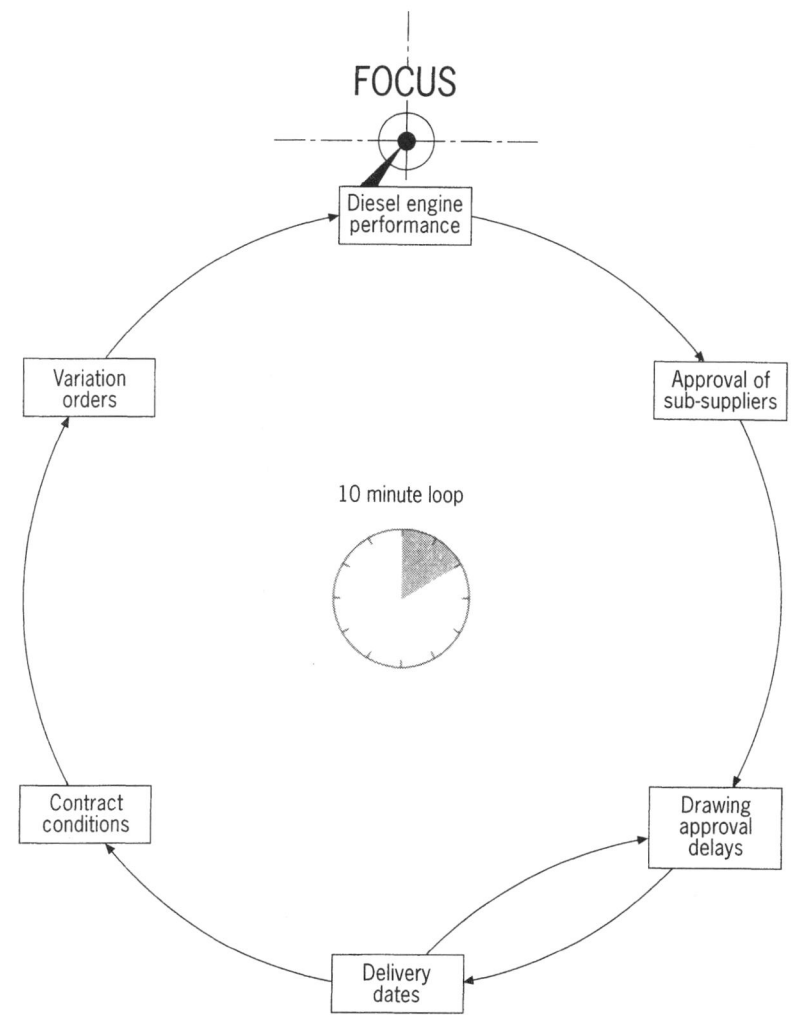

Fig 2.3 Keeping your focus on FFP

Asking and listening

The moderate view is that the activities of works inspection are those of an inquiry, rather than a trial. All parties start out neither innocent nor guilty and will remain that way, whatever the outcome of the inspection. It is also a fact, however, that a works inspection is about questioning the actions and technical knowledge of competent contractors and manufacturers and can therefore be a rather stressful activity for all parties. Perhaps the best general guidance I can offer a works inspector is to try and project a modest attitude of searching for the truth, rather than attempting to appear too confident, or authoritative, or even apologetic – I have seen some inspectors who continually apologize for being in the works. Granted, there are many different ways to do it, but over time the honest truth-seeking approach, albeit hard to maintain, probably works best.

Armed with this, you can start asking questions. A good technique to master, which will serve you in all kinds of inspection situations, is that of *chain questioning*. This is a very effective way of getting at the truth. The technique involves spending most of your time (upwards of 80 percent of the time you spend speaking) asking questions; not in a suspicious or confrontational manner (although these may have their place), but asking, just the same. The precision of the questions is important – they need to have two main properties. Firstly they should be capable of being answered by manufacturers or contractors in a way that allows verification. That is, manufacturers must be able to prove to you that what they say is true, and provide supporting evidence. This means you must be accurate in what you ask. Secondly the questions must be 'chained' – each one follows on from the last answer, developing progressively better levels of resolution on a connected subject in question, rather than hopping about from subject to subject. Note that it is essential you obtain verification of a previous answer before developing the next question in the chain, otherwise future questions and answers can get increasingly hypothetical and the integrity of the chain breaks down. In most situations, a mixture of closed and open-ended questions seems to work best.

It is at this point that we can see one advantage of the duplication of inspectors. With several parties using (knowingly or unknowingly) this technique, you get the benefits of cross examination, a situation which is widely accepted in legal circles as providing the best method of getting at the truth. You should find that a well-chosen chain of four or five questions, with tenacious perusal of the verification of each answer, will strip the veneer off most works inspection situations. Try it.

A final point on questioning. Remember that this is an inquiry. Once you have asked a question it is not wise to answer it yourself. Don't even start to, just wait. Thoughtful silence will normally bring the answer.

Making decisions

Making decisions is one of the things that you will have to do frequently during works inspections. Inspection is by its nature an adjudication process and even when working within large companies and complex projects the practicality of the situation is this:

THE MAIN DECISION MAKER IS YOU.

We are fortunate in that inspection decisions centre predominantly on technical matters, so a rational approach is possible, and more likely to be accepted, than in (for instance) more complex management situations. It is wrong to expect that all decisions will be simple accept-or-reject choices, many will involve deciding on rectification work that needs to be done, or the extent of actions required by the various parties present. The way in which you present your decisions is therefore of prime importance. Clarity is vital – make sure that you express decisions in full, but in simple terms, so everyone will know what you mean.

One of the pitfalls of decision-making during a works inspection is to be drawn into making qualified decisions based on something that is intended to happen in the future: i.e. if the manufacturer does this...then the equipment *may* be acceptable. This is a risky area from an inspector's viewpoint, and one which is responsible for many problems with fitness for purpose. The problem is that as an inspector you really only have the powers of scrutiny *when you are in the works*; once you leave, you relinquish much of the potential you have of influencing FFP. It is much better to make decisions whilst you are there. By all means consult specialists or managers during the inspection routine, as protocol demands, but don't defer decisions to them. Listen carefully, then *you* decide – there and then.

You must communicate your decisions to the other parties present. The only real way to do this is to put them in writing (see Chapter 15 for guidance on non-conformance reports). Verbal agreements, with the best intentions of all parties, can easily be misunderstood. Write it all down.

At this point it is useful to consider the influence that contractual agreements and protocol have on the way that decisions are implemented during an inspection. I have proposed that you should *make* the relevant decisions but this does not necessarily mean that you alone have to implement them. Normally, the various contractual agreements that exist between contractor and manufacturer, or between manufacturer and third-party body, are the best vehicles for getting things done. As, for example, an end-user's inspector, you would probably not have a direct contract with the equipment manufacturer, so the correct way to get your decisions implemented would be through the main contractor, who has placed the purchase order with the manufacturer. Keep the implementation chain short – and let the contractual protocol work for you.

The end game

It is at the end of a works inspection when things really start to happen. At this point ideas and decisions that have been accumulated by the parties during the course of the inspection will start to be put forward and implemented, bringing into focus any outstanding actions that are necessary. Effective inspection means planning carefully for this end game, understanding the mechanisms that are available, and using them to best effect. The main mechanisms available to you are twofold, the non-conformance report (NCR) and the corrective action (CA). In practice our objective is to link them closely together – this is a good way of making things happen.

Non-conformance reports (NCRs)

The purpose of a non-conformance report is to make a statement on fitness for purpose of the equipment that you have inspected. This definition is important; note that the purpose of an NCR is not to *reject* equipment. It can be contractually very difficult to formally reject equipment – you are in effect saying that it does not comply with the purchase order, with all the consequent contractual and legal implications. This is dangerous ground. You may feel that there is a paradox here; an inspector can *accept* equipment but cannot *reject* it? We could investigate this hypothesis but we would have to discuss legal liabilities and organizational responsibilities, and submerge ourselves in specific case law. This is another subject. We can do it more easily by putting to one side the accept versus reject argument and just look at what you are in the works to do. You have seen the first part of this before:

- You are there to get a guided consensus on FFP

and

- The purpose of an NCR is to help you get it

Figure 2.4 provides ten simple rules which I suggest you follow when issuing NCRs – if you look forward to Chapter 15 you can see a specific example. In essence, it is the content which is important. Be accurate. The layout will vary between companies and contracts but frankly this

Ten simple rules

1. Properly identify the equipment (serial Nos etc) so there is no confusion
2. Make specific technical observations; not generalized comments
3. Mention the evidence that you have seen
4. Be very accurate in your terminology – check with others so you get your description right
5. Refer to specific clauses of specifications and standards
6. Say why the equipment doesn't comply
7. Give the manufacturer (or contractor) the opportunity to add his comments
8. Watch your timing. Don't issue NCRs prematurely
9. You must issue an NCR whilst in the works, not as an afterthought when you have left
10. Make clear personal notes that you will understand later

Fig 2.4 Non-conformance reports

doesn't matter very much as long as you follow the basic rules on content and implementation. The next step is to link the NCR to an agreed corrective action. The message I have tried to convey throughout this chapter is that NCRs and CAs must be closely linked. In fact, they must come in pairs, it is not good practice to issue an NCR without an accompanying discussion and agreement on how the issue will be put right.

Corrective actions (CAs)

It is much easier to find faults with equipment than it is to put them right. You should not expect therefore, that implementing corrective actions is easy.

The objective of corrective actions is to bring the equipment to a condition where it meets fully the FFP criteria that have been set for it. Like any other action centred on technical matters, there will invariably be several views to be taken into account; for this reason you can think of CAs in a similar way to NCRs – i.e. you are looking for a guided consensus on what needs to be done. Consultation may be necessary, but procrastination is not, not if you stick to the premise of keeping extra time and costs to a minimum. Look for the single technical solution: it will be there – somewhere.

Here are the four practical steps involved in implementing a corrective action (refer to Fig. 2.5).

22 Handbook of Mechanical Works Inspection

Steps in implementing a Corrective Action (CA)

Practical guidance to help you

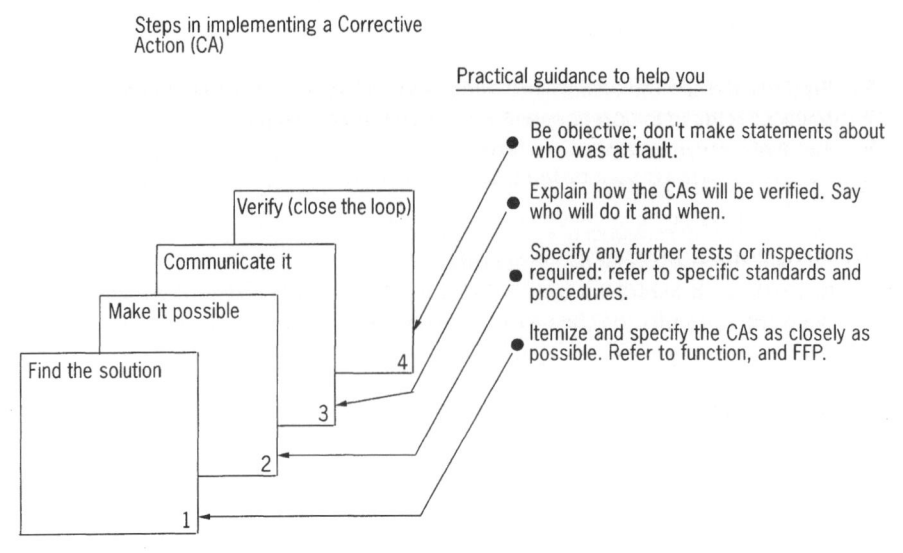

- Be objective; don't make statements about who was at fault.
- Explain how the CAs will be verified. Say who will do it and when.
- Specify any further tests or inspections required: refer to specific standards and procedures.
- Itemize and specify the CAs as closely as possible. Refer to function, and FFP.

Verify (close the loop)
Communicate it
Make it possible
Find the solution

Fig 2.5   Corrective actions

Finding a solution

Not only must you find the technical solution, you must *arrive* at it. The best way to do this is to revert back to the chain questioning technique introduced earlier in this chapter. If you choose your questions carefully, they will eventually draw out the real technical solution. There is one area to avoid at this stage, and that is the 'thinking aloud' type of veiled question. If you use this during the end game it will be seen as indecision, and you may be seen as 'fair game' for persuasion by some of the other parties present. Keep the questions precise and you can avoid this type of occurrence.

Making the solution possible

Whatever the technical solution is, someone is going to have to pay for it. Experience shows that the longer a party has to consider the costs of further work, the greater becomes the uncertainty and disagreement over who will pay, until commercial matters grow to predominate. One way to minimize this is to exert pressure to get corrective actions completed as quickly as possible. If it is an additional or repeat test that is necessary, have it done whilst all parties are still there – or first thing the next day. This will prove to be by far the most effective option, in terms of avoiding wasted time and cost.

Communicating the solution

Remember the principle of getting things done by using short lines of communication? It holds good for corrective actions. This means that the most effective instruction route for initiating the corrective action is that between those parties (for example contractor and manufacturer) who are linked by contract. Be prepared here to let others implement your decisions if necessary. As an inspector, your input can be expressed in the form of a statement of what needs to be done, without imposing on contractual links. Stand back – but make sure that the subsequent implementation is to your satisfaction.

Closing the loop

The important point is *verification*. It is not difficult, having reached the stage of issuing an NCR and agreed CAs, to think that the inspection job is complete. It isn't. One of the inherent weaknesses of the inspection business is that much is left unverified. The result is that some commitments made by contractors or manufacturers regarding actions to ensure the fitness for purpose of mechanical equipment are often not done in the works; the 'we can do that on-site' syndrome. In reality all this does is transfer the problem to somewhere else, where it is more expensive to solve.

The central message is that for non-conforming equipment you must develop a way of verifying that the corrective actions have been done. It is, frankly, poor practice not to do this, or to try and shift the verification activity to someone else. Closing the loop is to finish the job. Your client will see that the job is well done; clients want, after all, to buy *solutions* to their problems by employing inspectors. I will repeat this, for the sake of good order.

INSPECTION IS *ONLY* EFFECTIVE IF YOU FINISH THE JOB

Reprise : saying 'no'

Rarely; but sometimes, despite what everyone would like you to believe are their best efforts, the equipment that you inspect will just not be fit for purpose and making it so will be a lengthy and expensive task. In such situations, choose carefully the precise moment to start and issue your NCR, because as you reveal your decision you will be put under pressure to change your mind. You must anticipate this, and for the pressure to increase as the minutes pass, as the implications of extra cost start to be appreciated by the parties.

Expect criticism. You may be accused of being uninformed, pedantic or even unreasonable. It is normal for your technical knowledge in specialist areas to be questioned. You can now appreciate the importance of your timing. This is not the time to review the technical arguments, you should have *done this already* and therefore have your focus clear on FFP and the elements of your decision (from your notes) in logical order. You can explain your reasons again if forced, but don't go over old ground – and make sure you stop well short of being defensive.

Then just stick to it: say 'no'. Your final words should be to explain what your report will say, and to make sure that everyone at the inspection is well informed about your decision.

No one looks for this. It is not necessarily a failure of the manufacturer, or the contractor, or both, or the system, or of anything else. It *is* a failure, though, and *you* are part of it because you are there to ensure FFP and you haven't done it. The reality is, unfortunately, that you are not only the messenger, you are also a contributor to the failure and will be seen as such. So how can we stop this all happening? You have to start with the bedrock: specifications, standards, and plans.

KEY POINT SUMMARY : OBJECTIVES AND TACTICS

Objectives

1. The purpose of inspection is buyer confidence.
2. Multiple parties are involved. This is because of the rules of commerce.
3. Basically, all inspectors are looking for the same things, although their foci may differ.
4. Inspection is about determining fitness for purpose (FFP) or *utility*. The four elements of FFP are:

   - Function (will the equipment do the job?).
   - Mechanical performance (is it strong enough?).
   - Service lifetime (how long will it last?).
   - Quality.

Strategy

5. The plant end-user or purchaser is instrumental in defining overall inspection strategy. Duplication of inspection is unavoidable and you should plan for it. Too much inspection is as bad as too little. Try to understand the strategy behind what you are doing.

Tactics

6. Develop a tactical approach – it will help you become more effective and increase your business.

   Tactical elements include:

   - Managing information: make sure you have the information you need. Don't arrive at the works without it.
   - Focus: keep sight of your FFP criteria and priorities. Learn how to use the ten-minute loop.
   - Asking and listening: keep asking. The technique of 'chain questions' can be useful.
   - Decisions: the main decision maker is *you*. Aim to get decisions implemented by using the contract protocol to help you.

The end game

7. Make a plan as to how you will deal with the end game. Non-conformance reports and corrective actions belong together. You have to reach agreement. Make sure you finish the job. Expect frequent questioning and occasional criticism.

Chapter 3

Specifications, standards, and plans

The contract specification provides a structure to work to. You have to understand what it says – then use your judgement.

If the fundamental objective of works inspection is to verify fitness for purpose, it is safe to say that the contract specification is the way of expressing this objective in written form. It is still only a précis. Increasingly, plant purchasers are using functional or 'turnkey' specifications to try and obtain a predictable output (the plant that they want). The result is that inspectors often have only a very basic set of written requirements to work with. This is where technical standards, experience, and judgement fit in.

It is best not to put a negative perception on this for the (very good) reason that there is absolutely no logic in doing so – if you make a careful, and rather detached, study of the real difference in the works inspection role imposed by a detailed plant specification compared to a turnkey specification (for a power station for example) you should feel comfortable with this statement. Turnkey plant specifications involve similar inspection activities to more detailed specifications. They are just *expressed* differently.

You shouldn't accept the often expressed notion that works inspection activities are of less importance in turnkey contracts – that your role is somehow devalued by the fact that the contractor accepts all of the responsibility. Much of this is illusion – try and view the contract specification, however brief, as a firm justification for your works inspection remit.

This being said, it is always easier to work behind a *good* specification. It is not easy to write a good specification (and not many of us can honestly say we have never criticized other's attempts). One of the main problems is lack of feedback – particularly from 'shop-floor' and site activities. You can help this by developing a feel for what a good specification consists of. Delegate yourself the role of improving the content, and the clarity, of the contract specifications that you are involved with. Try to identify better ways to specify, as precisely as possible, technical requirements that will facilitate the achievement and verification of fitness for purpose. This is effective inspection. What of the position of the contract specification within the wall of technical information that we are faced with? Look at Fig. 3.1: I have attempted to show you here how it fits in. Broadly, the large box represents the totality of information sources relevant to the verification of FFP. You should find that the percentages shown correspond approximately to the relative importance of each component of information and, more importantly, the ratio in which you will find the categories of use to *you* as a works inspector. Note two interesting points:

- Recognize that statutory/safety requirements have an (often unwritten) input into each component.
- Note the overall significance of the contract specification (almost 50 percent if we include the various amendments and technical clarification that goes with it). Note particularly that the body of technical standards making up a further 45 percent are referenced directly or indirectly *by* the contract specification.

Contract specifications vary widely. Fortunately, it is the format that varies more than the content. Expect the content of a proven specification to be reasonably consistent – if only by virtue of the fact that contract specifications are rarely compiled from a zero base, they are invariably based on a previous consultant's, or end-user's, specification.

What you have to do is to identify those points of the specification that are directly relevant to the works inspection activity. There are five main groups of information, and they are normally well spread out within the body of the contract specification. An important word of warning – they are unlikely to be included under the five headings I have identified – be sure to look for *content* not title.

Figure 3.2 shows the groups. In the remainder of this chapter we will look in detail at these five areas, with the objective of deciding how best to deal with them. I will then propose to you that there is almost a formula solution to enable you to organize these activities in a way which maximizes the benefits of works inspection – we will adapt Fig. 3.2 to show how to start and put the structure in place. This will give us an effective inspection regime.

Specifications, standards, and plans 29

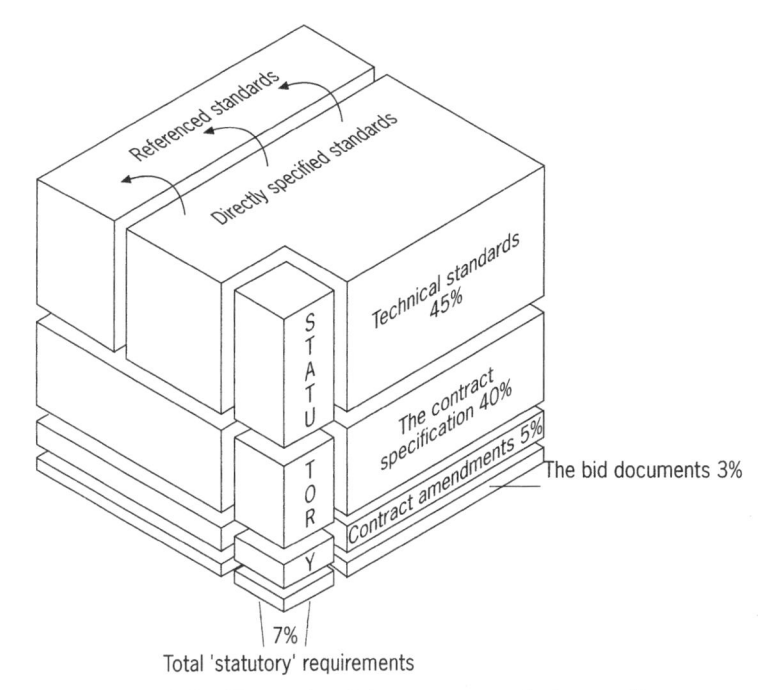

Fig 3.1 The ratio of document use in inspection

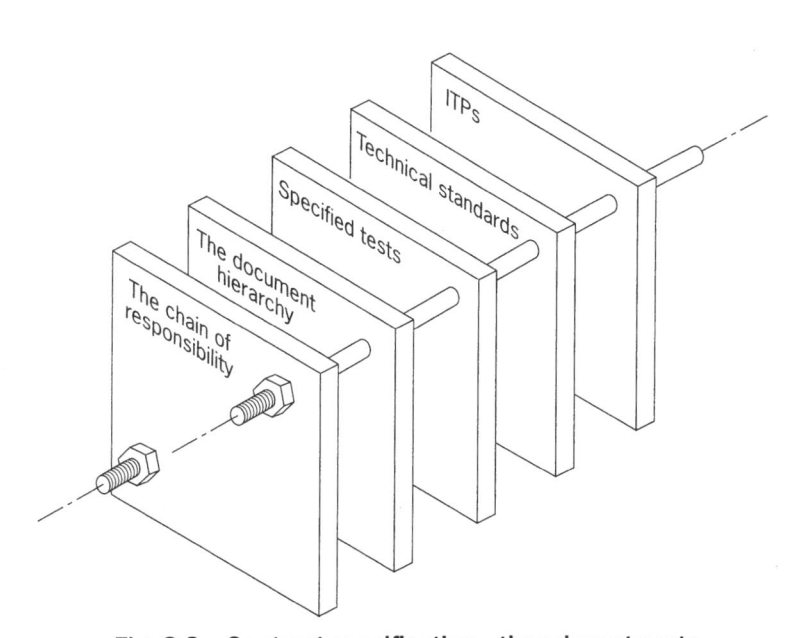

Fig 3.2 Contract specification – the relevant parts

The chain of responsibility

One of the purposes of the contract specification is to allocate responsibility for the procurement and construction of the plant. Even a rather poor technical specification will normally make clear the main responsibilities. The most common model used for power or process plants is that of a single contracting party (The Contractor) taking full responsibility for the fitness for purpose of the plant. For large projects, the contracting party is often a consortium, but the same principles apply. As simple as this may sound, there is much more to it than this – there is *always* an underlying structure of subordinated responsibility between the contractors, licensors and manufacturers.

It is an inspector's role to recognize and understand this sub-structure but, thankfully, not to become involved with it. The main questions of issue to an inspector are two-fold:

- Who is obliged to answer my questions? (although it may be the contractors' role to demonstrate FFP, it may be optimistic to suggest that they have to *prove* it to you).

and

- Who is responsible for carrying out any corrective actions that are needed?

These questions are easily answered by looking at the chain of responsibility as shown in Fig. 3.3. The bold line is the main responsibility link. If you are representing the plant purchaser (and that is the viewpoint I have used throughout this book) then you must only act on this line. You can only *directly* impose your requirements on the contractor. Now for the paradox. As I pointed out in Chapter 2 (take another look at Fig. 2.1 if you need to) the contractor, being remote from the activities of equipment manufacture, knows less than the manufacturer about the FFP of the equipment. This means that the best quality information for you, as the inspector, is obtained from as far down the chain of responsibility as you can comfortably reach. We can crystallize this as follows:

– Get your information near the bottom of the chain.
– When you act, do so at the top of the chain.
– If you don't do this the final result will cost too much. You will be partly to blame.

If you think carefully about this, the apparent paradox will fade. What

Specifications, standards, and plans 31

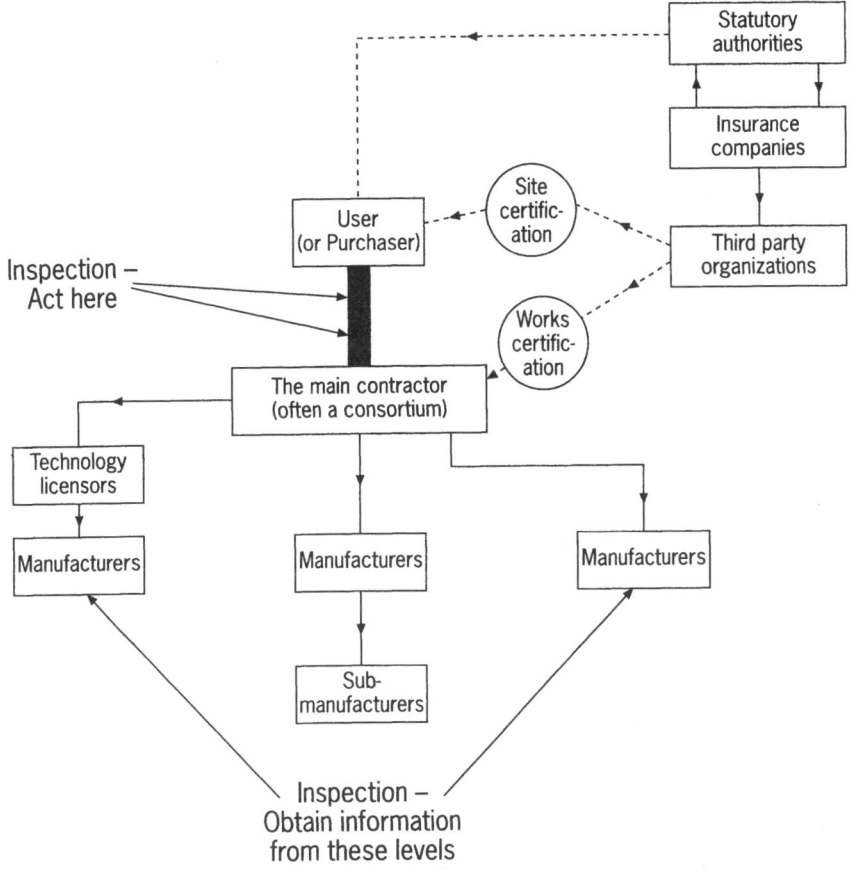

Fig 3.3 The chain of responsibility

you can actually produce by operating in this way is a mechanism for consensus.

Third parties

Third party certification bodies have a well-defined role in power and process plant contracts. What do they do?

Their job is to certify compliance with recognized standards for boilers, pressure vessels, lifting equipment and a few other equipment items that are governed by statutory regulations in the country where they are to be *used*. A minority of countries have stricter rules whereby equipment *manufactured* in that country must have statutory certification. A secondary driving force is provided by the requirements of the insurance industry who find comfort in the involvement of a nominally independent body. You would probably also not be wrong if you observed that third party bodies have history on their side – industry is used to them being there.

There is often a little confusion surrounding the exact role of third-party bodies. Basically, the following are true:

- Third parties *only* verify compliance with the standard or code that they are asked to. There is a link, of sorts, with safety requirements.
- Third parties do not assess fitness for purpose (i.e. will the equipment do the job for which it is intended?) even if some of their clients think they do.
- The level of inspection activity performed varies greatly between third-party bodies. There are no detailed listings in equipment standards or statutes that define *exactly* what they have to do. I concur with other people in the industry who believe that they do their best.
- Third parties are normally contracted directly by the equipment manufacturer (for boilers and vessels typically), rarely by the end user.

These roles tend to be clear – so it is not too difficult to formulate a basic tactical approach. Without going into much detail, there is a simple way to deal with the third-party aspect of works inspection. Build their role into the contract specifications where you can – as a minimum let their opinions weight your judgement in those tricky situations where you have to decide whether or not to issue an NCR for an important piece of statutory equipment.

I recommend that you:

- *Do* place your faith in the *integrity* of the well-known third party bodies. Accept their role as 'honest broker' organizations.
- Don't be so relaxed that you don't check their inspection reports. Make a real attempt to obtain the authority to review the scope of their work and how they are doing it. If you have the chance, build this into the contract specifications – then implement it.
- Assess how much of their work will duplicate your own. Probably around 40–50 percent of it (particularly on boilers and lifting equipment) will coincide with what you would be doing to check for FFP. Once you have built up confidence in the third-party inspector you can stand off, *ever* so slightly. Don't waste your budget.

Specifications, standards, and plans 33

Contractors

Contractors use several mechanisms to try and off-load technical and commercial risks onto the manufacturers that supply them with equipment. These include back-to-back guarantees, requests for extended guarantees and penalty clauses – I could go on. From a works inspection viewpoint we can safely avoid getting involved in this complexity. Only on the very rare occasions that subordinate guarantees from a major process licensor are contracted directly with the user, and are written into the contract specification, do you need to think about how the sub-structure of guarantees will affect you. Otherwise don't get involved. Keep your line of action simple, as shown in Fig. 3.3. It will help you to focus.

The document hierarchy

Put simply, the document hierarchy is there so everyone should know which documents to follow, when there is a choice. You will normally find the document hierarchy mentioned near the beginning of the technical part of the specification, following the commercial conditions. It is a useful tool in avoiding controversy over whether or not an equipment item complies with the purchaser's requirements.

The basic hierarchy

Figure 3.4 shows the basic document hierarchy. This is a common model; you may find a few variations but the principles of this hierarchy are well proven and accepted. The document hierarchy will impinge upon almost every works inspection that you do, so it is useful to understand some of the softer and often unwritten influences that operate here. Although the basic hierarchy is quite straightforward it takes a heavily simplified view of what is, in reality, a complex and nested structure of engineering requirements.

The inevitable result is that interpretation of the contract document hierarchy has to be tempered by practicality. Engineering experience of design and application, and an appreciation of how things are done, need to be added to make the hierarchy work.

Practical advice

In dealing with the contract hierarchy during inspections you have to be flexible. This approach is not a panacea, nor is it a substitute for

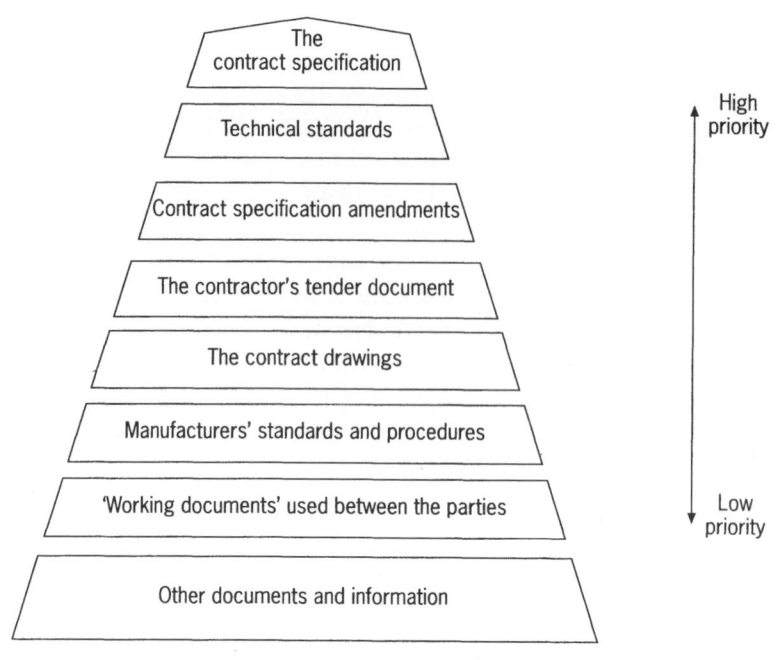

Fig 3.4 The basic document hierarchy

knowing what you want. Here are some practical points which you may find useful:
- *Standards.* Technical standards represent accumulated knowledge. Never underestimate the level of specialized ability and effort that has gone into compiling them. If in doubt about a technical point you should rely on their content. Does this mean that some technical clauses in the contract specification (which lies, remember at the top of the document hierarchy) may sometimes be wrong? The way to answer this is to make a quick check on the level of *detail*. Detail shows thought – so if a specification clause shows a level of detail *better* than a technical standard, there is a fair chance that the standard is outdated, and you should give the specification clause more weight in your judgement (gas turbine technology is a good example of how this can happen). Make a conscious attempt to comprehend this balance when you are carrying out works inspections. It is time well spent because you will find the same questions arising again and again.
- *Standards again.* Standards really are very difficult to argue against. If you are party to a technical, contractual or legal dispute with a contractor and the content of the relevant technical standards is not

on your side, then you will probably lose. Think carefully before you contradict standards.
- *Working documents.* Somewhere in the document hierarchy will be acknowledgement of the provision for working documents. Their status may be stated explicitly, but more often is just inferred. Working documents are those which act as an interface between manufacturer, contractor, and purchaser. They include:
  - design information and data-sheets
  - some sub-orders
  - a mass of technical information, including correspondence, which passes between the parties
  - the inspection and test plans (ITPs)
  - manufacturers' procedures and practices for manufacturing and testing.

You will see that many of these working documents are those which you will use during a works inspection – the ITP in particular is one of the core parts of an inspection programme. The difficulty is that these working documents occupy a low position in the document hierarchy. You have to be very careful, therefore, when you want to issue an NCR on the basis that a manufacturer has not complied with a working document. You can only do this if there is a clear linkage back to the upper elements of the document hierarchy. I find the best way is to draw reference to the document at the very top of the hierarchy (normally the contract specification); make explicit reference to a contract specification clause in an NCR, and you are on pretty firm ground. You may sometimes have to explain the document hierarchy to people from contractors and manufacturers whom you meet during inspections.

Specified tests

As if to compensate for the inferential approach of some technical standards, purchasers like to include specific works test requirements in their specifications. The approach is fairly common to power generation, offshore and process plant contracts, particularly where there are statutory certification or health and safety implications. You should find these test requirements divided into two clear groups; specific equipment works tests and general test requirements. Frequently, they are included in different areas of the contract specification.

Equipment works tests

Look for these in the parts of the contract specification describing the individual plant areas or pieces of equipment. Major performance tests are more likely to be specified in the guarantees section. Expect to see a schedule of works tests for major equipment items such as prime movers, pumps, compressors, fans, and cranes (these are in addition to the tests that are specified to be carried out after installation on-site).

It is important to look at these requirements carefully and try to understand why they are there. Some of them will be simply a duplication of the requirements of the referenced technical standard, included perhaps as a reminder to the contractor of what is accepted practice. There is no need to worry too much about these. The others will be there for very clear technical reasons linked to *special* engineering requirements of the equipment. By definition, you will find these in the more complicated technical areas. Don't be surprised if they raise apparent technical contradictions that you will be called upon to solve. Here are three common examples of special tests:

- Pumps which have specific q/H requirements for fire service, vacuum application or parallel operation.
- Lifting equipment. Cranes may have special classification or speed requirements. Some clients want extra functional tests or non-destructive tests.
- Gas turbines. We will see later in this book that gas turbine guarantee acceptance tests can be quite complex. Functional, no-load testing in the manufacturer's works is often specified.

General test clauses

In a part of the contract entitled 'General Technical Requirements' (or something similar) you will find the general test clauses. You may find that these cover quite a wide variety of engineering disciplines. This is good, in that they provide a defensive 'catch-all' to cover areas where standards or the technical specification are a little weak, but bad in that they increase the possibility of contradicting those standards that are very precise. You have to treat these carefully. They are there to use when necessary but do not force them into technical areas where they do not easily fit. There are five main sets of 'catch-all' requirements:

Materials choice

Some specifications impose the use of specific materials of construction. This is more common for large fabricated equipment such as boilers, desalination modules and process vessels than it is for discrete proprietary equipment items such as gearboxes or compressors. Better specifications leave the manufacturer some leeway by specifying suitable material 'groups'. An even better method is only to specify those materials which *should not* be used – this reduces significantly the amount of specification 'clash', whilst still giving a level of control.

Materials testing

For proven engineering components, I think this one rather unnecessary. Most materials standards are very advanced in their specification of material tests. Look at some of the ASTM (American Society for Testing of Materials) standards on tubing and castings if you need convincing of this. Very rarely, you will meet one of the few special technical areas where additional tests need to be specified. Watch for:

- Specific corrosion resistance tests performed in highly alkaline or acidic environments.
- Very low temperature impact (Charpy) tests for new materials which may not yet be recognized as proven for cryogenic use. Sometimes these are combined with fracture appearance transition temperature (FATT) tests.
- Creep tests for particularly *high* temperature 'experimental' alloys.
- Special tests on advanced high-temperature ceramic coating of alloy components.

Material traceability

This is a powerful tool to employ in a specification but you may not always find it used in the right place. Full traceability means that a complete and unique record exists of the origin of a material, its testing and its destination in an engineering component (we will look at this in more detail in Chapters 4 and 6). Full traceability is almost a prerequisite for equipment which is covered by statutory requirements, particularly boilers and pressure vessels, and the third-party certification bodies play a part in policing it. Outside the statutory area, however, material traceability is purely a technical option. I have seen it overstated.

My advice is that you look carefully at what the general material traceability clauses in the specification actually say. Look for the logic behind the statement – then use your findings to weight your judgement as to how hard you will police the stated traceability requirements. I have found that:

- on average, a general specification requirement for full material traceability will be complied with about 35 percent of the time

and

- you will find it almost impossible to police the other 65 percent on an economical basis.

I do not include statutory equipment in these findings. Compliance is much better for statutory equipment. One reason is that traceability has always been a requirement.

Non-destructive testing (NDT) and welding

It seems that most contract specifications make some provision for extra NDT of critical components. Once again, you have to decide whether these are well thought-out requirements. If they are, then the right thing to do is to impose them. If you have the chance to influence these clauses, try to get them organized in categories by primary production process – separate them into stock, forged, cast, and welded components. You should find them easier to use like this, rather than classified by equipment item. NDT requirements do not often transfer comfortably between types of *equipment*.

Welding is an area where it is difficult *not* to recommend the use of general specification clauses. Requirements for weld procedure specifications (WPSs), weld procedure qualification records (PQRs) and welder qualification records (WQRs) are, frankly, just good engineering practice. You should feel happy about having these overall requirements in the specification and be prepared to impose them. Poor welding means poor FFP.

Using standards

Technical standards form the backbone of a technical specification – I have made the point previously that without them, the cumulative experience of industry would be lost. If we exclude statutory requirements, technical standards only become part of the technical requirements of the contract if they are mentioned explicitly in the contract specification. Broad statements saying that 'industry standards' shall be applicable are not an effective way for purchasers to get what they want. It is useful for inspectors to see things from the purchaser's viewpoint – a good purchaser uses technical standards to:

- control the price

- bring objectivity
- minimize repetition
- utilize accumulated experience.

Inspection consists of involvement with many different technologies. Effective inspection requires that you develop a clear view of how a technology relates to the technical standards that exist in that area, and the way in which manufacturing industry interfaces with these standards. Within the normal scope of a large power or process plant contract you will find two trends: 'leading' standards and 'following' standards.

'Leading' standards

This is where technical standards take the lead in developing a technology. They are assembled by niche manufacturers and trade associations, who have often done their own quite entrepreneurial development work, and then disseminated quickly throughout the industry. The complexity and technical detail of the standards encourages manufacturers to strive to develop their own practices to meet these standards. The result is that the *standards* lead the industry. Disciplines such as material technology (particularly non-ferrous and corrosion-resistant materials), advanced fabrication, and some areas of pressure vessel technology are like this.

'Following' standards

In engineering disciplines where there is intense technical competition it is not unusual for technical standards to lag behind manufacturers' practices. One excellent example of this is gas turbine technology. It develops so quickly that standards do not keep pace. Most technical aspects of gas turbines are better described in manufacturers' own procedures than they are in the technical standards (see Chapter 7), which at best only form a framework for broad agreement on technical matters between the purchaser and manufacturer.

These two situations experienced with technical standards means that you have to tailor your approach carefully. An important aspect of works inspection is to make the best use of experience – it is worth trying to form an opinion on each technical standard, as you have occasion to use it, as to whether it is a leader or a follower. Use your opinions to weight your judgement when using these standards.

Here are some useful summary points regarding the use of technical standards.

- You should use standards to their full but you must understand their limitations.
- Match standards very carefully to their correct application. Make sure you know if you are using them loosely.
- Standards *are* an overt guardian of fitness for purpose but are not the full story. Remember the effects of the document hierarchy.
- Quote technical standards in your NCRs. If there is a clear non-conformance with a standard you are on very firm ground.

Inspection and test plans (ITPS)

ITPs are key working documents. Note that this is *only* what they are, they do not normally have high status in the document hierarchy. This can make things difficult for you. A large contract can have several hundred ITPs, using different formats and containing varying degrees of detail. They are almost the only possible mechanism for controlling and organizing the various inspection activities.

Purpose

Their purpose is to satisfy you, and the other parties' inspectors, by providing a mechanism for organizing inspection activities. For this reason some manufacturers see them as an imposition. Please don't be misled by the notion that ITPs are used by manufacturers as a method of production management or work instruction – they have better ways of doing this.

ITPs *are* a forum for discussion and agreement of technical matters between purchaser, contractor and manufacturer without the enforced rigidity (and emotion) of the purchase contract. They are also an excellent documentary record of the activities and commitments of the multiple parties involved. Inspectors like them because ITPs show them what to expect.

Essential content

You will not find any hard and fast rules about what should be included in an ITP – some standards include recommendations but any attempt at uniformity gets very difficult when faced with the variety of equipment, processes, and practices in a large engineering plant contract. The tendency is for many of the contractors and manufacturers to use differing types of content and format for their ITPs. Frankly, format doesn't matter that much, it is content that is

Specifications, standards, and plans 41

important. Good contract specifications will often include a specimen 'pro-forma' ITP to indicate the level of information that is required. See this as a valid attempt to introduce uniformity but remember the practical difficulties of this, as I have explained.

Here is an outline checklist of ITP content.

- A clear list of manufacturing and test steps for each manufactured component.
- Cross references to salient contract specification clauses.
- Detailed reference to show which acceptance standards (or technical standards) are applicable to each manufacturing and test step.
- Cross references to manufacturers' more detailed working procedures (so you can ask to see them if you need further clarification about a particular activity).
- Indication of the records and certification requirements applicable to each step. This will define conveniently the compilation list for the final documentation package (sometimes also called an 'end of manufacturing report') for that equipment. Documentation requirements should be comprehensive, but finite.
- Standards equivalence: a good ITP will always indicate where standards quoted are equivalent to others. This is especially important where less well known international standards are involved.
- A system of activity codes. These are useful for understanding which tests are being referred to (e.g. under general NDT categories).

Figure 4.8 in Chapter 4 shows a well laid out (but simplified) ITP – I have annotated it to illustrate how it complies with the checklist points. Make a point to guide contractors and manufacturers towards this type of ITP model. Remember that the figure concentrates on content, rather than format.

Using ITPs effectively

The secret of using ITPs effectively is *anticipation*. ITPs will be of much greater use to you if you can manage to exert an influence on their content at the earlier stages of a contract. To do this you have to anticipate what you want their content to be – this will then facilitate a more effective inspection process than if you are forced simply to act retrospectively once all the ITPs have been written. Using ITPs is not too difficult, but it is an area where concentration and some tenacious attention to detail pays dividends. It is easy to pass over some of the major issues, particularly in relation to complex plant (such as turbines and boilers) where the ITP will have a lot of steps. Try not to let this happen.

There are a few very well proven guidelines to follow – if you try to stick to these, they will provide a foundation for all the inspection activities in a contract (not just your own). They are:

- *First draft*. The best party to produce the first draft of the ITP is the main contractor, *not* the manufacturer or the end user. This is because the contractor is the party that will have best analysed and understood the purchaser's contract specifications. It is unfortunately all too common for contractors to let their manufacturers produce the first ITP drafts. I don't think this gives the best results.
- *Marking-up*. Aim for *very* early marking-up of the ITPs with the inspection witness points of all the parties. If you are the purchaser's (end user's) inspector, make sure you mark up your inspection requirements *last*. This way you can make your choices based on a knowledge of how much priority the other parties are giving to the various works test activities.
- *Timescales*. Make sure that you do not receive the ITP too late, when the material or equipment is almost complete. There is always plenty of pre-manufacturing contact between contractors and their manufacturers to discuss the purchase order so there are few credible excuses for not making the ITPs part of these discussions. It is in your interest to intervene sooner rather than later if you suspect this is not being done very enthusiastically.
- *Witness points*. If you are likely to perform the works inspections, then *you* mark up the ITP with your witness requirements. This gives continuity.
- *Focus*. Effective works inspection involves being decisive about very complex engineering issues. You will not learn to do this by spending lots of time in contract meetings discussing administration and document control (interesting though they may be). Leave this to the contract administrators. They are much better at it than you are. Put your effort into the engineering aspects of the ITPs.

I have tried to cover in this book most of the works tests that you will be expected to witness as a works inspector. Use the technical chapters, Chapters 4 to 14, as a broad guide to the tests that you should witness. Don't lose sight of the technical aspect of what you are doing.

Equipment release

The big questions is: when should you release equipment for despatch from the manufacturers works?

The mechanism by which equipment can be released for shipping is one of those areas that varies a lot between contracts. It is also the subject of a lot of discussion between the parties. This is because the release of equipment is inevitably linked in some way to payment stages, and forms one of the recognizable contract 'milestones'.

In theory the contract specification should define clearly the material release mechanism (material is a generic term covering all materials and equipment forming the scope of delivery within a contract). The release mechanism so described should match the chain of responsibility inherent in the contract terms. There is, however, not always a simple rationale in the design of this release mechanism – you may sometimes find it feels a little indistinct.

The release mechanism has a significant effect on the precision of the contract cash-flow. Hence you can expect real commercial pressures to show themselves as the various parties become involved in the release of equipment. My best advice on a general level is that you should firstly exert pressure for the release mechanism to be well defined. Figure 3.5 shows the chronological stages following the final works inspection. Stages A and B are two common locations of the material release point – bear these two possibilities in mind and then consider these general guidelines.

Guidelines

- In general, stage A in Fig. 3.5 is the most efficient location for the release mechanism. It fits in better with the practicalities of contracts and the way that manufacturers work. It allows adequate opportunity to verify fitness for purpose without involving risks of delays by the parties involved. I believe there is little benefit to be obtained by delaying release to stage B – it is almost all downside in terms of cost and time delays.
- You can make the release mechanism itself much more effective by linking your non-conformance reports firmly to corrective actions and then ensuring their implementation. I described this pro-active approach in more detail in Chapter 2. Remember that you lose most of your powers of verification once equipment leaves the manufacturers' works.

44   Handbook of Mechanical Works Inspection

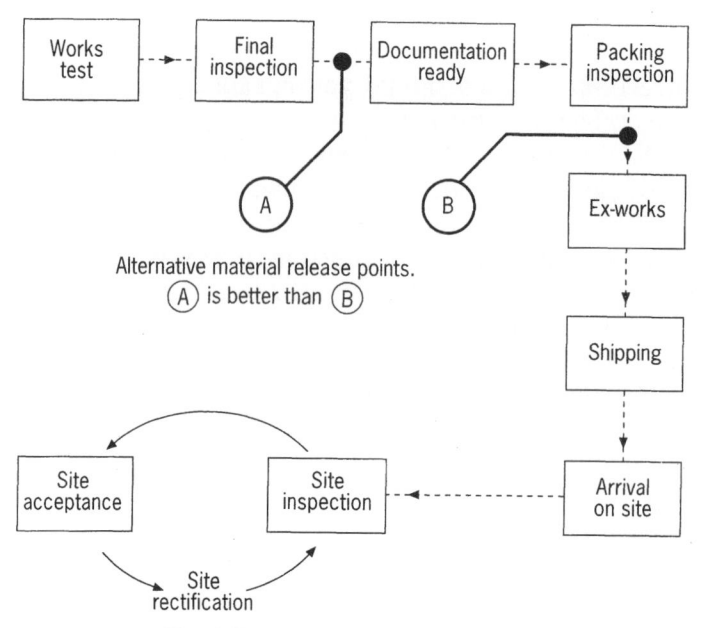

Fig 3.5   Material release points

- Manufacturers will sometimes want to carry out rectification work on site rather than in their works. You have to decide if this is acceptable. There are no hard and fast rules for this. The best advice I can give you is to think back to your FFP criteria – if any of the rectification work has a *direct effect* on FFP, make sure it is carried out in the works. Don't release a piece of equipment until you are convinced that it is correct.
- It sometimes happens that manufacturers or contractors will despatch equipment without formal release (expect them to explain this by reminding you that it is 'at manufacturers' risk'). Your main priority here is to make sure that any NCRs relating to the despatched equipment are awaiting it at its destination on site. Define the necessary corrective actions, and the responsibilities for their execution, with particular care.

In an 'average' power or process plant contract, perhaps ten or fifteen percent of equipment will leave the manufacturers' works in a condition which warrants the issue of a non-conformance report. Some will have major problems which influence FFP and some will only have minor faults. A contract which has an *effective* equipment release mechanism

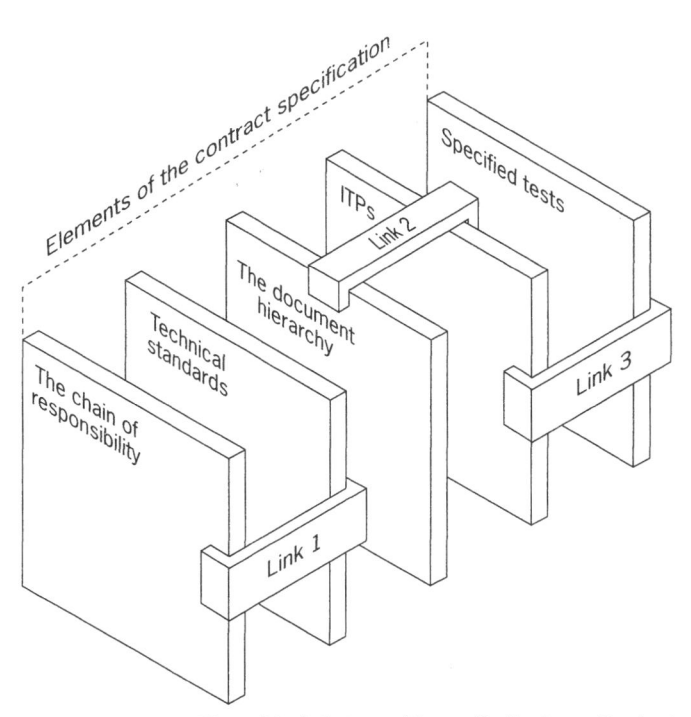

Specifications, standards, and plans 45

Fig 3.6 Contract specification – a 'formula solution'

These links help to provide an effective inspection 'regime'

can perform much better than this – the number of non-conformances can be reduced to a few percent, or zero, and the resulting administrative and management costs will be less.

Aim for clarity and objectivity in the equipment release mechanism.

The formula solution

I mentioned near the beginning of this chapter that I would propose a formula solution, suggesting the way in which the five areas of the contract specification with an influence on works inspection are best linked together. Strictly, this is the job of contract engineers rather than inspectors, but if you are interested (as perhaps you should be) and can influence future contract specifications, you should note the main principles. I know that there are a multitude of other factors to be taken into consideration – I would hope, however, that you will not find the solution to be in *conflict* with other aspects of contract management.

Refer to Fig. 3.6, which you will see is an amended version of Fig. 3.2, and look at the linkages I have shown:

Link 1

This is about responsibility. Make sure that the explicit chain of responsibility stated in the contract specification is used *properly* to pass the requirements for compliance with standards down to subcontractors – you are looking for an exact matching here, not just 'general statements'. In a practical works inspection situation you may need to apply pressure to contractors to ensure that they perform fully their responsibilities in this area, in particular their role in exerting control on manufacturers' technical practices. Monitoring alone is not enough. It is also important that you define the responsibility of third-party bodies correctly. Their role needs to be specified in some detail – if it is not, be careful not to under-estimate, or over-estimate, the validity of their input in verifying compliance with specified technical standards.

Link 2

Link 2 concerns decisions – more precisely decision *making*. It is necessary that the document hierarchy (which is one of the most prescriptive parts of the specification) has clear operational links to the ITPs. This helps to ensure that the acceptance criteria shown in the ITPs are closely controlled *and* are compliant with the actual contract requirements, not the manufacturers' opinion of what they may be, or those that the manufacturer usually uses. With acceptance criteria agreed in advance, you, the inspector, should not find too many contradictions when evaluating the results of tests in the works. It will be easier to make decisions, without a lot of wasted effort.

Link 3

This is to do with economy and money. It will help you if you can see link 3 in terms of the *costs* of verifying fitness for purpose. The protocol surrounding inspection activities governs which of the specified tests you will witness and how often. Strike a balance between the equipment tests that are notified to you (as defined in the ITP) and those which you make a decision to inspect. A ratio of *about* 2.5 to 1 is not too bad – it will allow good project monitoring but not involve too much administration.

Finally in this chapter, a brief review. We have looked at the important areas of documentation, how they fit together, and how they can affect your activities. Before proceeding to the technical chapters (which are the important parts of this book), reflect that *effective* inspection is one of those activities that appreciates good organization. This is the key.

KEY POINT SUMMARY: SPECIFICATIONS, STANDARDS, AND PLANS

1. The contract specification provides a structure to work to. You need to know how it works.

2. Important parts of the contract specification are:
    - referenced technical standards
    - the role of ITPs
    - specified material and equipment tests.

3. All contracts have a 'chain of responsibility'. Effective inspection involves obtaining information from near the bottom of the chain but *acting* near the top of the chain.

4. The contract document hierarchy will help to sort out some of the apparent contradictions in technical requirements that you will meet but:
    - it will not give you all the answers – you have to look to the logic behind it.

5. Published technical standards represent accumulated knowledge, but:
    - some are 'leaders' and some are 'followers': consider them with this in mind when you are using them during works inspections.

6. ITPs are working documents and lack contractual power. The secret is to make sure the technical requirements of the contract are pushed into them at an early stage.

7. *Effective* inspectors facilitate the equipment 'release' mechanism. You can help by understanding the contractor's responsibilities in this area.

Chapter 4

Materials of construction

Works inspection involves quite a lot of *verification* of the engineering materials used to construct equipment. It is not, strictly, about material *selection*, although this does come into the equation. Materials are one of the lowest common denominators of engineering practice. The common steels and non-ferrous alloys are used in many and varied types of equipment. There are also many specialized materials, each with its individual niche of application and designed specifically to resist a particular regime of stress, temperature or process conditions. It is a complex picture.

The place to start is with fitness-for-purpose (FFP) criteria. I introduced this concept to you in the initial chapters as a way to help cut down complexity to a manageable level. With materials of construction it is fortunate that there is a large body of technical standards that document and classify engineering materials. This does not mean, however, that works inspection of materials is reduced to a simplified exercise in verification – to do it well you need to understand a little of the basic thinking that lies within a material standard.

Materials inspection is not just technical. Whereas material technology is heavily standards-based you will also find that the issue of material *traceability* occupies a prominent position. This is a procedural aspect of engineering manufacture that forms a common feature of the works inspection role, particularly in relation to pressure parts and structural components. We will look at this in more detail later, but it is useful to keep this procedural side in mind when reviewing the technical aspects of FFP.

Fitness-for-purpose (FFP) criteria

There are many different tests and analyses that are carried out on engineering materials. The purpose of almost all such tests is to verify one or all of the three main FFP criteria: mechanical properties, temperature capability and positive identification.

Mechanical properties

The mechanical properties of a material are matched to its application by the equipment's designer, assisted by the accumulated experience of proven and accepted technical standards. The common mechanical properties are: tensile strength (this is truly a 'strength' measurement), ductility (measured by percentage elongation and percentage reduction of area), 'toughness' or resistance to shock (measured by impact value), and hardness (the ability to withstand surface indentation). You will see these parameters specified in material standards. Less commonly specified properties include fatigue, creep resistance, and resistance to corrosion. Such properties can be considered as complex functions of the other mechanical properties, rather than primary properties in themselves.

The mechanical properties of an engineering material only have meaning as a set, rather than individually. Tensile strength or hardness, taken alone, is inadequate to describe fully the way in which a material performs – measures of ductility and toughness must be added to complete the picture. It is better therefore to think of this FFP criterion as consisting of compliance with a *set* of mechanical properties, even if one of the properties does not appear to be predominant when you consider the application of the particular material.

I inferred earlier that one role of an inspector is to *verify* materials rather than to select them. You will not normally experience a situation where you are obliged to select a material for a specific design application. The nearest you should come to this is when you have to interpret the material requirements written in a contract specification. You need to be prepared, though, to widen the scope of your verification activity, to introduce a judgemental approach, if it is required. In practical works inspections, such situations will exist more frequently than you may think. This means that you need to develop your background knowledge of engineering materials – it is not always sufficient simply to check a set of test results against the listed values in a technical standard.

Temperature capability

The resistance of a material to elevated temperature does have some relation to its mechanical properties. It is easier to consider, however, the links between temperature capability and the elemental chemical analysis of a material – but it is still a complex subject. I have included temperature capability as a separate FFP criterion for the more practical reason that temperature capability is a parameter used to decide the scope of many of the material technical standards (both European and American standards differentiate clearly between low, intermediate and high temperature ferrous alloys). It is also a crude, but practical, way to evaluate technical risk. High temperature often means high pressure and the existence of less predictable failure mechanisms such as creep. Works inspection is a wide, almost cross-disciplinary technical subject, so a common denominator such as materials of construction needs a focus. Temperature capability is one useful point of reference that will help you keep things in perspective.

Positive identification

It is almost impossible to identify materials by visual observation alone. For this reason positive identification is an important FFP criterion. It is not strictly technical – it has a procedural basis – but the activities form a well-defined element of engineering manufacture. We will look at the detailed activities of material identification and the traceability chain later in the chapter.

Basic technical information

There is a finite amount of technical information that you need to know about materials of construction. This is because technical questions regarding the common engineering materials found in a power or process plant are generally repetitive. A clear framework of the type of information that is useful, coupled with some experience of the issues in practice, will give you what you need to carry out effective inspections in this area.

Take a global view of all the material standards that are available. There are several thousand of them. You can divide them rather neatly into four categories, based on the four predominant forms of material manufacture. These are: forgings, castings, plates (or sheets) and tubes. These categories can help focus your field of view – you will find that there are similarities in common technical areas such as material

FORGINGS

Forgings are used for:
- Pressure parts such as nozzles, rings and flanges on all types of vessels.
- Shafts on rotating equipment. Also gears.
- Some reciprocating components, such as engine connecting rods.
- Very high pressure vessel ends and cylinders.

Their application has the following points in common:
- They are high strength components.
- They can withstand shock loading.

ANALYSIS AND PROPERTIES

A straightforward Carbon-Manganese steel pressure-part forging (BS 1503) **(1)** will have properties broadly within the following ranges (depending on grade).

Chemical analysis		*Mechanical properties*
C $\leqslant$ 0.3%	Cr $\leqslant$ 0.25%	Yield (R_e) 200–340 N/mm^2
Si 0.1–0.4%	Mo $\leqslant$ 0.1%	UTS (R_m) 410–670 N/mm^2
Mn 0.8–1.7%	Ni $\leqslant$ 0.4%	Elongation (A) $\geqslant$ 15%
P $\leqslant$ 0.03%	Cu $\leqslant$ 0.3%	No impact requirement
S $\leqslant$ 0.025%		Typically such a material will have been normalized/tempered or quenched/tempered. It will be weldable.

A high strength alloy steel pressure-part forging (ASTM A-273) [ref (2)] will be slightly different (note the underlined areas).

Chemical analysis		*Mechanical properties*
C $\leqslant$ 0.4%	<u>Cr 0.8–2.0%</u>	Yield ($R_{p0.2}$) $\geqslant$ 825 N/mm^2
Si $\leqslant$ 0.35%	<u>Mo 0.3–0.5%</u>	UTS (R_m) $\geqslant$ 930 N/mm^2
Mn $\leqslant$ 0.9%	<u>Ni 2.3–3.3%</u>	Elongation (A) $\geqslant$ 14%
P $\leqslant$ 0.015%	The forged billet will have been vacuum degassed to remove undesirable gases.	Charpy impact $\geqslant$ 41J (average)
S $\leqslant$ 0.0.15%		

Traceability – you should find that:
- Small components are batch manufactured. Large ones are forged singly.
- Positive identification will be via a *heat number*. This will be linked to a record of the full material designation.

Large forgings will incorporate their own test-bars. There will be a sketch showing the orientation and location of these test-bars in the forging (see Fig. 4.10 for an example)

Fig 4.1   Basic information – forgings

CASTINGS

Castings are used for:
- Components which have an awkward shape or which are difficult to fabricate.
- Parts of equipment which are not subject to significant shock – loading.
- Components that have particularly thick sections.
- Valve bodies, turbine casings and many large, thick-walled components.

Major criteria of choice for a casting material are temperature, tensile properties and corrosion resistance. Castings for pressure-part purposes have special specifications such as BS 1504 **(3)** or ASTM A 487 **(4)** and A 703 **(5)**.

ANALYSIS AND PROPERTIES

An example of a 'high strength' Cr–Ni steel casting (BS 1504–425 C11) would be:

Chemical analysis	*Mechanical properties*
Cr 11.5–13.5%	Yield (R_e) $\geqslant$ 620 N/mm^2
Ni 3.4–4.2%	UTS (R_m) $\geqslant$ 770 N/mm^2
C $\leqslant$ 0.1%	Elongation (A) $\geqslant$ 12%
Mn $\leqslant$ 1.0%	Charpy impact $\geqslant$ 30 Joules minimum
P $\leqslant$ 0.040%	(average of 3 tests)
S $\leqslant$ 0.040%	Hardness 235-320 HB (Brinell)
Mo $\leqslant$ 0.6%	
	Note that this is stronger than an austenitic grade
Note how the trace elements contents are specified as *maxima*	

An austenitic grade would have approximately 17–19% Cr and 8–10% Ni (sometimes with 3% Mo eg. ASTM A 351–CF8)

Traceability–you should find that:
- There is a chemical analysis and mechanical test performed for each cast.
- Positive identification will be via a *cast number* and a *heat treatment* number; nearly all castings need heat treatment.
- The location of test-bars is an area for discussion between manufacturer and purchaser. Technical standards do not provide detailed guidance. It needs to be decided because the location and orientation of the specimens can affect significantly the results, particularly in large castings.

Fig 4.2   Basic information – castings

PLATES

Plates are produced by rolling and have wide application:
- Their most common usage is in sections > 600mm wide and < 16mm thick – for vessels and general fabrication.
- High-pressure applications can use plate up to 250mm thick – but this is a specialized area.
- Most plate is weldable and can be 'cold-formed' to shape in the fabrication works. It can also be 'spun' to make dished ends for vessels.

ANALYSIS AND PROPERTIES

Chemical analyses and mechanical properties vary significantly, depending on the type of plate. Basic plate is generally classified as 'Carbon' or 'Carbon-Manganese' plate and is covered by the EN 10130/BS 1449 **(6)** range of standards – this plate is for general use at non-elevated temperatures.

An indicative set of properties is:

Chemical analysis	Mechanical properties
C 0.08%–0.2% (depending on the grade)	Yield (R_e) 170–350 N/mm^2
Mn 0.45%–0.9% (the lower Mn levels give better formability)	UTS (R_m) 280–500 N/mm^2
	Elongation (A) $\geq$ 16%
S 0.03%–0.05%	Charpy impact – generally not specified.
P 0.025%–0.06%	Hardness 130-270HV (before hardening)

Higher-strength C–Mn steels have a higher Mn level, up to about 1.5%Mn. Note that C–Mn steels are available in a variety of conditions which are described by letter 'designations'. The main ones are:

R: a 'rimmed' steel: this has a skin almost free from carbon and impurities, produced by controlling deoxidation during manufacture.

K: a 'killed' steel: this has been fully deoxidized.

H: this is a temper designation – there are six grades H1–H6 denoting various temper hardnesses.

Steel plates for higher temperature (> about 120°C) applications are described by different technical standards. Boiler plate (which is an 'alloy steel') is covered by BS 1501 **(7)**.

Traceability – you should find that:
- Complete plates will be traceable to a mill certificate.
- For complete traceability, hardstamped identification marks should be transferred, whenever a plate is cut for manufacturing (see Fig. 4.6).
- Plate for pressure vessel use is normally ultrasonically tested to BS 5996 **(8)** to check for laminations (see Chapter 5).

Fig 4.3   Basic information – plates

TUBES

The terms 'tubes' and 'pipes' are almost synonymous – there is no easy definition of the difference. There are different types of tubes for each engineering duty – specific applications such as boiler tubes, heat exchanger tubes, and high pressure tubes have dedicated technical standards. Tubes for light duties are often formed from rolled strip and then longitudinally welded. High pressure and high temperature tubes are usually manufactured by forging, or a related 'rolling' process such as pilgering. A major criterion of the choice of tube steel is whether it needs to have elevated temperature properties.

ANALYSIS AND PROPERTIES

A typical carbon steel tube of dia 120mm for use in boilers would have the following characteristics:

Chemical analysis	*Mechanical properties*
C $\leqslant$ 0.16%	Yield (R_e) $\geqslant$ 195 N/mm^2
Mn 0.3%–0.7%	UTS (R_m) 320–480 N/mm^2
Si $\leqslant$ 0.35%	Elongation (A) $\geqslant$ 25%
P $\leqslant$ 0.04%	Tubes are also subject to a flattening
S $\leqslant$ 0.4%	test and a drift expanding test

This steel could be used for welded or seamless tube. Note how the low C level causes a relatively low Re but moderate ductility (A $\geqslant$ 25%). The common standard for this application is BS 3059 grade 320, which is similar to ISO 2604 **(9)**.

Tubes are subject to a range of tests during manufacture – such as eddy current testing for defects and a hydrostatic test.

Traceability – you should find that:
- Most tubes are manufactured in batches of up to 200 tubes so you cannot always expect tubes to have *individual* traceability. Test samples will be taken from tubes selected at random from each batch.
- Batch identification will be via a *heat number*, which should be quoted on the material certificate and hard-stamped on the tube itself.
- Large forged tubes should have individual traceability and their own test-bars.

Fig 4.4 Basic information – tubes

analysis and mechanical tests within each of the categories. The categories also reflect the way in which you will meet material documentation in a works situation, and the allocation of activities between material suppliers. I have used these categories to present some basic technical information in Figs 4.1 to 4.4. Accept that this is highly distilled information – you can use it as a 'lead in' to more detailed information via the technical standards that I have shown. Remember that works inspection is more concerned with material verification than material selection.

Acceptance guarantees

For the majority of the materials of construction included in a plant contract, there is no separate acceptance guarantee, as such. There will be, as explained in Chapter 3, a set of material standards raised directly, or inferred, by the contract specification and its attendant document hierarchy. The requirements of these standards effectively form the acceptance criteria for the materials of construction. Practically, the scope and clarity of material standards are very good. You would find it an arduous task to write a better and more sustainable set of 'controls'. Occasionally (perhaps 10 percent of the time) you *will* find special and overriding material performance requirements stated in the contract specification. These may take the form of explicitly stated values for chemical analysis or mechanical properties, *or* references to more specialized, less well-known standards. You will see industry-specific standards such as the NACE (National Association of Corrosion Engineers) series referenced like this. Some of the more common special requirements are:

- *Corrosion tests.* These are specialized tests to determine resistance to intercrystalline corrosion of austenitic stainless steel.
- *Compound distribution tests.* The most common one is the sulphur print, a chemical process used to obtain a visual print of the distribution of sulphur compounds in a material specimen. A similar type of 'macro-etch' test is carried out on cast ferritic and martensitic steels where there can be strict restrictions on total residual aluminium content. A test piece is taken from a 'heavy-section' location after heat treatment. The material surface is then etched with hydrochloric acid and inspected for aluminium compounds.
- FATT (fracture appearance transition temperature) test. This is an extension of the impact test to determine the relative amounts of brittle and ductile failure areas on a fractured specimen.

It is good practice always to check the contract specification for such test requirements. Concentrate on components that are subject to unusually high or low operating temperatures or are designed for corrosive process conditions (anything more corrosive than sea-water and you may start to see such special requirements).

Specifications and standards

I have mentioned that contract specifications tend to control materials of construction by referencing material standards. Material standards therefore have a more prescriptive influence on the final state of manufactured equipment than do some other less definitive types of equipment standard. This is not absolute – there are several areas of material usage where manufacturers like to develop their own materials, based on their experience. There is nothing inherently *wrong* with this but be prepared to put in a little extra effort to understand the situation. Note the following guidelines when faced with manufacturers' 'own materials':

- Try to identify the nearest published material standard to the manufacturer's own material. Use the elemental chemical analysis as a comparison first – then compare the heat treatment and any finishing processes.
- Make a careful comparison of the mechanical properties. Expect the 'own material' manufacturer to claim that that material is better than the nearest published standard. Ask how.
- For a material to be 'superior' to a published material, it is necessary that all the properties should be better. It is not always acceptable, for instance, to achieve better strength (yield or UTS (Ultimate Tensile Strength)) at the expense of ductility (percentage elongation or percentage reduction in area) or impact toughness, or vice-versa. Watch for this.
- Remember that there are other, less definable, aspects to material properties than those common ones that you see in material specifications and standards. Fatigue and creep are specific examples – corrosion and erosion resistance are general ones. You can ask the material manufacturer to explain how these have been taken into account.
- The areas where you can expect to meet manufacturers' own material specifications in a conventional power generation or process plant are reasonably well defined. Practically, they are:
  - Large castings that are subject to high temperatures, such as turbine casings and steam valve chests.
  - Cast materials that have been developed specifically for increased corrosion resistance, such as valves for sea-water and aggressive process service.
  - Forged materials that are designed to operate at temperatures

above 650 °C, particularly in gas turbine and process plant applications.

You are much less likely to find manufacturers' own materials in statutory boilers and vessels, stainless steel items, equipment which is used for non-specialized water service, low temperature engine components and proprietary 'catalogue' items of engineering equipment. A less judgemental approach is acceptable when dealing with these items as they generally rely on well-known published material specifications. Here, the acceptance criteria are well defined – most mechanical inspectors would concur that the level of definition of technical information is much better in published material standards than it is in many equipment standards. Perhaps this is because the technical scope is tighter for materials than for equipment. You can't afford to be haphazard about the way that you *use* published material standards. They are configured in a particular way, based on the results of many years development. Fortunately, most American and European standards that you will need follow similar structure and principles. A careful, perhaps rather deliberate, approach works the best. Try to incorporate the following steps:

- First get access to a good set of reference documents. There are two main ones I can recommend:
    - The *BSI Standards Catalogue* **(10)**. This covers British and corresponding international and European standards (ISO/EN, etc.). All standards are referenced, not just those dealing with materials. There is an invaluable subject index.
    - *The annual book of ASTM standards* **(11)**. The ASTM standards are published in a series of volumes so the best document to have is the *Subject Index.* This will enable you to identify quickly the relevant standard. The most relevant ones for works inspectors are Volume 1 (iron and steel products), Volume 2 (non-ferrous metal products), Volume 3 (metal test methods) and Volume 6 (paints). The ASTM standards are referenced in the ASME boiler and pressure vessel codes which impinge directly upon works inspection activities.
    - As a 'back-up' be aware of the German reference document 'Stahlschüssel' **(12)**. This lists and identifies the majority of ferrous materials manufactured throughout the world. It also shows equivalence between material designations (which can sometimes be confusing). I have used this book many times to identify obscure material references. It works well.

- Use the cross-references. If you look *inside* the front or back page of a material standard you will find a list of cross-referenced standards. Because of the way that standards have developed and 'spread' over the years it is almost essential to review these standards to see how they impinge upon your FFP verification. Check the 'nearest' ones carefully – watch for special requirements relating specifically to the four main material manufacturing processes (forging, casting, sheet, and tube) that I mentioned earlier. Only revert back to the 'main' standard when you are sure that you have not missed a key point that could compromise the accuracy of your FFP verification.
- Don't ignore superscripts, subscripts, and footnotes. They are a feature of material standards. It seems that many of them involve quite important changes in the 'sense' of the information – particularly chemical analysis and heat treatment. Be careful also about the notes concerning the way in which acceptable mechanical properties are allowed to vary depending on the size of section or thickness of the material – there are some significant differences.
- Note material *grade designations* slowly and carefully. One number or letter can make a big difference (for an austenitic Cr-Ni steel casting alloy, for instance, the difference in acceptable UTS and elongation values between BS 1504 grade 304C12 and BS 1504 grade 304C17 is around 10 percent and the carbon and chromium contents are very different. Heat treatment condition is also important – always double-check that you are referring to the correct table – and watch out for those footnotes.

It is a fact that the scope of published material standards is very wide. You will find some duplication between the American ASTM range and the European standards, and within different European standards themselves. I would also be surprised if you could not find, in some places, technical contradictions. You are almost forced to adopt a rather detached view of this – it *is* a specialized area, so elect to follow the guidance of a published standard, even if an expert tells you that it is wrong. For, I promise, the last time; works inspection is about material *verification*, not material selection. Don't get too involved.

Happily, in day-to-day works inspection activities you will not often get involved in areas of potential uncertainty relating to complex or unusual material standards. There is an informal 'top twenty' list of material standards which appear again and again in plant contracts. It is not entirely true to say that the list will be the same for power generation, chemical plant and offshore industry projects, but you will

60   Handbook of Mechanical Works Inspection

find commonality, especially for basic equipment such as pumps, pipework, boilers and general pressure vessels. Incidentally you may recognize (if you have read Chapter 3) that these tend to be the technical areas which have 'technology-leading' standards, i.e. where manufacturers' practice is led *by* technical standards, rather than the other way round. As a practical guide that you can use, I have shown a top twenty list for a typical combined cycle power generation project in Fig. 4.5.

These are typical standards used for a combined-cycle power generation plant.

Component	*Material reference*
Gas turbine	
• Turbine blading	'Nimonic 115' **(13)**
• Turbine discs	'Waspalloy' (Wk 2.4654) **(14)**
• Casing (hot end)	'Hastelloy 'X' **(15)**
• Combustion chambers	Inconel 617 (Wk 2.4663) **(16)**
Steam turbine	
• Rotor	ASTM A470 **(17)**
• Casing	DIN 17245 **(18)**
• Stop valve casing	ASTM A356 **(19)**
Waste heat recovery boiler	
• Drums	DIN 17 155 **(20)**
• Low temperature headers and tubes	BS 3604 **(21)**
• Intermediate temperature headers and tubes	BS 3605 **(22)**
• Superheater headers and tubes	BS 3059 **(9)**
High pressure steam system	
• Cast valve casings	ASTM A487 **(23)**
• Forged pipework	ASTM A430 **(24)**
Seawater system	
• Condenser tubeplate	BS 2875 **(25)**
• Cast valve casings	BS 1400 **(26)**
• Valve internals	AISI 316L **(27)**
• Pump casings	BS 3468 **(28)**
Gearboxes	
• Gear wheels, pinions, and shafts	BS 970 Part 1 **(29)**
General water service	
• Cast iron components	BS 2789 **(30)**
• Pump shafts	BS 970 **(31)**

Fig 4.5   The 'top twenty' material standards

Inspection and test plans

References to materials of construction are made in all ITPs for manufactured engineering equipment. With the possible exception of mass-produced items (i.e. catalogue products) the inspection and verification of materials is an important and commonplace activity. Strictly, those lines in an ITP concerned with material verification cover not only the technical aspects of compliance with published standards, but also the important *procedural* concept of traceability. Technical aspects aside, traceability is at the heart of material verification – it is fair to say that much of the content of a good ITP is configured with this in mind. Let us look at traceability.

Material traceability

You will need to guard against the whole issue of material traceability looking, to you, like a paperwork exercise. This is dangerous, because fundamentally it isn't. It is about *control*.

Traceability – to achieve what?

The purpose is to achieve positive identification, so that buyers know they are getting the material that has been specified and that industry has spent tens of thousands of work-hours studying, understanding, and improving. With this as the objective, the question becomes one of how, in the world of real manufacturing industry with all its pressures and personalities, can such positive identification be achieved? Here is the only answer that has developed and it is in three parts:

- Material is fully examined, tested and classified at the source of manufacture – in the foundry, forge or mill.
- The material carries *source* certification when it is sold.
- The system that follows this material to its final use is a *documentation* one. There *can be* some physical corroboration of what the documentation says.

Looking at this, you can see that it is essentially a documentation system which is given *solidity*, not exactly 'validity', by the existence of physical observations. Don't confuse this with an *administrative* system where documentation is more of an end in itself – in material traceability, the documents only form the means to the end. This demonstrates the absolute necessity of a predetermined schedule of witness and review points, to maintain the efficacy of the relationship between the documents and the pieces of material to which they apply. These witness points are shown in the ITP.

Figure 4.6 shows the 'chain of traceability' which operates for engineering materials. Note that although all the activities shown are available for use (i.e. to be specified and then implemented) this does not represent a unique system of traceability suitable for all materials. In practice you will find several 'levels' in use, depending both on the type of material and the nature of its final application.

Levels of traceability: EN 10 204

The most common document referenced in the material sections of ITPs is the European Standard EN 10 204 **(32)**. It is widely accepted in most industries and you will see it specified in both American and European based contract specifications. It provides for two main 'levels' of certification : Class 3 and Class 2 (see Fig. 4.7). Class 3 certificates are validated by parties other than the manufacturing department of the organization that produced the material – this provides a certain level of assurance that the material complies with the stated properties. The highest level of confidence is provided by the 3.1A certificate. This requires that tests are witnessed by an independent third-party organization. You should find however that the 3.1B is the most commonly used for 'traceable' materials.

Class 2 certificates can all be issued and validated by the 'involved' manufacturer. The 2.2 certificate is the one most commonly used for 'batch' material and has little status above that of a certificate of conformity. Although you will find increasing instances of customer requirements for such 'certificate of conformity' level of documentation, this is strictly not a part of the material traceability chain. It is a manufacturer's statement that may (or may not) be capable of validation under scrutiny. I think you should look for better evidence of traceability than certificates of conformity if you want to do a good job.

Retrospective testing

In cases where 'full traceability' of material (as in Fig. 4.6) is clearly specified, perhaps for statutory reasons, there is little room for manoeuvre if you find at an advanced stage of manufacture that the chain of traceability is incomplete. I will suggest that this is not too uncommon. One solution is to do retrospective tests.

Materials of construction 63

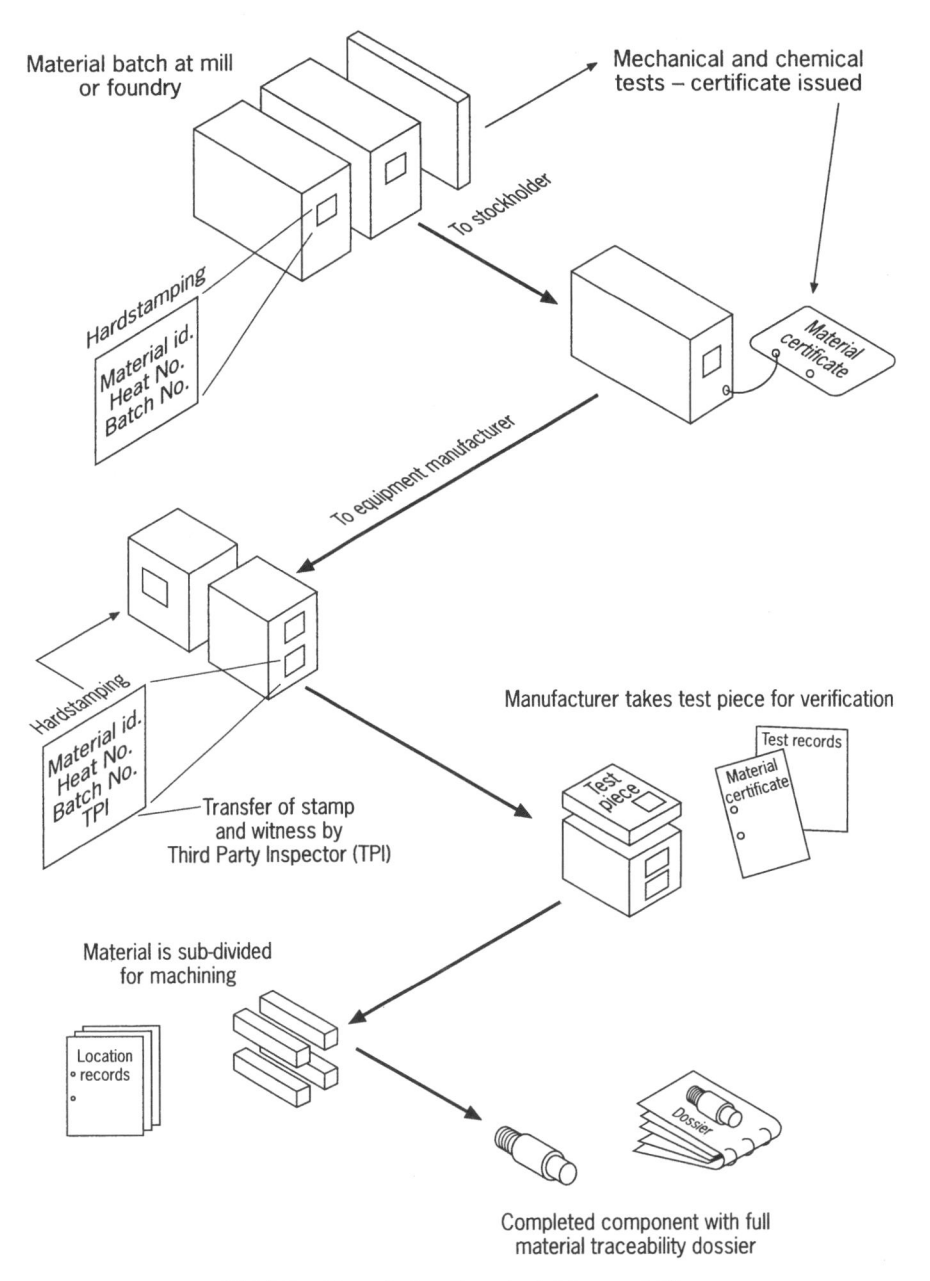

Fig 4.6 The chain of material traceability

EN 10 204 Certificate type	Document validation by	Compliance with: the order	Compliance with: 'technical rules'*	Test results included	Test basis Specific	Test basis Non-specific
3.1A	I	•	•	YES	•	
3.1B	M(Q)	•	•	YES	•	
3.1C	P	•		YES	•	
3.2	P + M(Q)	•		YES	•	
2.3	M			YES	•	
2.2	M			YES		•
2.1	M	•		NO		•

I – An independent (third party) inspection organization
P – The purchaser
M(Q) – An 'independent' (normally QA) part of the material manufacturer's organization
M – An involved part of the material manufacturer's organization
*– Normally the 'technical rules' on material properties given in the relevant material standard (and any applicable pressure vessel code).

Fig 4.7 Levels of certification

Some guidelines on retrospective material tests:

- The real objective is to identify and classify *fully* the material, not to carry out a few 'placebo' tests to save making proper decisions.
- In-situ tests such as 'metalscope' (an approximate method of elemental analysis) cannot differentiate accurately between grades of some materials. These tests give an indication only.
- To be really sure about a material probably means cutting a test piece from the component in question. This is often actually much easier than the various parties (particularly the equipment manufacturer) think it will be.
- If you are not convinced that the original ITP has been followed correctly, then you are justified in requesting retrospective material tests. Check that you are supported by the text of the contract specification: you should be.

Responsibilities

The party responsible for correct material traceability should be defined in the contract specification. It is normally the main contractor but, in practice, because of the structure of sub-contracts, the responsibility is frequently delegated. As an inspector you may be surprised by the way that material traceability requirements disappear or change within a 'chain' of two or three sub-contractors. I cannot start to offer an explanation as to why this happens, but it often does. Here is a useful mini-checklist of points related to the *responsibilities* of effective material traceability. I introduce them here because they underlie the way in which responsibilities are shown in the ITP.

- First, check the text of the sub-contract orders. This is normally where the traceability requirements have been 'mislaid'. There are few manufacturers that will do voluntarily what they are not contracted to.
- Check the role of the third party inspection (TPI) body. The TPI is the 'nearest' to the core witnessing activities at the foundry or mill before material starts being cut, welded, and machined. Help, and encourage, the TPI to pay extra attention to material verification at this stage.
- Monitor manufacture. The traceability of material during and after cutting and machining is the task of the manufacturer. The TPI has less capacity for observation and influence once manufacture has started. Effective works inspection involves managing these parties in the best way – at each stage of the manufacturing process.
- Investigate discontinuities. Discontinuities in traceability are nearly always a 'system fault' rather than an isolated incident (ask any BS EN ISO 9000 system assessor). You have to get to the root *cause* of discontinuities in the chain of material traceability, not just report that they exist.

ITP content

A relatively small number of entries are required in the ITP. They must be complete, concise, and absolutely accurate to provide the best vehicle for traceability. Figure 4.8 shows a model of the content that I recommend. In this case, it is for components of a waste heat boiler constructed to a pressure vessel code **(33)**. Format, as long as it is clear and unambiguous, is less important than content. I do not think that it is possible to omit any of the elements I have shown without reducing the effectiveness of this type of ITP. Look at the annotations that I have added to the figure. These should reinforce the 'guideline' points that I made earlier in this chapter. There are four levels of traceability shown, covering most of the situations you are likely to meet in your works inspection role. I have shown the highest 3.1A traceability level at the top of the table for emphasis – try to give extra attention to this category.

As a final point on 'material' aspects of ITPs, remember that the

* *It is essential to reference the applicable standard and acceptance levels* ↓

* *Note the role of TPI for traceable 'statutory' material* ↓

	Step No.	Operation	Inspection points M	C	TPI	Certificate requirements	Comments
High ↑ level of material traceability ↓ Low		**Boiler header tube**					
	1.	Transfer of marks	W	R	W	EN 10 204 (3.1A)	
	2.	Mechanical tests	W	W	W		
	3.	Chemical analysis	W	W	W		
	4.	Hydrotest	W	W	W		Note any
	5.	Tube NDT	W	W	W		NCRs in this
	6.	Documentation review	W	R	R		column
		Safety valve spindle					
	1.	Mechanical tests	W	W	R	EN 10 204 (3.1B)	
	2.	Chemical analysis	W	W	R	Certificates	
	3.	NDT	W	W	W		
	4.	Documentation review	W	W	W		
		Boiler structural Steelwork					
	1.	Mechanical tests (sample)	W	W	R	EN 10 204 (2.2)	
	2.	Chemical analysis (sample)	W	R	R	certificate for carbon steel	
	3.	Document review	W	W	R		
		Steam pipe expansion joints					
	1.	Document review	W	R	–	Certificate of conformity	

* *Non-traceable 'catalogue' item* ↑

* *Spell out clearly which level of certification is required* ↑

W = Witness point   M = Manufacturer   C = Contractor   R = Review
TPI = Third Party Inspector

↑ * *Include a key so there is no misinterpretation.*

Fig 4.8   ITP content: material tests and traceability

entries made have to link in with the clauses of the contract specification – this was discussed as a general principle in Chapter 3. You should check that this is the case, particularly for fully traceable 'statutory' materials. It makes the inspector's implementation role that much easier. Try to improve things at the draft or 'marking up' stage of the ITPs when it is still possible to make the changes you want.

Test procedures and techniques

Material testing techniques are well proven. They feature in many engineering reference books and are well documented in both American and European technical standards. Common destructive tests are used to determine the mechanical properties of a material – one of the principal fitness-for-purpose criteria (destructive tests of welds are covered in Chapter 5).

Effective inspection involves considering material tests in a particular way. There are two ways in which you may become involved, either to witness the tests themselves (and so satisfy yourself of the accuracy and validity of the results), or more simply in just a *reviewing* capacity – checking the test results. To witness the tests is fine, it may help to keep the test laboratory staff alert. The problem, however, is that it is rarely the test procedures or activities themselves that reveal any potential problems with FFP. This is because most of the technical *risk* is not in the test, it is in the choice of specimens. It is perfectly feasible for a set of well specified, executed and documented mechanical tests to be unrepresentative of the properties of the component from which the test piece was taken – if the specimens are badly chosen. It will pay benefits if you can bear in mind this rather slanted viewpoint when considering your involvement in material testing. Partly for this reason, I have provided only outline details of the actual methodology of the common destructive tests (you can easily look them up in the relevant standards) but more detail on the choice of specimens and a few other peripheral issues. This is where you will find most of the latent problems.

I doubt whether you will find many non-conforming results when watching the instruments during destructive tests in the laboratory – good manufacturers will have satisfied themselves of the test results before you arrived. That said, let us look at the way you can expect to see test results presented to you.

Test results – presentation

What should you expect to see on a material certificate? Normally this is quite straightforward, and Fig. 4.9 shows a typical example, in this case for an austenitic steel casting: a material broadly known as a 2½ percent molybdenum steel. Although it may appear simple, a few points are worthy of note:

- *Acceptance levels.* The field of materials science is so wide that it does not make sense to try and anticipate specified values for either the chemical analysis or mechanical properties. Surprisingly, you will see a lot of 'material certificates' that do not show specified values, instead they may just quote a material specification number (or sometimes not even that). Unless you are very experienced with a particular material you should make a point of checking with the material standard itself – compare the data carefully, making sure that the test results meet the specified acceptance values for the *particular* grade and heat treatment condition that you want. Be careful not to confuse minimum and maximum specified values.
- *Units.* The majority of material certificates will use the SI system of units. You will see tensile strengths expressed in MPa, MN/m^2 or N/

This is a '2½ Molybdenum steel' austenitic casting material

Chemical analysis

	C%	Si%	Mn%	P%	S%	Cr%	Mo%	Ni%
Specified (BS 1504 316 C12)	0.03 max	1.5 max	2.0 max	0.040 max	0.040 max	17–21	2–3	10 min
Actual	0.025	1.35	2.0	0.038	0.036	18	2.8	11

Mechanical properties   *Note these marginal results*

	$R_{p\ 1.0}$ (N/mm^2)	R_m (N/mm^2)	A (%)	Impact (J)	HB
Specified (BS 1504 316 C12)	215 min	430 min	26% min	–	–
Actual (room temp)	220	430	51%	–	–

Fig 4.9   Typical material certificate content and presentation

mm² – they all mean the same. Elongation and reduction of area are always expressed as a percentage figure. Note that tensile results are only relevant if the correct specimen size has been used. A good material test results document will show this; look for a symbol like '$L_o = 5.65 \sqrt{S_o}$' which indicates the specimen gauge length and diameter. Impact values are always given in joules (J).

- *Temperature classification.* Check the temperature at which the mechanical tests were carried out and compare it with that required by the standard. The material designation is normally a clue, for instance, BS 1504 castings with special low temperature properties have the suffix 'LT' after the material code. A series of digits may also be included to specify the actual temperature at which tests have to be carried out. If there is no explicit temperature information given, you are safe in assuming that all the tests and acceptance levels are referenced to ambient temperature (20 °C).

- *Impact tests.* Not all materials have a specified impact value. Materials which are chosen for temperatures applications above 150 °C rarely need it, unless they are subject to shock loading conditions. For the alloy shown in Fig. 4.9, the standard only requires impact tests if a low temperature (LT) grade is specified.

- *Marginal results. Always* look carefully at marginal results – this means chemical analysis or mechanical test results that are *just within* the acceptance limits, or fall exactly 'on the limit'. Whenever you see a marginal result (look at the Mn content, and the $R_{P1.0}$ and R_m results in the example), it is time to dig a little more deeply. In practice you are almost obliged to accept the validity of a marginal chemical analysis result (unless you are willing to specify a re-test). For marginal mechanical test results, however, there are several steps that you can take.
    - Do a Brinell hardness test on the component (this can be carried out in-situ using a portable tester) and make an approximate conversion to tensile strength using the rule of thumb shown under the hardness testing part of this chapter. Be wary if you see results which indicate the material is significantly softer (lower strength) than the marginal test results suggest.
    - If it is a graded material, look at the next grade down (lower strength) and compare the strength and the elongation values. Is there a big difference in the elongation values? – this can give you a 'pointer'. Note in the example how the marginal $R_{P1.0}$ and R_m are accompanied by an elongation of almost twice the minimum level. Whilst the material appears within specification

there is still a shadow of uncertainty. You can often see this pattern of results on steel sheet and strip material. This material is a good candidate for a re-test.

Tensile tests

This is the main destructive material test. European Standard BS EN 10002-1 **(34)** (replaces BS 18) provides a detailed technical explanation of the technique. An equivalent American Standard is ASTM A370 **(35)**.

From an inspection viewpoint one of the main points of concern is the origin and *orientation* of the test specimens. It is a fact that steels worked by all the main processes; forging, casting, rolling, and their variations, have some directional properties. This is due mainly to the orientation of the metal's grain structure. Large castings, which should in theory be homogeneous, still exhibit directional properties due to the complex mechanisms of cooling and shrinkage. Tensile properties can vary significantly (perhaps up to ± 25 percent), depending on the orientation of the test specimen with respect to the local grain direction. The shape of the load-extension characteristic can also vary. Some specifications, such as boiler and pressure vessel standards, specify clearly the orientation of tensile specimens but many others do not. Figure 4.10 provides a broad summary of the situation for the major types of engineering component you are likely to meet during works inspection – good manufacturers will have a diagram showing where test specimens have come from. A few other guidelines may be useful to you:

- *Gauge lengths.* There are a number of different specimen sizes and gauge lengths. Make sure the correct one has been used for the material in question. There is a standard on this, ISO 2566/2 **(36)** but you shouldn't need it very often, if at all.
- *Defined yield points.* Make sure you know whether a material can be expected to show a defined yield point (note that there may be two, R_{eL} and R_{eH}, but the lower yield strength, R_{eL}, is the important, commonly used one). It is almost impossible to tell this from the chemical analysis alone – look at the mechanical properties section of the material standard to see whether it expresses a yield value, or resorts to a proof stress measurement.
- *Don't forget traceability.* Tensile specimens should be identified by documentary records and hardstamping. Piles of unidentified specimens in the test laboratory do not inspire confidence. If in doubt ask for a re-test.

Materials of construction 71

Castings

- Small castings tend to have a homogeneous structure so a unidirectional test coupon is acceptable.
- Large castings (e.g. turbine casings) do have directional properties so X, Y, and Z axis specimen orientations are needed.

A cast-on test block

- A separately-cast test block is often used

Plates

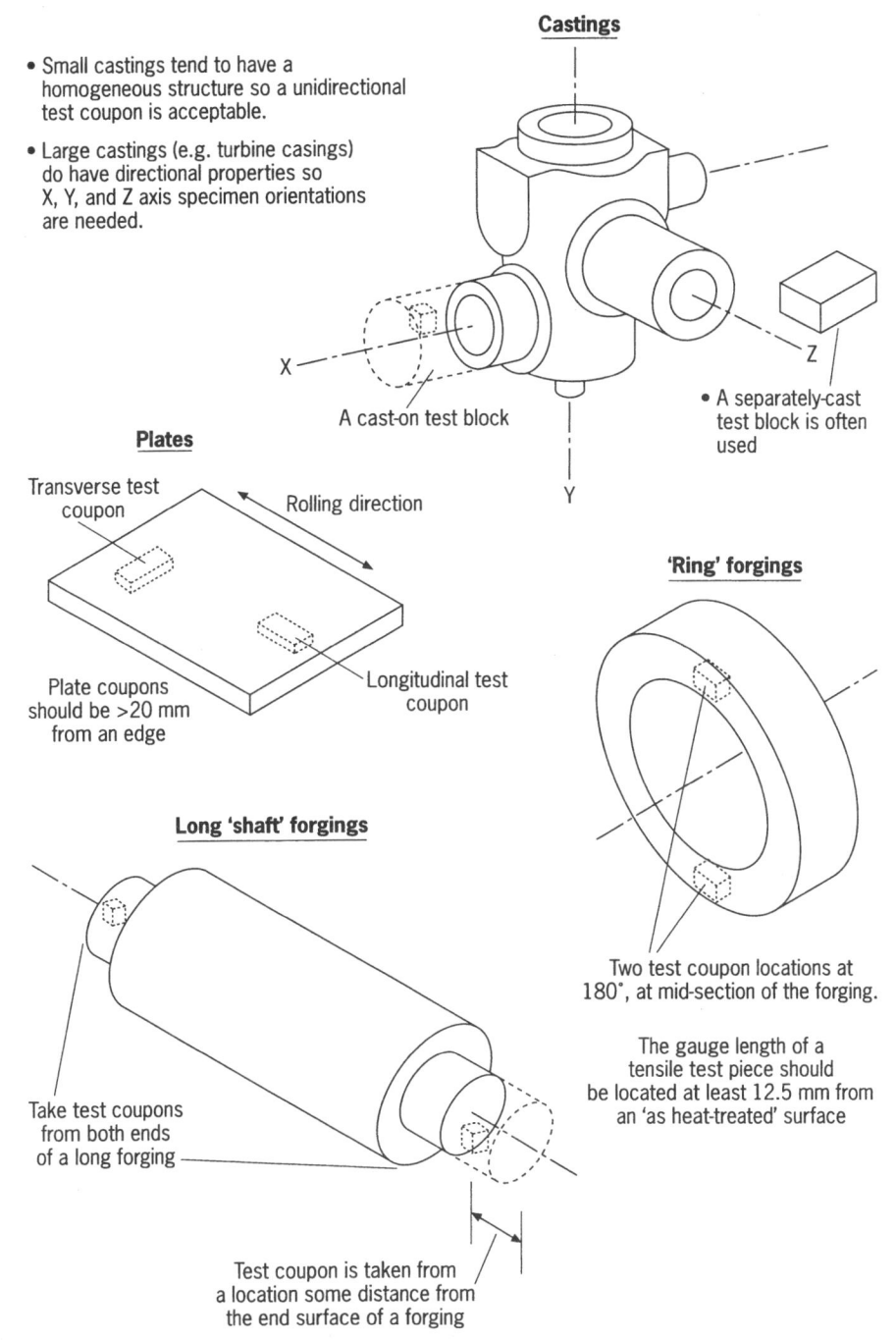

Transverse test coupon

Rolling direction

Plate coupons should be >20 mm from an edge

Longitudinal test coupon

'Ring' forgings

Two test coupon locations at 180°, at mid-section of the forging.

The gauge length of a tensile test piece should be located at least 12.5 mm from an 'as heat-treated' surface

Long 'shaft' forgings

Take test coupons from both ends of a long forging

Test coupon is taken from a location some distance from the end surface of a forging

Fig 4.10 Test samples – a quick guide

Tensile strength: easy rules of thumb?

Are there any easy 'rules of thumb', or convenient and concise tables of comparative material properties that can be used as a quick reference list to compare tensile strengths during a works inspection? Unfortunately the field is just a little too wide, at least as far as ferrous materials are concerned. Because of the large number of different grades in most material categories, there is a large, sometimes total, degree of 'overlap' between material types. Grades of cast iron, for instance can have tensile strengths ranging from 200 N/mm^2 to 800 N/mm^2. Carbon steels and stainless steels extend throughout a similar range. I introduce this point only to demonstrate the complexity of materials science and to try and justify why there are so few prescriptive rules. Unless you have a real tendency towards metallurgy (and most inspectors do not) I can only reinforce my earlier advice that you should learn to rely on published standards. Express mild surprise when you meet inspectors that don't.

Impact tests

Impact tests are specified for materials that experience low temperature, shock loadings or both. There are several types of test, but the one that you will meet nearly all the time is the Charpy V-notch test. This is described in the standard BS EN 10045-1 **(37)** 'notched bar tests' (there is no direct ISO equivalent) and ASTM E812 **(38)**. As with tensile tests the choice of specimen location and orientation is important. Figure 4.10 is equally valid for the location of impact test pieces for common material forms. It is important to realize that impact test results are inherently *less* reproducible than are tensile test results. This is due to the nature of the test itself – it relies upon a very accurately machined notch and is affected by inherent effects of material structure. Impact values have great sensitivity to minor variations in grain size and precipitates, hence small variations in heat treatment can cause different results. For these reasons, impact tests are always carried out in groups of three and the results averaged. Note two key points:

- *Test designation.* Although you will meet the Charpy V-notch test more than 90 percent of the time you may occasionally see Charpy U-notch or Izod tests specified. Note that there is *no conversion* between these impact results. If the wrong test has been used, the only option is to re-machine correct test pieces and repeat the test.
- *Units.* The Charpy impact value is given in joules (J). This is the energy absorbed by the specimen in breaking and and it is specified as

a minimum acceptable value. There may sometimes be a maximum quoted, mainly for alloys where ductility (as represented by reduction of area and elongation) can be falsely represented because of the propensity of the material to rapidly work harden during a tensile test.

Fracture appearance transition temperature (FATT)

The FATT test is an extension of the activities involved in the impact test and is rather simpler than it sounds. The objective is to determine the temperature at which the material will become brittle and break by a brittle fracture rather than a ductile fracture mechanism. This temperature is quoted as the 'transition temperature' – it is essentially an indirect measure of the low temperature impact properties of the material. The test is performed by doing a series of impact tests (usually Charpy) at several temperatures, typically 0 °C, −20 °C and −40 °C. The impact values are recorded for each specimen.

A visual (and often microscopic) examination is made of all the fracture faces with the objective of describing the fracture surface in terms of the relative percentage *areas* of ductile fracture and brittle fracture. As the test temperature gets lower, the percentage of brittle fracture surface will increase. The FATT is defined (some would say approximated) as the test temperature at which these percentages are equal, i.e. 50 percent ductile fracture/50 percent brittle fracture. You will see that there is room for some uncertainty in this test. This is precisely what happens in practice – do not expect the FATT test to give precise results. It is often specified for highly loaded rotating equipment operating at near ambient temperature. As a guide, you should find that steel alloys which have high carbon, sulphur, silicon, or phosphorus levels tend to have a relatively high transition temperature. Manganese and nickel additives have the opposite effect – they lower the transition temperature.

Hardness tests

Hardness tests are very quick and can be carried out in-situ on a finished component as well as on a machined test piece. For in-situ tests it may be necessary to lightly grind the metal to provide a good surface. The hardness test consists of pressing a steel ball, diamond, or similar shape into the surface of the metal and then measuring either the force required, or the size of indentation for a particular force. Both are a measure of hardness of the material. There are several hardness scales, depending on the method used. The most common ones are Brinell, Vickers and Rockwell (B and C). Unlike impact tests, it is possible to convert readings from one scale to another. Figure 4.11 shows the approximate comparison scales for steel – note the abbreviations which designate the method used. More detailed data are given in the standard ISO 4964 which is similar to BS 860 **(39)**. Specific standards are available which describe the individual test methods – these are BS EN 10003 (for Brinell tests), BS EN 10109 (Rockwell), and BS 427 (Vickers).

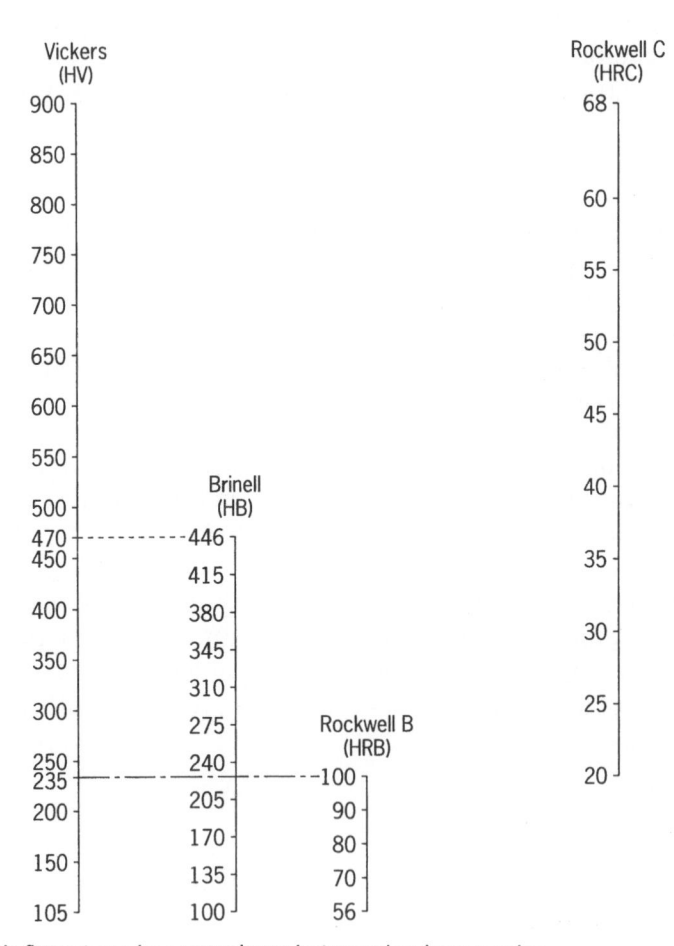

- Use this figure to make comparisons between hardness scales
- Treat the comparisons as approximate rather than exact
- It is often best to quote the HV hardness in your report – it is the most commonly used and understood

Fig 4.11 Comparison of hardness scales (for steel)

The most frequently used hardness test for common forgings and castings is the Brinell test. The Vickers and Rockwell C tests are more suitable for harder materials. Note the following specific points on the Brinell test:

- *Terminology.* You should see a Brinell result expressed like this:

    226 HBS 10/3000

    This is more easily read from right to left.

    - 10/3000 shows the ball size used (10 mm). The '3000' is a 'load symbol', which is expressed as a factor (0.102) × the test force in Newtons. The force used depends on the expected hardness of the material.
    - HB denotes the Brinell scale. S (or W) shows that a steel (or tungsten) ball was used.
    - 226 is the actual hardness reading. In laboratory tests it is determined by measuring the diameter of the indentation using a microscope and a special eyepiece incorporating a measuring aperture. With portable hardness testers, the result is recorded as a function of the indentation force and displayed directly on a digital readout.

- *Accuracy.* For a laboratory test with a machined specimen the test should be accurate to approximately ±2 percent (of the Brinell number). If you are using a portable tester it should be calibrated on a strip of test material before use. Even so, expect the accuracy to be a little less, perhaps ±3–4 percent. It is not advisable to rely on a single reading – many portable meters have an 'averaging' facility which will calculate the mean of 10–20 readings.
- *Castings.* For in-situ tests on cast components it is essential to grind off the surface layer where you plan to do the hardness test. If not, you will obtain misleading (harder) readings due to contamination and other surface effects.
- *A quick approximation to tensile strength.* For the Brinell test you can make an approximate conversion to the tensile strength using:
    - UTS (N/mm^2) $\simeq$ 3.39 × Brinell hardness number (HB)

This can be useful as an 'order of magnitude' check when the tensile properties of a material are questionable – it is a quick test, with no machined test piece required. Be wary of the accuracy though, hidden factors such as work-hardening rates and grain structure can distort the conversion. It is not very accurate for cast iron. If in doubt do a full tensile test.

Retests

Retests of material specimens are commonplace. Despite the efforts of technical standards to specify closely the methodologies of material testing to reduce uncertainties, the majority do make provision for retests if a material fails to meet its acceptance criteria during the first test. In a works inspection situation you may have to remind manufacturers of these provisions, in the unlikely event that they do not inform you first. For components with complex metallurgy such as high temperature castings and turbine rotors, manufacturers often have their own retest practices – this is acceptable as long as they exceed the requirements stated in the relevant standard.

The philosophy of the acceptance criteria for retests is different for tensile and impact tests. Tensile tests have better reproducibility and most acceptance decisions are made on the basis of one or two results, whereas impact tests use more of an averaging approach. Figure 4.12 shows the way (in this case for BS 1504 casting material) in which additional test specimens add cumulatively to the average. I suspect that this approach is based more on statistics than metallurgy. So, for retests:

- Check the material standard for the number of retests that are allowed.
- It is acceptable for components to be heat-treated again to try and improve poor mechanical properties. Pay special attention to the way in which the manufacturer plans to obtain the new test pieces (there may be no test bars left). It is advisable to positively identify test pieces with hardstamping before any repeated heat-treatment. As the material has already failed a test it is *unacceptable* to rely on unidentified test specimens. Make doubly sure that you ask the manufacturer about the origin of retest specimens. What you do not want to hear is uncertainty. Note the agreed retest piece location and orientation – as well as traceability details – on your non-conformance/corrective action report.
- Always issue an NCR if you find an incorrect material test procedure or an unacceptable test result. Let the retests or repeated heat-treatment follow later:
- Figure 4.13 is a useful 'aide mémoire' when witnessing material tests.

Materials of construction

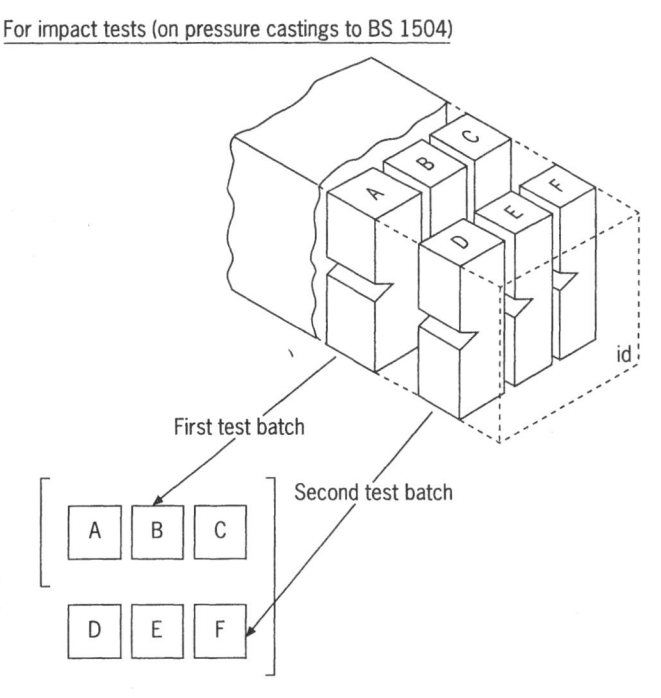

For impact tests (on pressure castings to BS 1504)

- If acceptance level = (say) 35J minimum. The material has failed the first test if the average (A+B+C/3) < 35J or if any one of A, B, or C < 70% of (35J). A retest is allowed; take 3 more test pieces DEF from the sample: the material has still failed if the average {(A+C+D+E+F)/6} < 35J or if more than two are < 35J, or if more than one is < 70% of (35J).
- Then the component may be re-heat treated a maximum of twice and the above tests repeated. If it still fails, no further activities are allowed.

For tensile tests

If the first specimen fails the test   Two more specimens are allowed

If any one still fails

Two re-heat treatments allowed then retest

Fig 4.12   Re-tests on castings

1. *For all tests.* Make sure you know the location, orientation and number of test pieces that were taken.

2. Tensile tests. Standard symbols should appear on material test certificates. These are:
   - R_e Yield strength (R_{eL} is the 'lower' yield strength, R_{eH} is the 'higher' yield strength)
   - $R_{P0.2}$ 0.2% proof strength.
   - R_m UTS or 'tensile strength'
   - $A\%$ Percentage elongation of gauge length

Check that:
- The correct test temperature has been used. Some materials are tested at elevated temperatures
- The correct R_e or $R_{P0.2}$ measurement has been taken (look at the relevant standard).

3. Impact tests
   - Check the specimen size and notch configuration. There are several different types and it is not possible to convert the results.
   - Try to examine the broken specimens so you can describe the fracture surfaces in your report.
   - Check the contract specification to see whether it specifically asks for a FATT test.

4. Hardness measurements
   - Make sure the correct scale (HV, HB, HRB or HRC) is used.

Fig 4.13   Material tests – some useful 'aide mémoire' points

Materials of construction 79

KEY POINT SUMMARY: MATERIALS OF CONSTRUCTION

Fitness for purpose (FFP)

1. The three FFP criteria for materials of construction are:
   - Mechanical properties
   - Temperature capability
   - Positive identification

Material standards

2. Material standards often relate to a particular material *form* – forging, casting, plate, or tube.
3. Acceptance guarantees are heavily dependent on technical standards. The situation is complicated by manufacturers who use 'own specification' materials – I have provided some simple guidelines for dealing with this.
4. Some material standards contain complicated tables and footnotes – these merit careful reading because they affect significantly the technical content.

Traceability

5. Verifying material *traceability* is an important part of the works inspection role. Try to see traceability as a control mechanism, rather than a documentation exercise. Expect, at times, to find this difficult.
6. There are various *levels* of traceability. EN 10 204 is the accepted standard covering the type of material certificates corresponding to the different levels.

Material tests

7. For tensile and impact tests the location and orientation of the test piece in the component is important. This is because material properties are *directional*.
8. You can convert between hardness scales (Brinell, Vickers and Rockwell) but not between impact strength scales (Charpy and Izod).
9. *Retests* are allowed if materials fail their first mechanical tests. Check for positive identification of the test specimens.
10. Inspection is about material *verification* – not material *selection*. Don't get too deeply involved.

References

1. BS 1503: 1989. *Specification for steel forgings for pressure purposes.* A related standard is ISO 2604/1.
2. ASTM A-273/A273M: 1994. *Specification for alloy steel forgings for high strength pressure component application.*
3. BS 1504: 1984. *Specification for steel castings for pressure purposes.*
4. ASTM A487/A487M: 1993. *Steel castings suitable for pressure service.*
5. ASTM A703/A703M: 1994. *Steel castings, general requirements for pressure containing parts.*
6. BS EN 10130: 1991. *Specification for cold-rolled low carbon steel flat products for cold forming: technical delivery conditions.*
7. BS 1501: Part 3: 1990. *Specification for corrosion and heat-resisting steels, plate, sheet and strip.*
8. BS 5996: 1993. *Specification for acceptance levels for internal imperfections in steel plate, strip and wide flats, based on ultrasonic testing.* A related standard is Euronorm 160.
9. BS 3059: Part 1: 1993. *Specification for low tensile carbon steel tubes without specified elevated temperature properties.*
   BS 3059: Part 2: 1990. *Specification for carbon, alloy and austenitic stainless steel tubes with specified elevated temperature properties.* Related standards are ISO 1129, ISO 2604/2 and ISO 2604/3.
10. The BSI *Standards Catalogue*, ISBN O 580 25370 8, published annually by the British Standards Institution.
11. The annual book of ASTM standards
12. '*Stahlschüssel*', 1995. Verlag Stahlschüssel Wegst GmbH (key to steel).
13. 'Nimonic 115' is a trademark of the INCO family of companies.
14. 'Waspalloy' is a trademark of United Technologies Corporation. The nearest equivalent is material referenced as Werkstoff No. 2.4654.
15. 'Hastelloy X' is a registered trademark of Haynes International Inc. The Hastelloy range is a family of high-nickel alloys.
16. 'Inconel 617' is a registered trademark of INCO Alloys International, Inc. The nearest equivalent is material referenced as Werkstoff No. 2.4663.
17. ASTM A470. *Vacuum treated carbon and alloy steel forgings for turbine rotors and shafts.*
18. DIN 17 245: 1987. *Ferritic steel castings with elevated temperature properties; technical delivery conditions.*
19. ASTM A356/A356M: 1992. *Specification for steel castings, (carbon low alloy and stainless steel, heavy wall) for steam turbines.*
20. DIN 17 155: 1989. *Steel plates and strips for pressure purposes.*
21. BS 3604 Part 1: 1990. *Specification for seamless and electric resistance welded tubes.*
22. BS 3605: 1992. *Austenitic stainless steel pipes and tubes for pressure purposes.*

23 ASTM A487/A487M. *Specification for steel castings suitable for pressure service.*
24 ASTM A430/A430M: 1991. *Austenitic steel forged and bored pipe for high temperature service.*
25 BS 2875: 1969. *Specification for copper and copper alloy plate.*
26 BS 1400: 1985. *Specification for copper alloy ingots and copper/copper alloy castings.*
27 316L stainless steel is described in BS 970: 1991 (see reference 29).
28 BS 3468: 1986. *Specification for austenitic cast iron.*
29 BS 970 Part 1: 1991. *Specification for wrought steels for mechanical and allied engineering purposes.*
30 BS 2789: 1985. *Specification for spheroidal graphite or nodular graphite cast iron.*
31 BS 970 (in various parts): 1991. *Specification for wrought steels for mechanical and allied engineering purposes.*
32 EN 10 204: 1991. *Metallic products, types of inspection documents.*
33 The TRD boiler code (Technischen Reguln für Dampfkessel)
34 BS EN 10002-1: 1990. *Method of [tensile] tests at ambient temperature.*
35 ASTM A370: 1992. *Mechanical testing of steel products.*
36 ISO 2566/2 (equivalent to BS 3894: Part 2: 1991. *Method of conversion for application to austenitic steels.*
37 BS EN 10045-1: 1990. *The Charpy V-notch impact test on metals.*
38 ASTM E812: 1991. *Test methods for crack strength of slow bend precracked Charpy specimens of high strength metallic materials.*
39 BS 860: 1989. *Tables for comparison of hardness scales.*

Chapter 5

Welding and NDT

Introduction: viewpoints

Almost every works inspection involves non-destructive testing (NDT) in some form or other and much of this is related to welds in ferrous materials. Welding (and its metallurgy) and NDT are well-developed engineering disciplines, with the result that levels of staff competence and qualifications are high. Experts and their opinions abound during works inspections – it just seems that they prefer to wait for the inspector to make the decision. If you can develop a good working knowledge of NDT and apply it effectively, you undoubtedly have something useful to sell.

From an inspection viewpoint I think it is an advantage to consider welding and NDT as closely linked, almost synonymous disciplines. The main fitness-for-purpose criteria involved, and the inspector's role in their verification, are dovetailed closely together – there are strong technical and procedural links between them. A discussion about welding usually involves mention of the NDT aspects, and vice versa. The labour-intensive nature of welding and NDT raises the inevitable questions about the need for *supervision* of such processes. Is the inspector a supervisor? Should she interact with the manufacturer's staff, guide them, as well as assess them? I think probably not. You can find at least partial justification for this view by looking at the way in which manufacturing activity is itself structured. Despite BS EN ISO 9001, a manufacturer's own shop-floor inspectors or quality control co-ordinators still play a vital role in supervising welding and NDT activities – a type of closed system operates, with its own checks, balances and control functions. Within such a system (and most of them are like this), the division of labour is clear – the operatives or technicians weld, or test, whilst the inspectors and supervisors do not. It is wise to fit in with the way that this system works. It is easy, with only a small amount of knowledge, for an external inspector to intervene. Avoid doing this: try hard to develop a conceptual grip on what is happening then stand back. Keep your energies for activities that help you to *assess* and therefore verify FFP.

Fitness-for-purpose (FFP) criteria

The keyword is *integrity*. Weld integrity is always the overriding FFP criterion. It is an explicit objective of the relevant statutory codes and standards for all types of engineering equipment incorporating the jointing of metals. It is also an implicit requirement for non-statutory equipment because of its impact on the safety aspect – there is a direct link to manufacturers' duty of care to provide inherently safe equipment, and to their formal product-liability responsibilities. Your job as an inspector is to *verify* this integrity – using it as a focus in your inspection activities, in your reports, and in your non-conformance reports.

The difficulty with verifying integrity is that you can't see it – not directly. There are three parts to a good verification process. Treat them as sub-elements of FFP but remember that they are not 'stand-alone' items, they are all part of the picture. Consider them as having approximately equal weight and refine this (with care) as you develop experience of particular welding processes and applications. Here are the three sub-elements:

The correct welding technique

Every welding application has a welding technique that is the most suitable one for it. There is little that is 'new' in this field – the information is well documented and understood by manufacturing industry. For this reason a well established methodology has developed for documenting and controlling welding techniques. This comprises the use of weld procedure specifications (WPSs), weld procedure qualification records (PQRs) and the testing and approval of welders themselves. The purpose of this methodology is to ensure the correct 'design' of welding technique and then to control its implementation. We will look at this in more detail later.

The correct tests

There is only a limited amount that you can conclude from visual examination of a weld. Destructive tests and non-destructive tests are required to look properly for defects that affect integrity. Non-destructive techniques are essentially *predictive* tests. Because each technique can only detect certain types and orientation of defects, the choice of technique is critical – it always depends on the application. As an inspector you need to pay attention to the suitability of the NDT technique being used in order to understand the accuracy and validity of the results that will be achieved. This can be slightly tricky.

Acceptable levels of defects

Note the reference to *acceptable* levels of defects. The inference here is that it is necessary to compare the defects you find with the correct defect acceptance criteria for the equipment in question. Acceptance criteria are documented in many engineering standards but not always in the component-specific way that would be most helpful. Be prepared to exercise some judgement on the applicability of defect acceptance standards to the component you are inspecting – as a preparation for this it is best if you can develop a fair working knowledge of those acceptance criteria that are available – then apply your own experience.

I have tried to make it clear that these three FFP sub-elements, even when taken together, cannot provide an absolute verification or 'guarantee' of fitness for purpose. Any non-destructive technique based on predictive principles will inevitably leave an element of risk. Works inspection is about reducing this risk to the absolute minimum.

Basic technical information

There is a broad spread of information applicable to the field of welding and NDT. There is also a lot of depth to it – almost every aspect of welding technology has been studied extensively and there are literally thousands of technical papers available if you care to look. I introduced in Chapter 2 the principle of limiting your information requirements by only looking for what you need to know. This is a key feature of effective inspection and there is perhaps no better example to use than that of welding and NDT. Poor focus and then confusion await you if you try to go too far. There are a number of key subject areas, however, that you do need to look at – their purpose is to give you a feel for the basic processes and engineering activities involved, so take a broad view of them.

Welding techniques (and why defects occur)

No welding activity is perfect. Forget the notion that a closely controlled welding procedure backed up by a series of 'proving' tests will produce a good weld every time. If we exclude the complex welding techniques used in nuclear plant components and some very advanced offshore pipeline applications, most general power and process plant welding of ferrous materials fits into one of four main categories. These are manual metal arc (MMA), submerged arc welding (SAW), metal inert gas (MIG/TIG), and a more general group classified as 'automatic techniques', which includes advanced techniques such as plasma and electron-beam welding. The techniques themselves are extensively documented in other books.

Together, these techniques will account for perhaps 95 percent of the welding methods that you will meet during works inspections of engineering plant. Each technique has its own set of uncertainties that can (and do) result in defects. Frankly, it is difficult to place the techniques in order of the *risk* of producing defects – I have seen no firm evidence to convince me that manual techniques produce more defects because of the existence of a human operator. Sometimes the opposite is the case. Butt welding techniques for small diameter heat recovery boiler tubes, for instance, are often really only semi-automatic techniques: critical weld variables may be automatically controlled but the process still requires an operator and defects are still produced. You may find that automatic does not necessarily mean *better*.

So why *do* defects occur in the common 'proven' welding techniques? Fortunately, the character of defects provides some information as to their origin. Figure 5.1 shows an overview of how weld defects occur, or more specifically *what* causes them. On balance, I think the information in Fig. 5.1 holds good for any of the main ferrous welding technique groups, whether automatic or manual. This figure has key implications for the way that you should organize your priorities during works inspections involving welding. Note the following guidelines:

- The *main* cause of weld defects is poor process conditions during the welding activity.
- Operator error is directly responsible for nearly one third of defects, even when the process conditions are all correct.
- Errors caused by incorrect material and electrodes, or an unsuitable

technique, are not too common. Those that do occur are likely to be due to poor manufacturing practices rather than technical uncertainties.
- Although documentation is involved at various stages of the control of a welding process, the chances of a pure documentation fault being the root cause of a defect are actually quite small (note the shaded areas in the figure). Most of these consist of straightforward errors of communication where the wrong work instruction has been passed to the shop floor – almost all will have occurred *before* an inspector became involved.
- Figure 5.1 can only be an approximation because it applies to all four main technique groups, which are quite wide. Expect the proportion of technical (i.e. non-operator) problems to be slightly higher for those techniques that have a greater level of inherent technical risk. These are:
  - materials with heavy wall sections (greater than 75 mm).
  - a parent material of ferritic stainless steel (greater than 16 percent Cr) or unstabilized (no Ti or Nb) austenitic stainless steel
  - any structural butt weld involving dissimilar materials, particularly carbon steel to stainless steel
  - any welding activity where the parent materials (or the electrodes) are of uncertain origin or do not have positive identification.

Be assured that there is a fair degree of reproducibility in weld defects, you *will* see the same problems repeating themselves. It should not take long for them to fall into an approximate pattern, broadly in line with that shown in Fig. 5.1.

Types of welds

There is a surprisingly small number of fundamentally different types of weld although there are of course many variations of each type. Weld type is of key interest to an inspector mainly because of the way it affects the NDT requirements (or lack of them) that attach to particular types. Some of these linkages are specified in engineering standards, BS 5500 and ASME pressure vessel standards being good examples – whilst for other types of equipment greater reliance is placed on what is generally accepted as 'good engineering practice'. The most important weld type, which has the most stringent NDT requirements, is the *full penetration*

88   Handbook of Mechanical Works Inspection

▨ These areas represent documentation-based errors

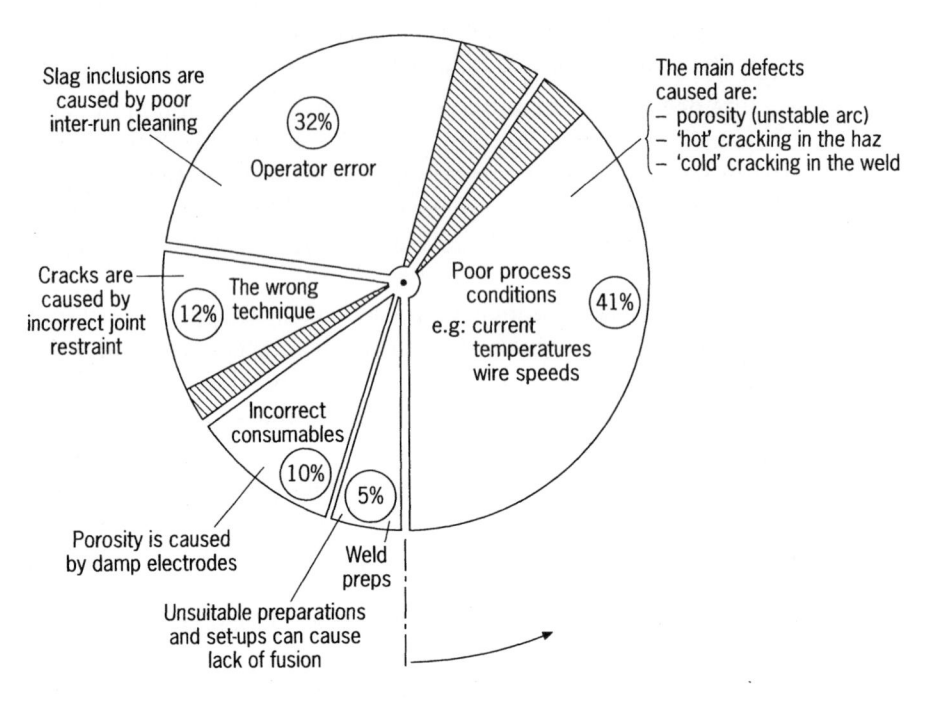

Slag inclusions are caused by poor inter-run cleaning

(32%) Operator error

The main defects caused are:
- porosity (unstable arc)
- 'hot' cracking in the haz
- 'cold' cracking in the weld

Cracks are caused by incorrect joint restraint

(12%) The wrong technique

Poor process conditions (41%)
e.g: current temperatures wire speeds

Incorrect consumables (10%)

5%

Porosity is caused by damp electrodes

Weld preps

Unsuitable preparations and set-ups can cause lack of fusion

Fig 5.1   How weld defects occur

weld. This is in common use in pressure components and other similar equipment.

The full penetration weld

The most common definition of a full penetration weld is a weld in which:

- The fusion faces are completely joined by the addition of a third (filler) metal *and* the weld is only accessed from one side. You may see other, possibly different, versions of this definition – the definition itself, however, is less important than the engineering principles which lie behind it. Note the inferences contained in this definition:
  - *The existence of 'fusion faces'*. In practical terms this means the area that will become the weld root – there will nearly always be a 'land' and often a *root gap* to help penetration of the filler material.

– *The reference to the weld being accessed only from one side.* This statement encapsulates the concept of a full penetration weld as one in which the root cannot be ground back on the reverse side. It is this aspect that results in the need for the stringent NDT requirements for full penetration welds – there is the risk that defects will remain in the root. To prevent this they must be identified by a volumetric NDT technique. Welds that have the reverse side of the root ground have had a major source of weld defects eliminated. Strictly then, any type of 'double-vee' butt or fillet weld is *not* a full penetration weld.

Coincidentally, it is often the case that full penetration welds are the most important ones in an engineering component or structure. This is a further reason why they should be subject to the most stringent levels of NDT, in order to keep risk to a minimum. You should always be on the lookout for full penetration welds. Figure 5.2 shows some common examples, and their application. You will see later in this chapter and in Chapter 6 the types of NDT activities that are specified for them, and the links with the various weld 'categories' specified by the BS 5500 and ASME pressure vessel standards.

Other types

Other types of weld which do not meet the 'full penetration' criteria normally have less stringent NDT requirements (fillet, lap, and seal welds are the most common). There are many other types, some of which are often mistaken for full penetration welds. Figure 5.3 shows some examples that you will meet. Don't be misled by any of these.

Weld heat treatment

Heat treatment is an important activity in obtaining desirable properties in an engineering material. You can think of welds in much the same way – because of the high local temperatures imposed during the welding process, some type of heat treatment is often required. There are basically three sorts: pre-weld heating can be performed on the prepared edges immediately before welding; post-weld heat treatment (PWHT) can be either the local type which essentially just reduces the cooling rate after welding, or a fully controlled stress relief carried out in a furnace.

The purpose of all these activities is to prevent cracking in the weld material and the heat affected zone (HAZ) – an important consideration, whether manual or automated techniques are used. It is a consideration not just at the initial welding stage but also during repair welding, when a

Note: A full penetration weld is welded from *one side* only

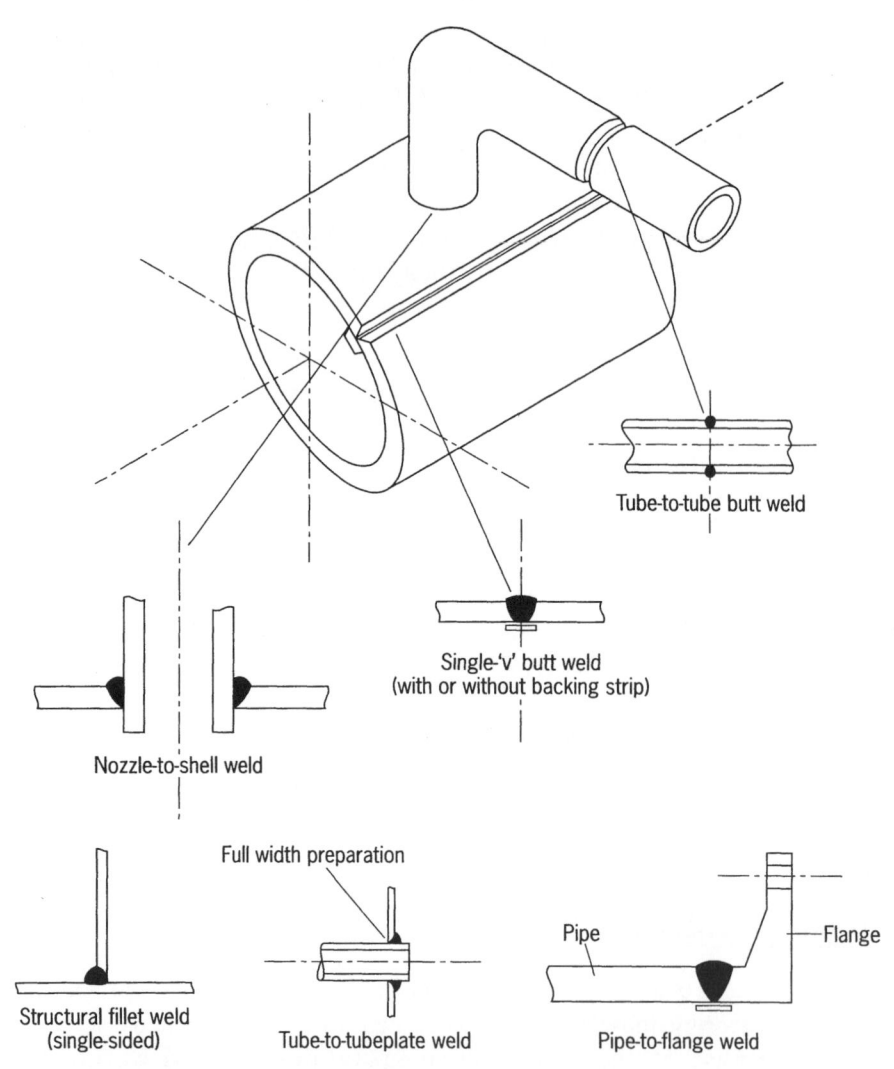

Fig 5.2 Full penetration welds – common applications

defect is found at a later stage of manufacture. Expect to see pre- and post-weld heat treatments shown on the ITP for fabricated items. If they are not, you should check to confirm that they are indeed not required. You can do this by checking in the applicable material standard. I have given some broad guidelines to follow in Fig. 5.4.

Verification that precisely the correct weld heat treatment has been carried out is part of your inspection role. In terms of *priority* I think it

Welding and NDT 91

These are not <u>always</u> considered full penetration welds

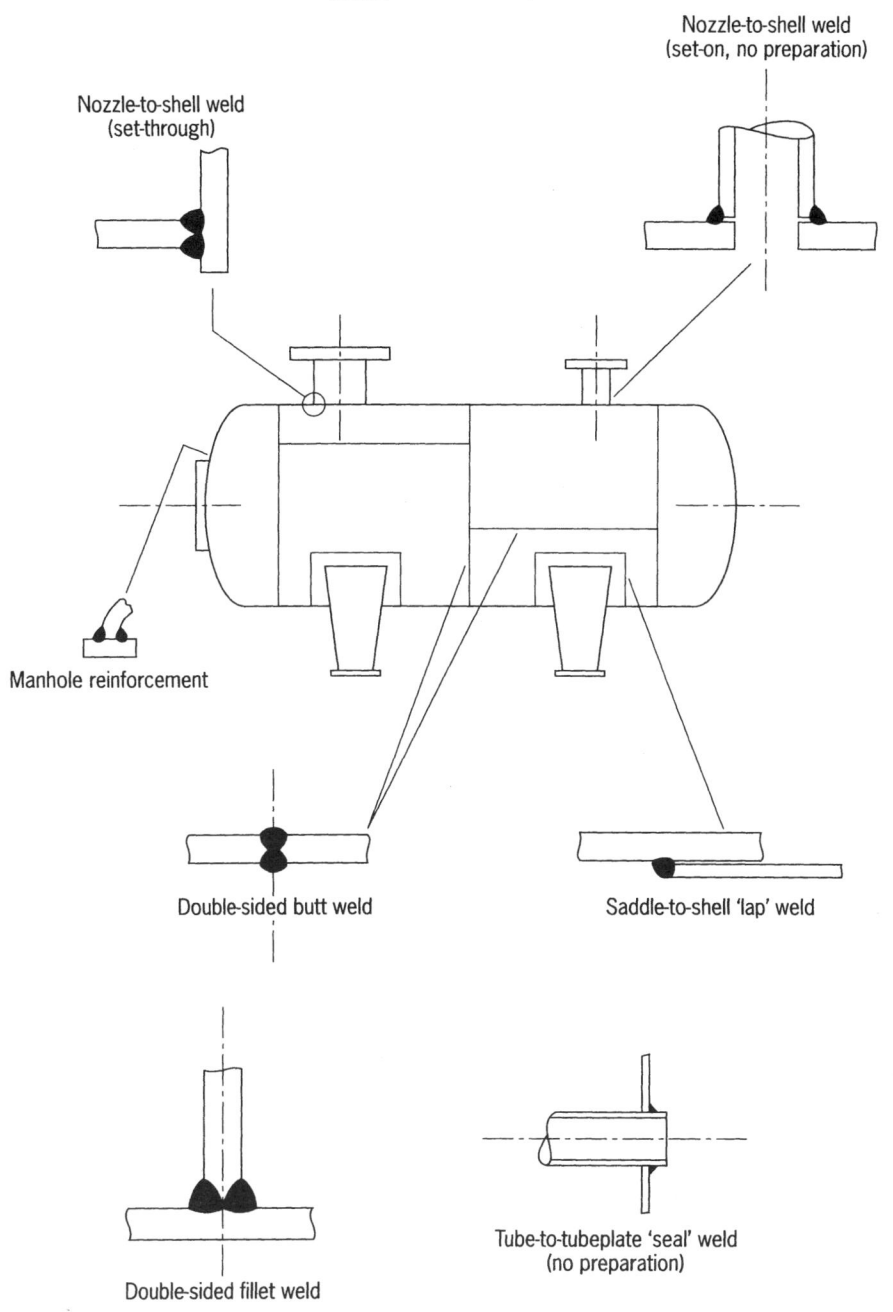

Nozzle-to-shell weld (set-through)

Nozzle-to-shell weld (set-on, no preparation)

Manhole reinforcement

Double-sided butt weld

Saddle-to-shell 'lap' weld

Double-sided fillet weld

Tube-to-tubeplate 'seal' weld (no preparation)

Fig 5.3 Other weld types

is best to concentrate on whatever post-weld heat treatment (PWHT) requirements are specified. Bigger and better weld defects are caused by incorrect PWHT than by poor pre-heating. We will see later in this chapter the importance of weld test pieces in verifying the final properties of the weld and HAZ.

Controlling documentation

Welding is associated with a well-defined set of documentation. The term *controlling* documentation is a slight misnomer – the documentation cannot *control*, as such. Perhaps it is better to think of this documentation as comprising a part (albeit an essential part) of the mechanism of controlling and recording. The documentation set also fits neatly in with the relevant requirements for statutory equipment – it provides a common-sense way of working which has become generally accepted for all types of fabricated equipment.

The objectives of this documentation set are to:

- Specify a particular weld 'method' to be used.
- Confirm that this weld method has been tested and shown to produce the desired weld properties.
- Ensure that the welder who performs the welding has proven ability.

Terminology can be confusing. Technical standards such as BS 5500 and ASME pressure vessel standards, which are used as the basis for meeting statutory requirements, refer to the need for the review of weld *procedures* and of the need for weld *testing*. These are fine as general definitions (they are well understood) but it is better if you refer specifically to those individual documents that make up the set. This reduces the chances of misinterpretation. Figure 5.5 shows the documents, and how they are related to the weld in question, and each other. The example is for a simple single-sided butt weld of a type commonly found in pressure vessels. Note that it has a backing strip – an acceptable configuration where one side of the weld is inaccessible. This is defined as a full penetration weld and could be added therefore to the examples shown in Fig. 5.2.

The weld procedure specification (WPS)

The WPS describes the weld technique. It is a technical summary of the relevant weld parameters and is detailed enough to act as a work

Welding and NDT 93

1. Pre-weld heating

Temperature–indicating point

Test plate

200°C

Typical preheat temperature is 200°C

75 mm

Note:
- Austenitic steels do not often need preheat
- Ferritic steels normally do need preheat

Gas burner for local pre-heating

Thick-walled vessel

2. Post-weld heat treatment (PWHT)

Furnace

Typical 'soaking' temperature is 600°C

600°C

Temperature/time recorder

Welded test plate in furnace

Note: plate thicknesses >25 mm normally need PWHT

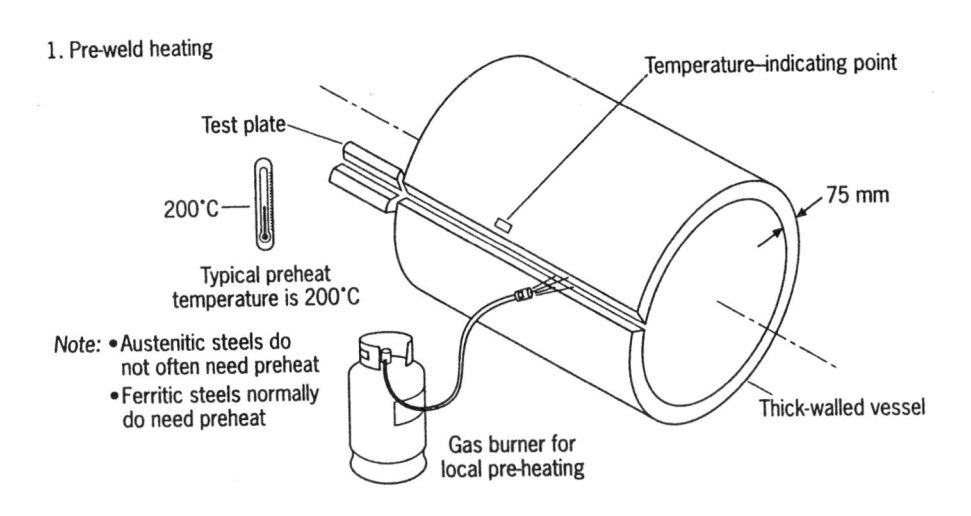

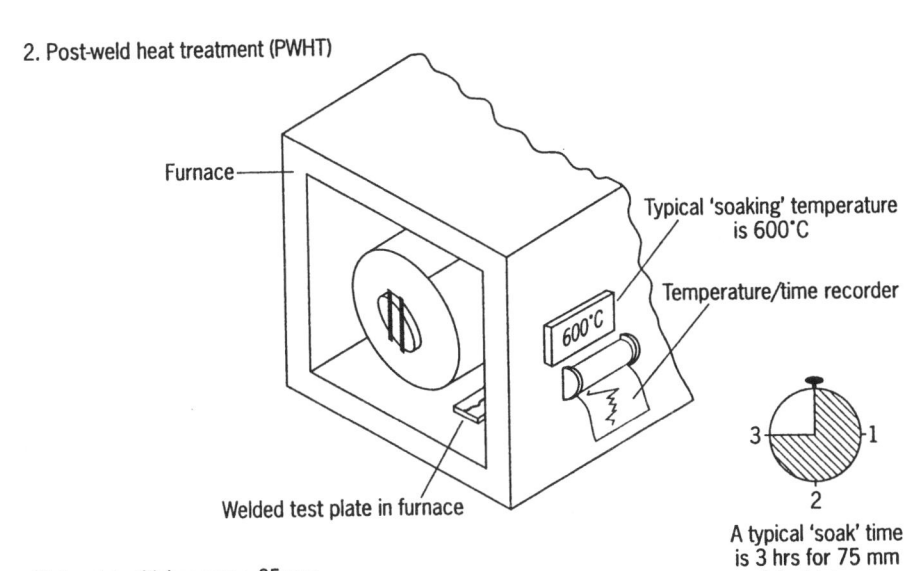

A typical 'soak' time is 3 hrs for 75 mm plate thickness

Always check the material specification to confirm pre- and post-weld heat treatment

Fig 5.4   Weld heat treatment: some guidelines

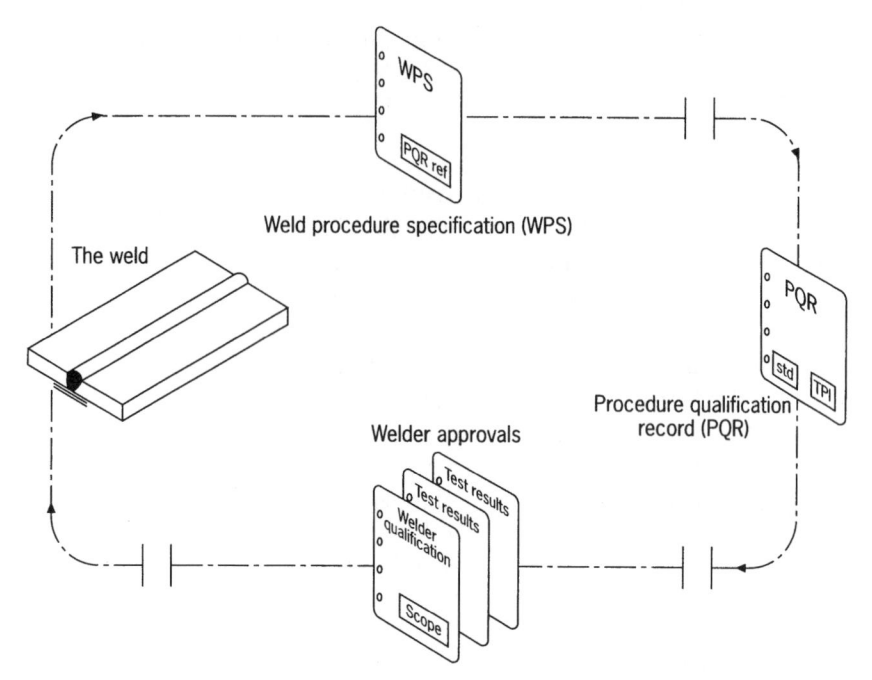

Fig 5.5 Welding: controlling documentation

instruction for the welder. The WPS is prepared by the manufacturer and the general principle is that each weld type should have its own WPS. It will include details of:

- parent material
- filler material/electrode type
- the weld preparation
- fit-up arrangement
- welding current, number of passes, orientation, and other essential variables
- back-grinding of the root run
- pre- and post-weld heat treatment
- the relevant procedure qualification record (PQR).

The procedure qualification record (PQR)

Different standards use different terminology for this. You may see it referred to as a weld procedure qualification (WPQ), or sometimes as a weld procedure approval – the term PQR, however, is commonly used as a generic meaning. The PQR is the record of testing of a specific weld, analogous to the 'type-test' carried out on some items of equipment. A test weld is performed to the specified WPS and then subjected to an extensive series of non-destructive and destructive tests to determine its physical properties. These include:

- Visual and surface crack detection.
- Ultrasonic or radiographic examination.
- Destructive tests (normally by bending). Look at BS 709 **(1)** if you need details. These test the fundamental strength characteristics of the welded joint.
- Hardness tests across the weldment and HAZ regions (done on a polished and etched macro-sample). Hardness gives an indication of heat treatment and metallurgical structure. The range of hardness readings across a particular sample can identify factors that lead to an increase in the risk of cracks developing in the *future*.

We will look at the other tests later in this chapter. A weld test leading to the issue of a formal PQR should be witnessed by an independent third-party organization. This does not have to be one of the well-known classification societies but usually is – this is because of their experience and capability in witnessing such activities. If you look at a typical set of welding documentation and the relevant technical standards you will see that not every weld (or every WPS) *necessarily* needs its own dedicated PQR. Note also that PQRs are unlikely to be 'job-specific' (whereas WPSs are). Most common weld types have been tested over the years and, as long as the essential weld variables do not change, PQRs do not go out of date.

Welder qualifications

Conventional wisdom is that welding ability is *non-transferable*. This means that the ability of welders to weld correctly, for instance, low carbon steel does not (rather than *may* not) mean they can do the same on stainless steel. Weld technique, position, and material type all have a significant effect on ability. This is a harsh concept, but one which you are obliged to follow. As with the WPS situation, the requirement for qualified or 'coded' welders has spread from its origin in the manufacture of pressure vessels and other statutory equipment. You will find it a generally requested concept in most power or process plant contracts.

A welder is tested to a specific WPS (or a range of them depending on the welder's scope) and the weld is then tested in a similar, or identical, way to that used for the PQR. Approval is confirmed by the welder being awarded a personal certificate, which includes a photograph. The test results and certificates are normally certified by a third-party organization.

The problems of 'mismatch'

It would be nice if the technical scope of every particular WPS was matched exactly by a corresponding PQR, with the welder qualified to precisely the same weld arrangement (and the tests) used in the PQR. In practice it does not happen quite like this. Both EN and ASME standards allow a certain leeway in the essential variables of a weld, both between the WPS and the PQR, and between the PQR and a corresponding welder qualification. It is a little misleading to think of these 'gaps' as a mismatch – there is certainly no *conflict* between the technical requirements of a WPS and its nearest corresponding PQR. What it does do though, is introduce a certain level of risk. This is one reason why the common set of technical documentation accompanying welding activity cannot provide *full* control of what is happening.

These gaps between WPS, PQR and welder approval requirements can cause you problems. You will not find it easy to recognize instantly whether a WPS and PQR correspond or not. Unfortunately the 'ranges of approval' are quite large – the standards contain complex matrices explaining which parent material groupings and material thickness differences are allowed. I don't think you have much to gain by worrying about the philosophy of these gaps I have described. Accept that they are the result of extensive study of the essential variables of a weld, their effect on the integrity of the welding process and the resultant mechanical performance of a completed weld. Make sure that you do *check* the matching of the WPS, PQR and weld approvals during your works inspections, though – it is still an important part of the mechanism for ensuring fitness for purpose. You will find further details given in the technical standards mentioned in the following section.

Specifications and standards

It is comparatively rare for contract specifications to say much about welding and NDT despite the significant impact these have on FFP.

Instead, they make specific reference to published standards, thereby imposing indirectly a raft of technical requirements. This is supported by a few carefully selected general statements, often in the form of 'fall-back clauses' which allow the purchaser to specify additional testing if it is felt necessary. We have discussed this type of approach before when considering materials of construction. Technical standards therefore are very much in evidence. This means that a lot of your work as an inspector in this field will be standards-based.

The technical coverage of welding and NDT standards is extensive. There is reasonable uniformity of approach between European and American (ASTM) sets of standards but expect also to find some duplication. The ASTM range, many of which are referenced directly by the ASME pressure vessel codes, are very practical 'doing' standards. The European ranges seem to adopt a more technological approach. In large contracts you are likely to find a mixture of both systems.

The role of acceptance criteria

There are a group of technical standards covering defect acceptance criteria – although some put more effort into describing defects rather than telling you when they are (or are not) acceptable. These documents are spread around the standards range (see Fig. 5.6). Be wary: although technical coverage is good in principle, you will find that many of the defect acceptance criteria are not *equipment specific*.

You should not expect that the definitive and prescriptive defect acceptance criteria quoted for pressure vessels and statutory items will be available for other equipment. This is an important point. As an inspector you have the opportunity (indeed duty) to apply your interpretation to these acceptance criteria. Normally this will be interpretation not of the *content* of a published standard but to the items of equipment to which it *precisely* applies. It you look at some of these standards I have listed in Fig. 5.6 you will see that this is not always perfectly defined. Judgement is required.

If a set of published acceptance criteria fit an application well, then you have a valuable and proven standard for fitness for purpose of that particular type of weld configuration. This is invaluable assistance for you – and you can relax your judgemental role just a little.

Try to follow these general guidelines on acceptance criteria.

- Make sure you look carefully at their content – they do contain good accumulated knowledge.

- For pressure vessels and statutory equipment you are fully justified in applying rigorously those acceptance criteria laid down by published pressure vessel standards as long as the *applicability* of the standard is clear.
- For other plant, put some effort into checking the applicability of the criteria to your particular piece of equipment. Don't fight the *content* of the acceptance standards.
- There will always be room for judgement and interpretation. Be prepared for this.

The main standards

The standards that you will need fall broadly into three groups. Technically these groups are fairly discrete, although you may often see them all quoted together.

Welding standards

You will not find a lot of detailed guidance in welding standards – at least not in those that address the pure *technique* of welding. Welding practice tends to be industry-led rather than standards-led (remember the discussion on these concepts in Chapter 3), with the result that technical standards present only a broad spread of information. The clearest examples of such standards are: BS 2633 **(2)**, BS 2971 **(3)** and BS 4570 **(4)**. These standards provide a general technical coverage of the subject, including heat-treatment, materials and explanation of welder approvals, but they do not all contain a lot of direct inspection-related information.

Moving on from the overtly technical standards on welding, there are two important standards that are in common international use. You will meet the implications of these two in almost all works inspections involving fabricated equipment.

BS EN 288 **(5)** is divided into six parts and covers the subject of weld procedure specifications (WPSs) and weld procedure qualifications (PQRs or 'tests'). The part you will meet most frequently is BS EN 288: Part 3 which specifies how a WPS is approved by a weld procedure test, to produce an 'approved' or 'qualified' weld procedure. The real core of this standard is to enable you to determine the range of approval of a particular PQR – by using a set of tables. These are detailed but well set out and relatively easy to follow. Useful information is provided in the form of Table 1', which lists the extent of NDT and destructive testing required on test pieces taken from butt, tee-butt, branch, and fillet welds.

BS EN 287 **(6)** relates to the approval testing of the welders themselves. The approval system is based on a set of uniform test pieces and welding positions. As with BS EN 288, there is a range of approval involved – the principle is that by demonstrating competence on a particular type of weld joint a welder becomes *qualified* for that weld type and all those weld joints that are 'easier' than the test joint. The standard provides a series of tables to determine where the limits of the ranges of approval lie. The welder performs the test welds under supervision and then various NDT and destructive tests are carried out. Note there are specific standards referenced for evaluating the results. BS EN 26520 (ISO 6520) **(7)** is used to *classify* the imperfections (defects) whilst BS EN 25817 (ISO 5817) **(8)** provides guidance on *quality levels* of these imperfections (this standard does not recommend FFP levels that you can apply to equipment – so be careful not to use it out of context).

BS EN 287 and 288 are intended to be used as a 'matched set'. They are well written and clearer than the standards they replace. Try and see them as a part of the control mechanism operating in a manufacturer's works. Their acceptance has spread from statutory pressure vessels to general fabrication. You can think of them as providing access for inspectors to the details of manufacturers' welding techniques, and in the case of BS EN 287 as a management and training tool.

NDT technique standards

Figure 5.6 shows the commonly used standards covering NDT techniques. Note the clear split between volumetric methods – radiography (RG) and ultrasonics (US) – and surface methods – magnetic particle inspection (MPI) and dye penetrant (DP). NDT standards tend to be very comprehensive – there are few of the problems of 'applicability' that reduce some of the usefulness of the acceptance criteria standards, for instance. In spite of this you may sometimes come to feel that manufacturers do not understand these standards very well. I do not think that manufacturers always appreciate the level of technical guidelines that these standards contain – possibly this is due to overfamiliarity caused by using NDT techniques on a daily basis without needing to refer to a standard in detail. As a key part of FFP, inspectors must not let NDT standards slip. This is an area in which to expect strict compliance with the spirit and the letter of the standards I have listed. It is best to allow no leeway at all. I think the standards themselves are detailed and clear enough to support this.

NDT TECHNIQUES

Radiography
- BS EN 444: 1994. NDT General principles for RG examination of metallic materials
- BS 7257: 1989. Methods for RG examination of fusion welded branch and nozzle joints.
- BS 2600 Parts 1 and 2 (ISO 1106). RG examination of fusion welded butt joints in steel.
- BS 2910: 1986. RG examination of fusion welded circumferential butt joints.
- DIN 54111: Guidance for the testing of welds with X-rays and gamma-rays
- ISO 4993: RG examination of steel castings.
- ISO 5579: RG examination of metallic materials by X and gamma radiography.

Standards concerned with image clarity are:
- BS 3971: 1985. Specification for IQIs for industrial radiography (similar to ISO 1027)
- DIN 55110: Guidance for the evaluation of the quality of X-ray and gamma-ray radiography of metals.
- ASTM E142: 1992. Methods for controlling the quality of radiographic testing.
- BS EN 462-1: 1994. IQIs (wire type). Determination of image quality value.

Ultrasonic testing
- BS 4124: 1991. Methods for US detection of imperfections in steel forgings.
- BS 3889: Part 1: 1990. Automatic US testing for imperfections in wrought steel tubes.
- BS 3923: Part 1: 1986. Manual examination of fusion welds in ferritic steel.
- ASTM A609: 1991. Practice for US examination of castings.
- BS 6208: 1990. US testing of ferritic steel castings – including quality levels.
- ASTM A418: US testing of steel rotor forgings.

Surface crack detection
- BS 6072: 1986. Method for MP flaw detection.
- BS 5138: 1988. MP inspection of solid forged and drop forged crankshafts.
- ASTM E709: 1991. MP testing of castings.
- ASTM E1444: MP detection practice for ferromagnetic materials.
- ASTM A275: MP testing of steel forgings.
- ASTM E165: Dye penetrant examination.
- ASTM E433: 1993. Reference photographs for DP examination.
- BS 6443: 1984. Methods for DP flaw detection.
- ISO 3452: 1984. Dye penetrant examination general principles.

Acceptance criteria/reference levels
- BS 5500: 1994. Specification for unfired pressure vessels (see tables 5.7(1), (2) and (3)).
- ASME Section VIII: 1995. Rules for construction of pressure vessels (see UW-51 and UW-52).
- ASTM E71 and E446: 1989. Reference radiographs for steel castings up to 2″ thickness.
- ASTM E186: 1991. Reference radiographs for steel castings $2 - 4\frac{1}{2}″$ thickness.
- ASTM E280: Reference radiographs for steel castings $4\frac{1}{2} - 12″$ thickness.
- ASTM E99: Reference radiographs for steel welds.
- BS 2737: 1995. Terminology of internal defects in castings as revealed by radiography.
- AD Merkblatter HP 5/3.
- BS 4080 Parts 1 and 2. 1989: Severity levels for discontinuities in steel castings.
- BS 5996: 1993. Acceptance levels for defects in ferritic steel plates.

Fig 5.6 NDT: relevant technical standards

Acceptance criteria standards

I have mentioned earlier the stoicism with which you should approach and use published defect acceptance criteria standards. Figure 5.6 shows the main ones in use. By far the most commonly used ones are the relevant sections of the BS 5500 and ASME VIII pressure vessel codes. The ASTM range shown in the figure gives particularly comprehensive coverage of defects in castings – the pictures of reference radiographs can be a useful basis on which to help decide fitness for purpose. Because of the method of manufacture, castings do tend to be prone to defects.

Inspection and test plans

It is rare to find an ITP without some welding and NDT content. There are normally only a few lines for each type of welding activity but these may be repeated many times through the ITP. The best example of this is for a pressure vessel, in which individual seam and nozzle welds have their own lines of the ITP. Figure 5.7 shows a typical ITP extract for a simple welded joint, in this case for a pressure vessel seam. This type of layout and content is very common. You will see a similar layout used for other applications such as welded forgings, castings or pipes (a more detailed outline of inspection stages for statutory vessels is discussed in Chapter 6).

The apparent simplicity of Fig. 5.7 can be misleading. For an effective inspection, the 'review' activities indicated need to be much more comprehensive than they appear. The ITP in Fig. 5.7 is not *wrong*, it's just that there are a number of activities that are inferred, rather than stated explicitly. These activities are best considered as slightly more than good engineering practice, one or two are directly standards-related and standards, we remember, are a cornerstone of FFP.

Figure 5.8 is a more detailed interpretation of the simplified ITP steps shown in Fig. 5.7. I have stated explicitly here the activities you should carry out during these 'review' parts of a weld inspection. Look at these carefully because they are quite specific and detailed. For best effect you should do these whilst in the works – not by reviewing the documents at your office later. Note that there are few witness (W) points shown in the ITP in Fig. 5.7. This is often the case for non-statutory fabricated equipment. It is not an ideal state of affairs – some witnessing is probably necessary, if only to stop 'a documentation culture' taking over. When you visit a manufacturing works, make a point of trying to witness some key welding and NDT activities. Aim to include these:

Step No.	Operation	Reference documents	Inspection points M	C	TPI	Certification requirements	Record No.
1	Weld procedures	WPS/PQR	R	–	–	BS EN 288	XX/Y
2	Welder approvals	BS EN 287	R	–	–	BS EN 287	XX/Y
3	10% RG	BS 2600	R	R	–	Record sheet	XX/Y
4	100% MPI	BS 6072	R	R	–	Record sheet	XX/Y
5	Visual inspection	BS 5289	R	W	W	Record sheet	XX/Y
6	Document review	–	R	R	R	–	XX/Y

W – Witness point
R – Review
M – Manufacturer
C – Contractor
TPI – Third Party (or client's) inspection organization

Fig. 5.7 The welding part of an ITP

- Weld preparations, particularly for thick steel sections (say greater than 20 mm). Do a visual inspection *then* compare the set-up with the WPS.
- Circumferential (mis)alignment on head-to-shell joints before welding for all types of vessels. BS 5500, ASME VIII and all other recognized vessel codes quote maximum limits.
- 'Grind-back' of double-sided butt welds.
- MPI examinations in progress (because it is easier to get these wrong than dye penetrant techniques).
- Repairs of any type.

Above all, when you are in the works *inspect the hardware*. The fact that you are reading this book shows that you are an *inspector not an administrator*. Then report (take a quick look forward to Chapter 15).

Test procedures and techniques

Witnessing tests is not the same as having to carry them out. As an inspector your objective should be to possess just the right level of knowledge to enable you to verify fitness for purpose without wasting time or money (either your own or the manufacturer's). As a result, you

Welding and NDT

STEP 1: WELD PROCEDURES

Means
- Check the main WPSs against the relevant drawings (look at the edge profile and set-ups);
- Check the WPS *content* complies with BS EN 288;
- Check the links between WPSs and PQRs;
- Check the essential variables and PQR test results;
- Check all TPI authorizations.

STEP 2: WELDER APPROVALS

Means
- Look at the qualification certificates for individual welders, checking scope and approval dates, check compliance with BS EN 287;
- Cross-check each welder's identification stamp with welds that have been completed.
- Check that the 'leeway' allowed between the welder's approval scope and the actual weld is within the scope of BS EN 287/288.

STEP 3: 10% RG

Means
- Check the radiography procedure and technique for compliance with a standard (see Fig. 5.6).
- Look at the films and their report sheet: check against the acceptance criteria.
- Check the RG location map and markers on the weld itself.
- Sign the report sheet (if you are satisfied).

STEP 4: 100% M.P.I.

Means
- Check the MPI procedure (and techniques) for compliance with a standard shown in Figure 5.6;
- Check that the correct areas have been tested;
- *Witness* a further sample of MPI tests if you have any doubts;
- Sign the report sheet when you are satisfied.

STEP 5: VISUAL INSPECTION

Means
- Open up any access hatches etc. to get access to load-bearing welds;
- Make a close visual inspection in line with BS 5289 (even if this was not quoted in the ITP – it is good practice).

STEP 6: DOCUMENT REVIEW

Means
- Check there is an index of documents, preferably cross-referenced to the step numbers on the ITP;
- Check the document package itself for completeness, and for any necessary TPI approval.
- If anything is missing, issue an NCR.

Fig 5.8 The correct interpretation of Fig. 5.7

cannot expect to achieve the level of 'feel' for welding and NDT techniques that an experienced technician or operator will possess. Your job is to concentrate on *verification*, not to act as unpaid technical advisor to those performing the tests.

We will look therefore at the main inspection and test activities relating to welding and NDT that you will be called to witness. Figure 5.9 shows them in the approximate order in which they are carried out. I have concentrated on a 'how to inspect' viewpoint of these tests rather than providing rigorous technical descriptions of the activities. Check the relevant technical standards if you need more detailed information.

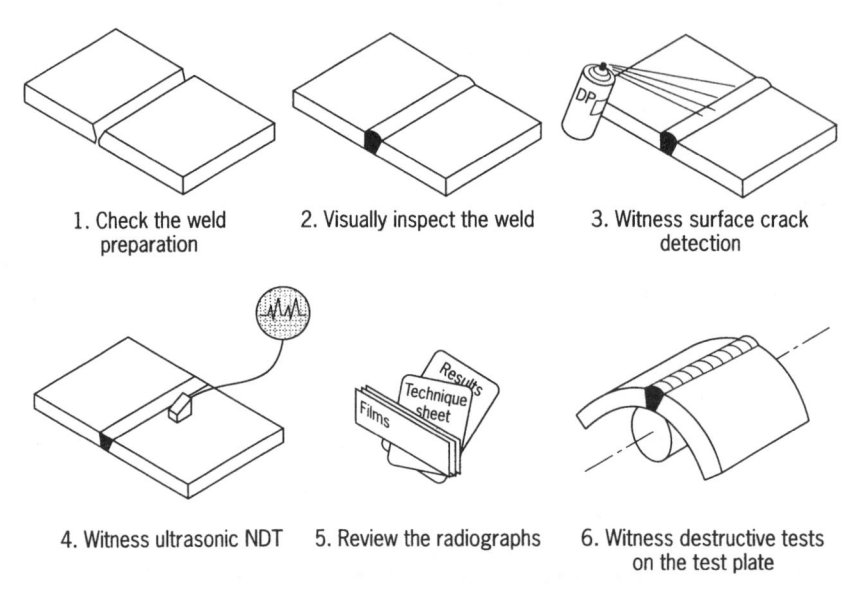

1. Check the weld preparation
2. Visually inspect the weld
3. Witness surface crack detection
4. Witness ultrasonic NDT
5. Review the radiographs
6. Witness destructive tests on the test plate

Fig 5.9 The main welding/NDT tests

Checking weld preparations

Poor weld preparation results in poor welds. An inspection programme for equipment containing load-bearing welds should make provision for inspection of various stages of 'welding in progress' – make a point of being in the works to inspect some of the major welds at the preparation stage. The required checks are quick and easy – Fig. 5.10 shows the checks that you should make. The same principles apply for butt and nozzle welds.

Start with *design*. Check that the preparation design complies with the fabrication drawing, the WPS/PQR and any applicable standard (normally BS 5500 or ASME) – in that order. Pay particular attention to the weld preparation angles, and to the root gap or backing strip, as applicable. Pressure vessel codes show typical weld preparations that are acceptable for coded vessels but don't expect these standards to be fully prescriptive as to whether *your* particular joint design is allowed. Sometimes you will have to use a bit of interpretation. The situation is normally clearer for butt welds than for nozzle welds. If in doubt you can issue an NCR, asking the manufacturer to demonstrate compliance with the relevant standard.

The next step is to inspect the prepared plate *edges*. It is important that defects are removed before welding begins. Any defects will remain in the weldment, often at the edge of the heat-affected zone, and be a prime source of initiation for cracks. You will find that these are often planar defects, so they may not be easily found by radiographic examination. Make sure that:

- Flame-cut plate edges and sheared edges have been ground back (before the preparation angles are machined) by at least one-quarter of the plate thickness ($t/4$), preferably more, to remove burnt and work-hardened areas.
- MPI or DP checks are performed on the prepared edges, to look for cracks or inclusions, on both sides of the plate, up to a distance of at least twice the plate thickness ($2t$) back from the prepared edge (see Fig. 5.10). No crack-like 'indications' are acceptable at this stage – the only solution is to grind out and re-machine the preparation, then repeat the check.
- For nozzle welds, be wary of poor surface finish on the outside of the nozzle. It may be a forging, if so all mill-scale should be removed to obtain a good clean surface before welding.

Finally check the weld set-ups. Physical alignment of the joint set-up can be a difficult task, particularly on large fabrications. The proper time to do the check is *after* all the necessary tack-welds, braces, and joint 'dogs' have been fitted: that is, when the joint is fully ready for the first weld seam. Check the soundness of the tack welds and that all slag has been cut out. Look for a uniform root gap all along or around the weld. For nozzle welds, check that adequate bracing has been fitted to prevent the nozzle 'pulling over' during welding. You will have to use experience here, or ask the manufacturer, it will not be on the WPS but it might be on the fabrication drawing. Don't lose sight of your objectives – you are looking for a weld set-up which will give a sound full penetration weld. It is not necessary to witness the welding activity itself, you can concentrate instead on a thorough examination between weld runs, and then again when the seam is finished.

Visual inspection of welds

It is safe to say that all welds should be subject to visual inspection. Strictly, visual inspection is a part of the NDT process – it plays a part when checking the location of radiographs and during the ultrasonic tests on the parent plate or completed welds. The way in which you approach visual inspections is more important than first appears. This is largely because the visible appearance of welds is easily commented upon by users, site engineers or inspectors after the equipment has arrived at its construction site. You can expect all manner of opinion – some of it informed, and some not – on how the welding looks. For this reason, I suggest that you adopt a fairly thorough approach to the way that you *report* your visual inspections of welds, and any defects that you find.

The best way is to use a *checklist*. This will provide a clear and simple way to present your findings in your report. It will help others (and occasionally yourself) to differentiate visible observations that truly influence FFP from those that are merely cosmetic. Figure 5.11 shows common visible defects and Fig. 5.12 is a simple checklist that you can use. The commonly used standard for visual inspection of welds is BS 5289 **(9)** – note though that this does not include details of acceptance criteria, instead it cross-references the relevant 'application standard' (pressure vessel code or similar).

Surface crack detection (and its limitations)

A disproportionate amount of cracks and defects are to be found on the surface of cast, forged and cold-worked materials, compared to those that are concealed in the body of the material (with the exception of the weld root, which *is* a common source of defects). This is due mainly to the way that the component material is formed or worked and it seems that little more can be done, technologically, than is done already, to stop such defects occurring. Although much study and work has been done to develop specialized, often esoteric surface crack detection techniques, the majority of manual methods used in manufacturing works still use simple dye penetrant (DP) or magnetic particle inspection (MPI) techniques.

In a practical works inspection situation these techniques provide *only* an enhanced visual assessment of the surface. Frankly, they will show

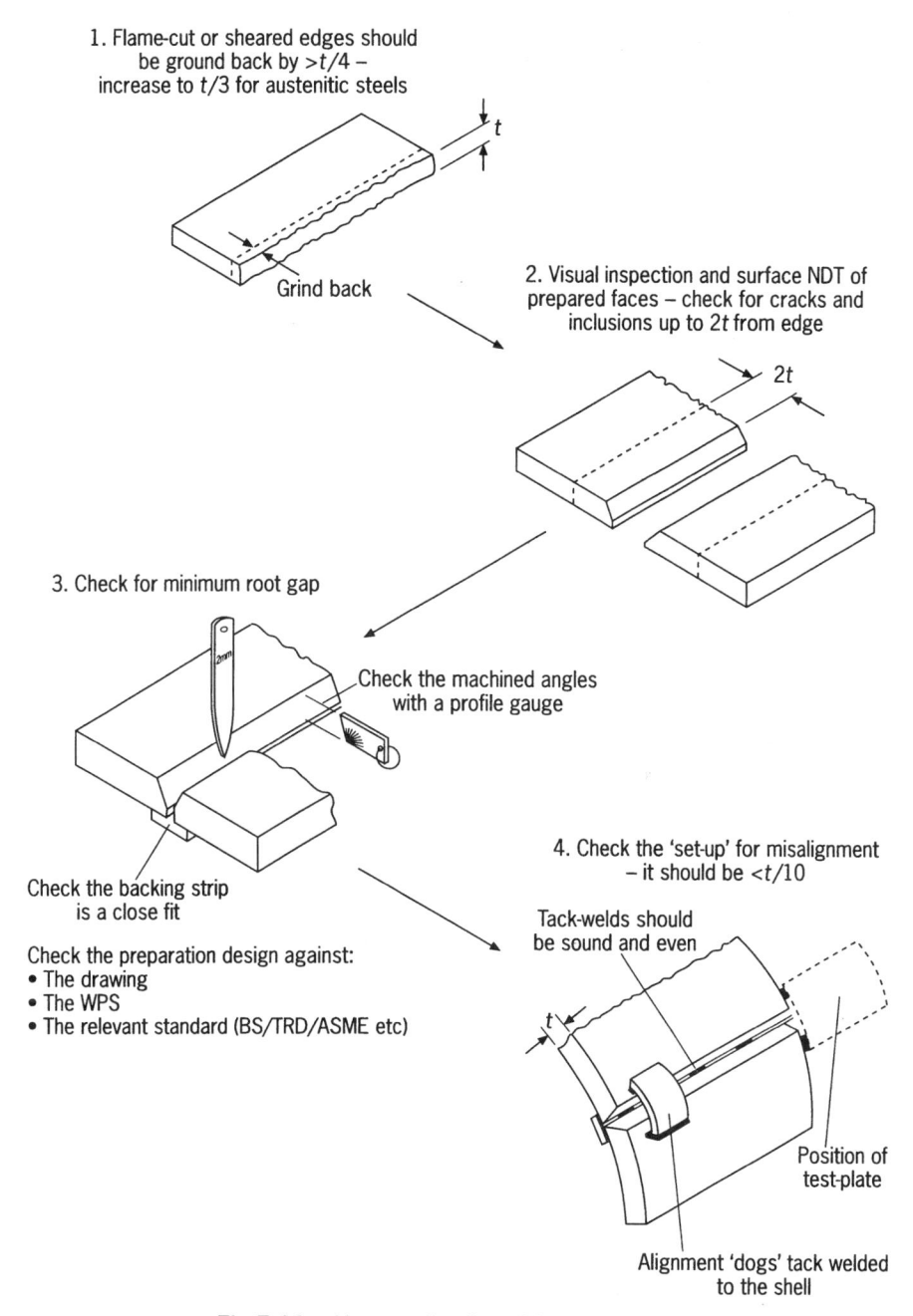

1. Flame-cut or sheared edges should be ground back by $>t/4$ – increase to $t/3$ for austenitic steels

Grind back

2. Visual inspection and surface NDT of prepared faces – check for cracks and inclusions up to $2t$ from edge

3. Check for minimum root gap

Check the machined angles with a profile gauge

Check the backing strip is a close fit

Check the preparation design against:
- The drawing
- The WPS
- The relevant standard (BS/TRD/ASME etc)

4. Check the 'set-up' for misalignment – it should be $<t/10$

Tack-welds should be sound and even

Position of test-plate

Alignment 'dogs' tack welded to the shell

Fig 5.10 How to check weld preparations

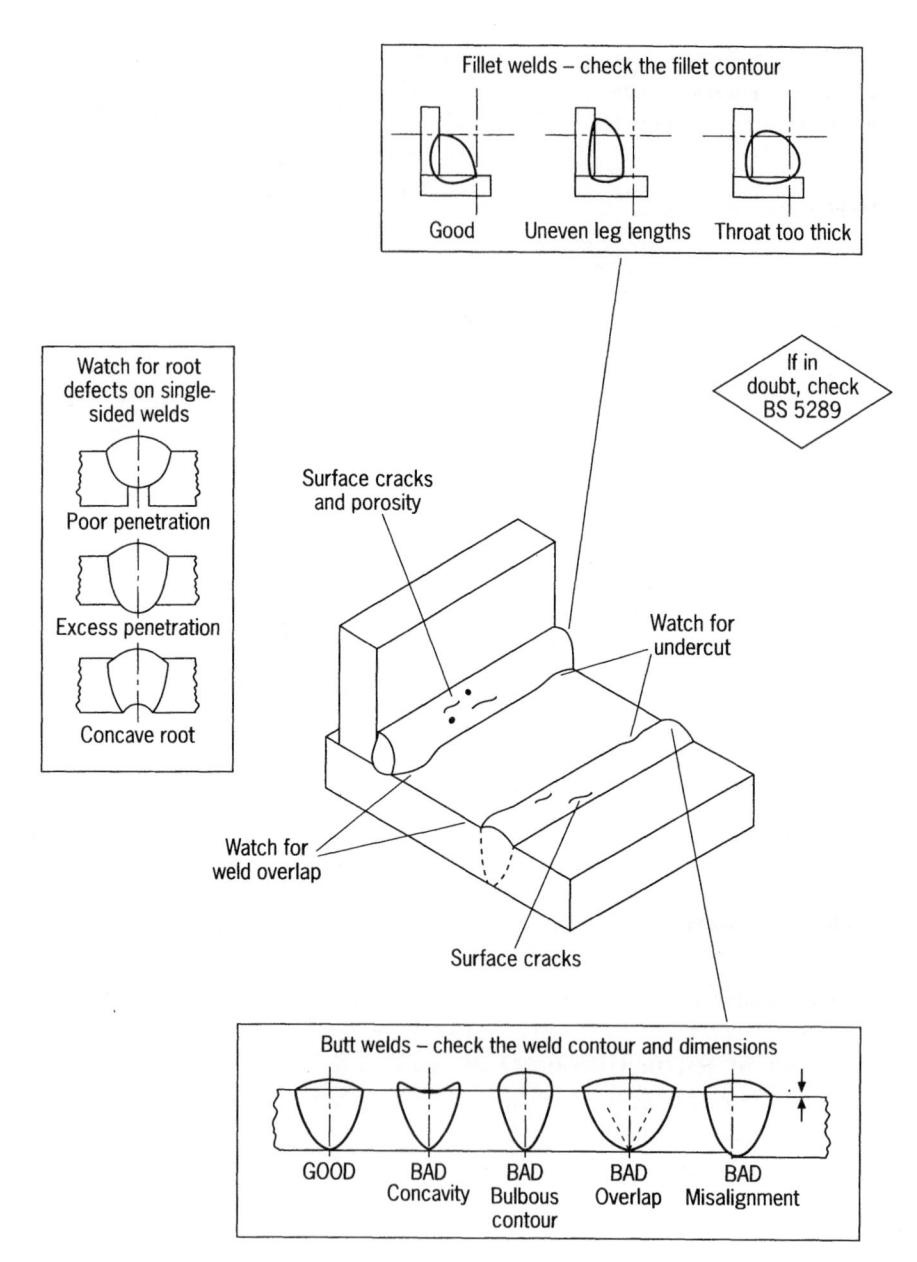

Fig 5.11 Visual inspection of welds

CHECKS

Cleaning and dressing
- All slag removed
- Arcing marks removed
- Weld spatter removed

Weld contour and dimensions
- Equal leg-lengths (fillet welds)
- Correct throat dimension (fillet welds)
- No throat concavity (fillet welds)
- Regular surface pattern (fillet or butt welds)
- Weld cap size correct (butt welds)
- 'Blended' weld toes (fillet welds)

Defects
- Undercut (fillet or butt welds)
- Overlap (fillet or butt welds)
- Surface cracks (fillet or butt welds)
- Surface porosity (fillet or butt welds)
- Notches at weld toes (fillet welds)
- Visible root defects (single-sided butt welds)

If you do find defects then:
- Describe them carefully.
- Identify the weld by location or seam number.
- Take a photograph (or do a location-sketch)

Fig 5.12 **Visual inspection of welds – a checklist**

you very little that a close visual examination under bright illumination will not. They are useful, however, for identifying surface 'indications' and generally making the job of finding defects that much easier. They are also quick and relatively simple to perform. As with all NDT techniques their effectiveness depends on factors such as the surface finish of the material, orientation of the defects, and a certain amount of skill and familiarity.

As an inspector, you have to treat surface NDT results with some caution. Granted, DP and MPI techniques will detect cracks of visible size which, if left unattended, can cause failure of the material – but there is a limit of *very approximately* 0.1 mm (100 microns) minimum defect size that will be detected when using these methods in a practical shop-floor situation. This 100 microns is *larger* than the 'critical crack size' for many materials. (Critical crack size is that which is considered large enough to propagate under normal working stresses and hence cause failure.) Treat surface NDT seriously, certainly, but try to complement the results with an understanding of the metallurgical realities of the mechanisms by which materials actually fail. This means 'take care'.

Dye penetrant (DP) inspection

This is the simplest and best-known test for surface cracks. There are several standards explaining the technique and showing typical defects (see Fig. 5.6), and DP has the advantage that a visual record is available – indications can be photographed. Despite it being such a simple technique you will still see it done incorrectly. Figure 5.13 shows the correct method. Three aerosols are used – the cleaner is a clear liquid, the penetrant is red and the developer is white. Note the following guidelines:

- DP is equally effective on all metallic materials, but it *is* sensitive to surface finish. Sharp edges and weld spatter will give false indications. As a guide, a normal 'as-cast' finish should be acceptable after fettling all rough edges – anything rougher may need light grinding.
- Make sure the correct cleaning steps are carried out as shown in Fig. 5.13. It is not advisable to scrub or rinse the surface to remove the red penetrant before applying the developer. This may wash it out of surface defects and give poor results.
- The most common error is not allowing the red penetrant enough time to penetrate. At least 15 minutes is required.
- Surface cracks will show as hard red lines on the surface of the white developer. Surface porosity will show as a group or chain of small red dots.
- Edge laminations on weld-prepared or sheared steel plate will show up as thin, often indistinct, crack-like indications orientated along the cut edge. On sheared plate, you may also see transverse cracks caused by excessive work-hardening of the plate. It is not possible to do a successful DP test on a flame-cut edge.
- If you find defects, take photographs to include in your inspection report.

DP techniques are finding increasing use on complex components such as heat recovery boiler headers where the number and arrangement of tube-stubs and nozzles makes MPI difficult. It is hard to identify any areas where DP could not be used as an alternative to MPI if required.

Magnetic particle inspection (MPI)

MPI works due to the magnetic properties of the material being examined. A magnetic flux is passed through the material and the surface sprayed with a magnetic medium or ink. An air gap in a defect

Dye penetrant testing – the correct method

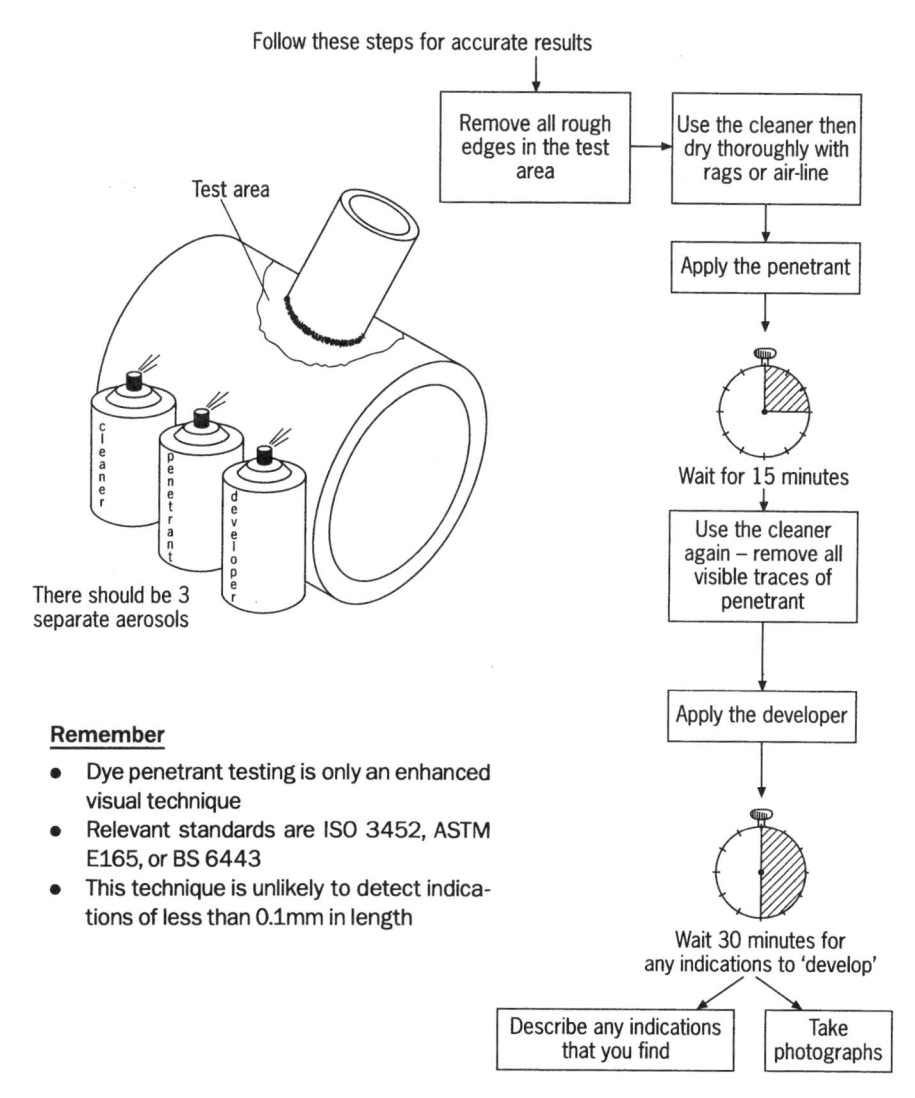

Follow these steps for accurate results:

- Remove all rough edges in the test area
- Use the cleaner then dry thoroughly with rags or air-line
- Apply the penetrant
- Wait for 15 minutes
- Use the cleaner again – remove all visible traces of penetrant
- Apply the developer
- Wait 30 minutes for any indications to 'develop'
- Describe any indications that you find
- Take photographs

Test area. There should be 3 separate aerosols: cleaner, penetrant, developer.

Remember

- Dye penetrant testing is only an enhanced visual technique
- Relevant standards are ISO 3452, ASTM E165, or BS 6443
- This technique is unlikely to detect indications of less than 0.1mm in length

Fig 5.13 Dye penetrant testing – the correct method

forms a discontinuity in the magnetic field which then attracts the magnetic filler in the ink and makes the defect visible. MPI enables large areas of material or weld to be tested without too much preparation. For this reason it is commonly used on large vessels and fabricated structures.

Theoretically, because the magnetic field penetrates some distance below the surface of the material, this technique can indicate the presence of sub-surface defects, as well as true surface defects. In practice this is difficult – an experienced MPI operator is required to spot anything less distinct than a pure surface defect. It is best to take the view that MPI, like DP inspection, can only provide an enhanced visual assessment.

There are several types of magnetic medium that can be used; dry red or blue powder, black magnetic ink, and fluorescent ink viewed under ultraviolet light are all in common use and do basically the same job. There are also several methods of applying the magnetic field to the component under test. Figure 5.14 shows the common arrangements. Note the following guidelines:

- MPI cannot be used on materials that are non-ferromagnetic. These include austenitic stainless steel and non-ferrous materials and alloys.
- MPI is better at detecting non-metallic surface inclusions than DP – for other defects its resolution is about the same.
- It is difficult to take good photographs of defects found by MPI. Make sure you identify any defect during the test itself and note its position on a sketch for inclusion in your inspection report. *Describing* the defect is important. There are three specific categories of defects that can be detected by MPI: cracks (you will often see these described more generally as 'crack-like flaws'), rounded indications, and linear indications (see Fig. 5.14). Your report should use these specific terms.
- There are two ways of applying the magnetic field; either permanently (the current is kept on while the ink is applied) or temporarily (the field is applied, removed and *then* the ink is applied); both are acceptable. In the second method the field is maintained by residual magnetism for sufficient time to apply the ink and identify any defects. If in doubt look at BS 6072 or ASTM A275/E1444.
- The technique is most sensitive when a flaw is orientated at 90 degrees to the magnetic field direction. When the angle falls below 45 degrees the sensitivity becomes poor.
- The most common technique error is not using *two* perpendicular field directions. This totally negates the usefulness of the test. Two directions are essential in order to locate defects in various orientations (see Fig. 5.14). Use the field tester, which contains 'test defects' to ensure that the field strength and ink properties are good enough.
- Beware of 'phantom indications' in corners and around sharp changes of section – particularly in castings. The correct terminology for these is 'non-relevant indications'.

- A paint coating of < 50 µm will not affect significantly the sensitivity of the technique – a thin coating of white contrast paint is often used to help the indications 'stand-out'.

Volumetric NDT

Volumetric NDT is the generic name given to those non-destructive techniques that identify defects in the body of a material or component rather than on the surface. You cannot properly find and identify surface defects using volumetric techniques. They are also more complicated than surface techniques and have greater variety. As an inspector your objective is *familiarity* with the two main techniques, radiography (RG) and ultrasonics (US), and how they are applied during works inspections, in particular on structural welding of pressure vessels and other statutory equipment.

Which technique – radiography or ultrasonics?

Ultimately, this is a question for you (and maybe your client). I will try and explain why. Radiographic and ultrasonic techniques are very different methods. There is little commonality in their method of use, the conditions under which they work best, or their capability in finding specific types of defects. Broadly, however, they both have do the same job. It is safe to say that for most applications they can be *considered* as being alternatives, whereas strictly they would be better described as being complementary. In practice you will rarely find them used together.

Much of what can be considered conventional wisdom relating to volumetric NDT comes from pressure vessel practice. The BS and TRD vessel standards are frequently used as a design basis for other types of fabricated equipment where a high standard of integrity is required. The content of these standards changes gradually by a process of amendments and addenda – they are documents of sound and proven practice, but also of consensus. A consequence is that these standards *do not like* to specify one of the radiographic or ultrasonic techniques in preference to the other. It is accepted that both exist, and are effective if used correctly and within their technical limitations, but the *choice* is deferred to 'by agreement between manufacturer and purchaser'. This means that the standards will not help you as much as you might like. Figure 6.8 in Chapter 6 shows what the vessel standards actually say about volumetric NDT on welding, plates, forgings, and castings.

114 Handbook of Mechanical Works Inspection

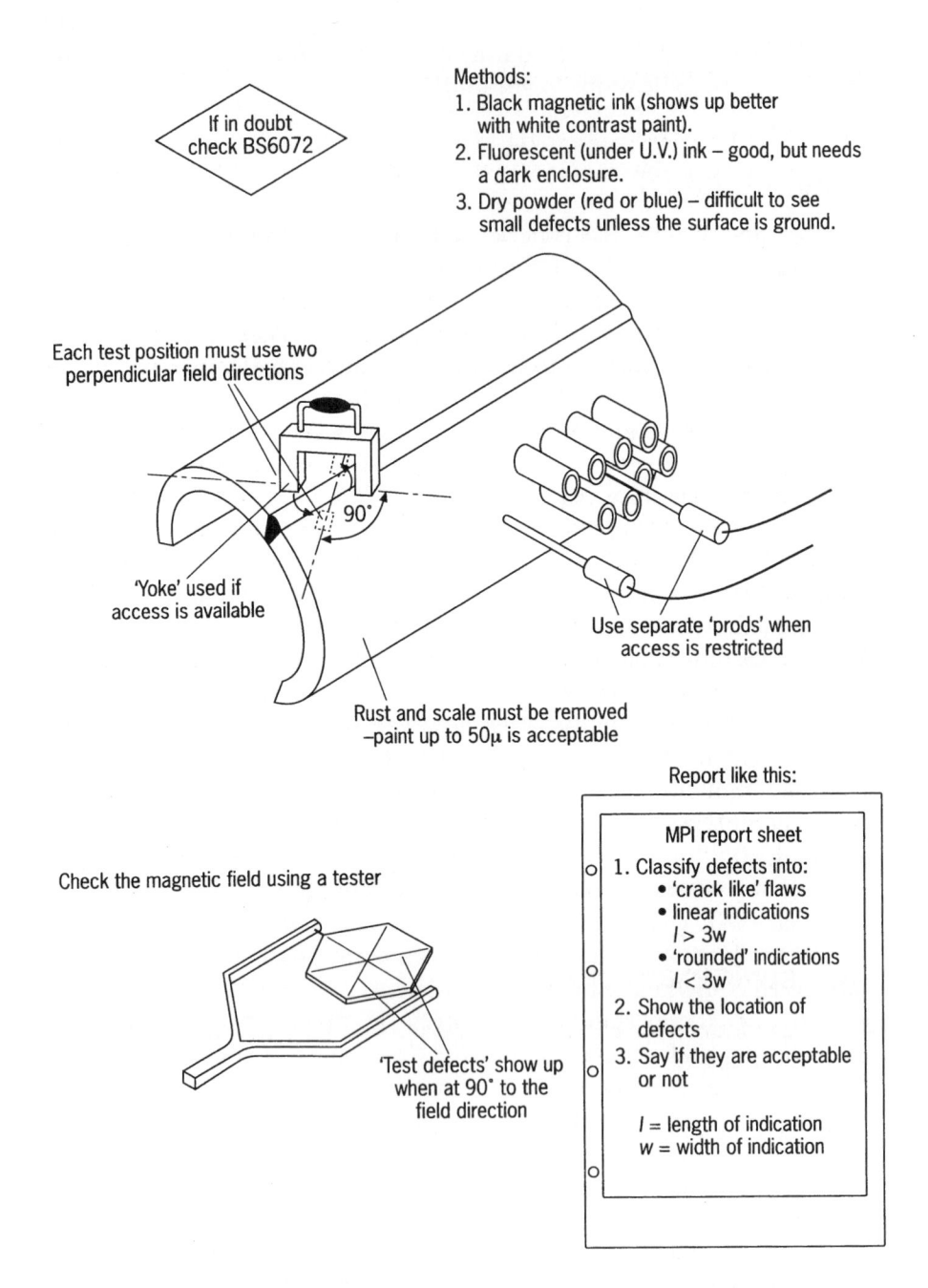

If in doubt check BS6072

Methods:
1. Black magnetic ink (shows up better with white contrast paint).
2. Fluorescent (under U.V.) ink – good, but needs a dark enclosure.
3. Dry powder (red or blue) – difficult to see small defects unless the surface is ground.

Each test position must use two perpendicular field directions

90°

'Yoke' used if access is available

Use separate 'prods' when access is restricted

Rust and scale must be removed – paint up to 50µ is acceptable

Check the magnetic field using a tester

'Test defects' show up when at 90° to the field direction

Report like this:

MPI report sheet
1. Classify defects into:
   • 'crack like' flaws
   • linear indications
     $l > 3w$
   • 'rounded' indications
     $l < 3w$
2. Show the location of defects
3. Say if they are acceptable or not

l = length of indication
w = width of indication

Fig 5.14 **Magnetic particle crack detection – guidelines**

Against this background, manufacturers are increasingly using ultrasonic techniques for NDT on structural welds, with the cautious approval of purchasers and classification societies. As an inspector you should follow the *agreement* made between the manufacturer and your client (the purchaser). Frankly, this often only manifests itself in the form of several levels of requirements, including various technical specification clauses, the unwritten requirements of the classification society, and broad contractual references to 'correct standards' or suchlike.

There are often technical contradictions between these requirements. What do you do in such a situation? Much depends on the precision with which the various clauses have been written. First, refer to the document hierarchy – look at the discussion of this in Chapter 3 – to bring the relative importance of the written requirements into focus. A key purpose of this document hierarchy is to reduce the incidence of contradictory technical requirements – use it now. In practice, there will be times when the document hierarchy will not help you, it may be indistinct (it shouldn't be), or it may subordinate tried and tested published engineering standards to some vague 'catch all' NDT-related clause in the contract specification. Welcome to the end of the line. What remains is engineering judgement – mainly yours, but also that of the other parties concerned: the manufacturer, contractor and classification society. I will not go into detail as to how you act with the other parties to reach the right conclusion (look again at Chapter 2 if you need reminding of this). We can look at the technical issues that are important – they are shown in Fig. 5.15. This is an area where you can always learn more. Devote a section of your inspection notebook to it and document your experiences for future use.

Ultrasonic examination

There are four main areas where ultrasonic NDT techniques are used: plate, castings, forgings, and welds.

Ultrasonic examination of plate material

Ultrasonic testing is used to check plate for laminations, which are extended flat discontinuities between layers of the rolled material. Plate for general structural purposes is not normally subject to a full lamination check, mainly because most plate is manufactured using a vacuum degassing method which reduces significantly the occurrence of

For plate materials	• US is the best. It is difficult to detect laminations using RG.
For castings	• For castings and thick weldments, the main issue is *access*. RG needs access to the reverse side. US can be used on very thick material but needs a good surface finish.
For forgings	• Most forged shafts are thicker than the practical limit of X-rays (about 150 mm). US is used in such situations.
For butt welds (full penetration)	• The main procedural difference is that RG provides a permanent record. • US is better for finding planar defects. RG is better for defects having volume, such as inclusions and lack of penetration.
For fillet welds	• RG can be used but US is better (and easier).
For tube-to-tube butt welds	• US is not suitable for very thin walled tube.
For tube-to-header and nozzle welds	• US is the best technique. RG is difficult due to problems of film 'creasing'.
For welds in stainless steel	• US is unreliable for welds in austenitic stainless steel because transverse waves cannot be reliably used.

Fig 5.15 Radiographic (RG) versus Ultrasonic (US) testing – technical issues

laminations. The relevant technical standard is BS 5996. Note how it treats the subject of acceptance criteria – it specifies various 'body grades' and 'edge grades' corresponding to different numbers of imperfections per unit area. The principle is that a plate is tested to see if it complies with the particular grade specified – if it fails, it may be downgraded to the next lower grade. A typical pressure vessel steel would be specified as body grade B4 and edge grade E3. Special high frequency probes are required for austenitic steels. Material that is 100 percent checked is often certified and hard-stamped by a third-party organization for use in statutory pressure vessels and cranes. Figure 5.16 shows points to check when witnessing this technique.

Examination of castings

Ferritic and martensitic steel castings can be examined using straightforward ultrasonic methods. The main technique is known as the A-Scope (or A-Scan) pulse echo method – the same technique is used for forgings. Figure 5.17 shows details. Note that it is a reflection

Welding and NDT

Material 'edge'-grades

Acceptance 'grade'	Single imperfection Max length (area)	Multiple imperfections Max no. per 1 m length:	above min length
E1	50 mm (1000 mm^2)	5	30 mm
E2	30 mm (500 mm^2)	4	20 mm
E3	20 mm (100 mm^2)	3	10 mm

- Use a 'pulse echo' A-scope technique (2–5 Mhz)
- Imperfections are delineated by area using a 'db drop' method
- If in doubt, look at BS 5996

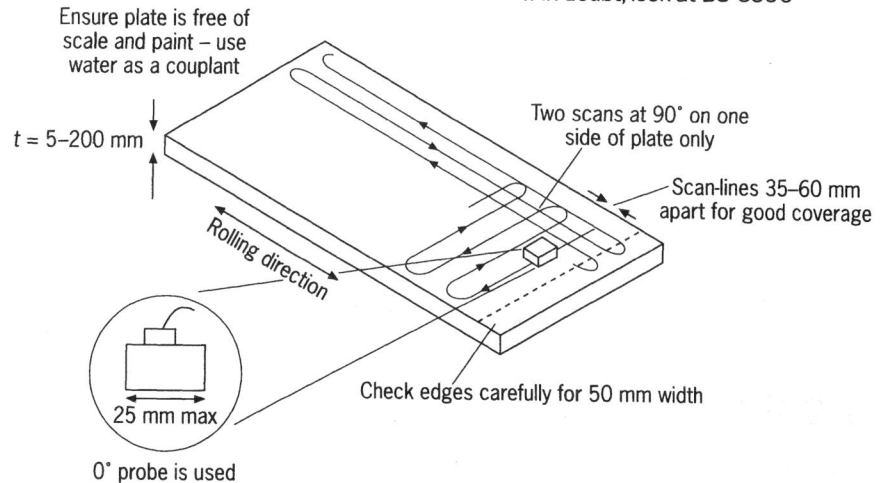

Ensure plate is free of scale and paint – use water as a couplant

$t = 5$–200 mm

Two scans at 90° on one side of plate only

Scan-lines 35–60 mm apart for good coverage

Rolling direction

25 mm max

0° probe is used

Check edges carefully for 50 mm width

Material 'body'-grades

Acceptance 'grade'	Single imperfection maximum area (approximate*)	Multiple imperfections Max no. per 1 m square	above min size (area)
B1	10 000 mm^2	5	100 mm × 20 mm (2500 mm^2)
B2	5 000 mm^2	5	75 mm × 15 mm (1250 mm^2)
B3	2 500 mm^2	5	60 mm × 12 mm (750 mm^2)
B4	1 000 mm^2	10	35 mm × 8 mm (300 mm^2)

* these are approximations—if results are marginal, look at BS 5996

Fig 5.16   Ultrasonic testing of steel plates

118 Handbook of Mechanical Works Inspection

Note these points:

- A 'pulsed' wave is used – it reflects from the back wall, and any defects.

- The location of the defect can be read off the screen.

The probe transmits and receives the waves

Signal amplitude

Transmission pulse

Couplant

d

Defect

Back wall

The A-scope screen looks like this

Defect echo — Backwall echo

The horizontal axis represents time – i.e. the 'distance' into the material

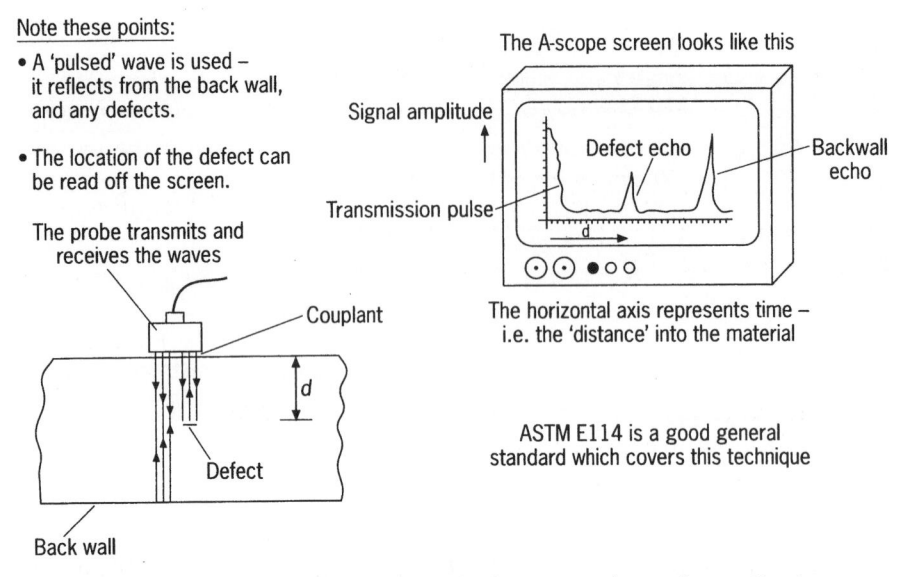

ASTM E114 is a good general standard which covers this technique

Fig 5.17   Ultrasonic testing – the A scope pulse-echo method

technique (the waves are transmitted and received by the same probe, which has two crystals) and that it uses compression waves; each wave oscillates longitudinally along the axis of the beam. The time scale, which represents distance into the object under test, is always displayed on the horizontal axis. The so-called B-Scope method uses a vertical time base but is not in common use.

Ultrasonic examination of castings is a much simpler technique than that used, for instance, to examine welds. Its main application is for castings with wall thickness greater than 30 mm and the techniques are the same whether the casting is designed for general or pressure service. Because of the complex geometry of many castings, full examination by ultrasonic means can sometimes be difficult. Together with the *nature* of the defects that occur in castings this leads to a certain amount of 'freedom' being incorporated into the applicable technical standards (such as BS 6208 and BS 4080) – they provide guidelines, rather than definitive rules to follow. This is particularly the case with acceptance criteria: standards are oriented towards describing 'grades' of castings, rather than defining absolute accept/reject criteria. Such decisions are left to agreement between manufacturer and purchaser – a concept that we have met in previous Chapters. Check the contract specification for any clauses that constitute this agreement, as there is often a section entitled 'general requirements for castings', or similar. If there are no clearly specified requirements, the correct thing to do is to follow the guidance on technique and defect assessment given in the relevant standard, then apply your judgement. As always, FFP is the key point.

The technique

The examination technique itself is relatively straightforward but there are a few particular requirements because of the characteristics of cast components (see Fig. 5.18). When witnessing the test, check the following points:

- A 0 degree (normal) probe is commonly used. It is difficult to use an angle-probe technique because of variations in the wall thickness of many castings (although you may see the angle-probe technique specified for turbine castings).
- The casting *wall thickness* should be divided into 'zones' as shown in Fig. 5.18.

It is of key importance, in order to make a proper assessment of the results, that the ultrasonic test of a casting follows a well-defined set of steps. You need to liaise closely with the test operator here – to keep track of what is happening. Look for these essential steps:

- *Calibrate* the equipment – first using test blocks and then on the area of greatest wall thickness of the casting itself.
- Do a *preliminary scan* over 100 percent of the casting surface. Do this at a frequency of 2 MHz, changing to smaller probes if necessary to accommodate tight radii. The purpose of this preliminary scan is to locate both planar *and* non-planar discontinuities, but not to investigate them fully at this stage.
- Before proceeding further *agree the technique* that will be used to 'size' both planar and non-planar discontinuities – they must be sized if the test is to have any real relevance in verifying FFP of the casting. The most common method is known as the '6 dB drop' technique. This is simply a method by which the operator defines the edge of a defect as that position which causes a 6 dB (50 percent of screen height) drop in the back wall echo. There is also a more sensitive method termed the '20 dB drop' technique.
- Assess the planar discontinuities for size by finding the location of the edges of each discontinuity.
- Assess the non-planar discontinuities for size – there are two parts to this; finding the upper and lower bounds of the discontinuity, and then finding (delineating) the edges, in order to calculate the overall volume.

- Record all the details carefully – it is necessary to describe the sizes and orientation of discontinuities before they can be properly compared with the grades shown in Fig. 5.18 for the purpose of assessment.
- It is normal to define critical areas of castings that will be subject to the most rigorous levels of assessment. These will be areas which will experience particularly high stress (such as blade roots on cast impellers) or where even small discontinuities would cause FFP problems such as areas that will be subsequently welded (note the valve weld-end shown in the figure). These critical areas will only be defined in the manufacturer/purchaser 'agreement' – you will not generally find them in technical standards.
- Castings with an austenitic grain structure can be very difficult to test. A general check for sound absorption should be carried out to see if a casting is suitable for being tested ultrasonically. Check with the operator – if permeability is greater than 15 dB, results may be poor and radiography is probably better.

Evaluating the results

In ultrasonic terminology, the purpose of testing castings is to identify *discontinuities*. There are two distinct types: planar (in one plane only with no thickness) and non-planar (having multiple dimensions and hence an area or volume). A key principle is that planar and non planar discontinuities are treated *separately* – each type having a different set of criteria by which the casting can be allocated a 'grade'. I have shown these criteria in Fig. 5.18. Note how the sizes of discontinuities are relatively large, compared to the very small defects that are important when examining welds. It is difficult to over-emphasize the importance of accurate, descriptive reporting in this area. Castings *will* contain discontinuities and it is not good enough just to refer to 'minor' or 'insignificant' defects in your report – such terms have no proper meaning in this context. Practise describing accurately what you find – and then apply your FFP judgement to it.

It should be clear from these explanations that ultrasonic testing of castings is far from being a perfect technique – some types of small planar discontinuities are difficult to find and the technique itself can be restricted by the complex geometry of the component being examined. Officially, ultrasonic testing of castings is probably still regarded as complementary to radiographic examination but you may find this

Welding and NDT

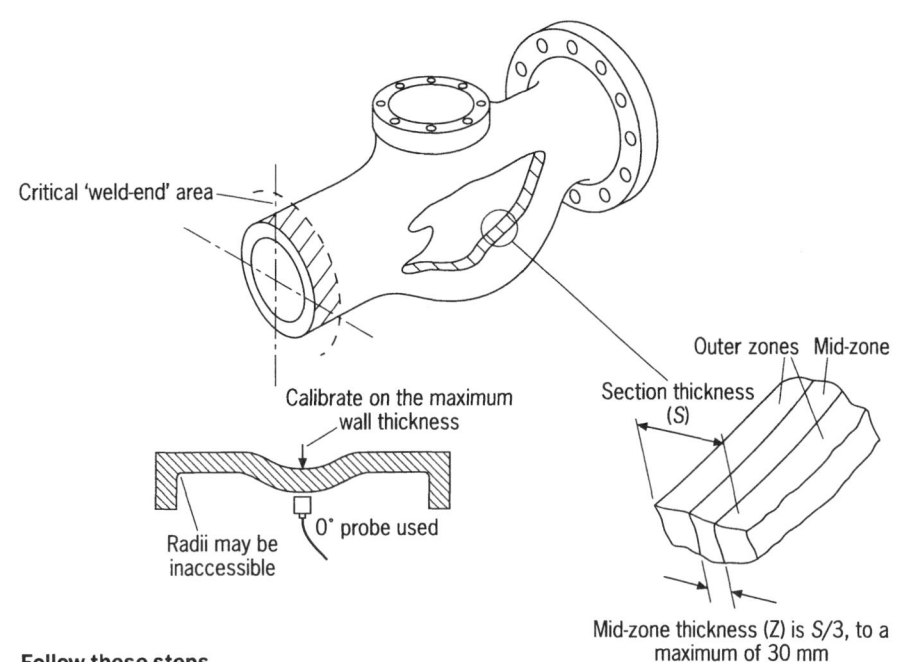

(Figure labels:)
Critical 'weld-end' area
Calibrate on the maximum wall thickness
Radii may be inaccessible
0° probe used
Outer zones   Mid-zone
Section thickness (S)
Mid-zone thickness (Z) is S/3, to a maximum of 30 mm

Follow these steps
- Check the suitability of the casting for an ultrasonic technique (the sound permeability)
- Do a visual inspection
- Do a preliminary scan – looking for both types of discontinuity
- Assess the planar discontinuities
- Assess the non-planar discontinuities (find their upper and lower 'bounds' and then delineate their area)
- Classify the results into grades, using the tables shown
- Report accurately
- If in doubt, check BS 6208 and BS 4080

Planar discontinuities	Grade 1	2	3	4
Max 'through-wall' discontinuity size	0mm	5mm	8mm	11mm
Max area of a discontinuity	0mm	75mm^2	200mm^2	360mm^2
Max total area* of discontinuities	0mm	150mm^2	400mm^2	700mm^2

Fig 5.18   The correct ultrasonic method for castings

continued on next page

Figure 5.18 — continued

Non-planar discontinuities	Grade			
	1	2	3	4
Outer zone max size	0.2Z	0.2Z	0.2Z	0.2Z
Outer zone max total area*	250mm^2	1 000mm^2	2 000mm^2	4 000mm^2
Mid zone max size	0.1S	0.1S	0.15S	0.15S
Mid zone max total area*	12 500mm^2	20 000mm^2	31 000mm^2	50 000mm^2

* All discontinuity levels are per unit (100 000mm^2) area

increasingly disproved in practice. Many experienced manufacturers use *only* ultrasonic methods for castings which have both high design pressures and are of complex shape. Steam turbine casings and high pressure steam valves are good examples that you will see.

Ultrasonic examination of forgings

For forged components that contain thick (greater than 150 mm) material sections, ultrasonic examination is the only type of volumetric NDT that can be performed. Its most common use is to examine rotors for turbines, generators, gearboxes, and pumps (smaller rotors will be solid and the larger ones hollow), as well as specialized forged components such as globe valves and high pressure thick-walled vessels. The NDT procedure for all such components normally comprises 100 percent ultrasonic and 100 percent surface crack detection by DP or MP – it is rare to find 'critical areas' defined as is often the case with castings.

Although both general and component-specific technical standards are available, you will find that much of the standards' content is devoted to the detailed techniques of ultrasonic examination rather than to the key aspect of defect acceptance criteria. The main 'general' standard BS 4124 **(10)**, does not address acceptance criteria at all, instead referring to the necessity for agreement between manufacturer and purchaser to define what is acceptable. From an inspector's viewpoint it can therefore be a difficult task to interpret any discontinuities found in terms of their effect on FFP, even though the examination technique itself is relatively straightforward.

Welding and NDT

The technique

Forgings are normally ultrasonically tested twice; before machining (when only a rough assessment is possible) and then after final heat-treatment and machining. Figure 5.19 shows the basic techniques that are used for the most common types of forgings that you will see. Note the following points:

- An A-scope presentation is used, as shown in Fig. 5.17.
- Some specialized forgings may have an additional activity, the 'near surface defect test' specified. This needs a high frequency probe

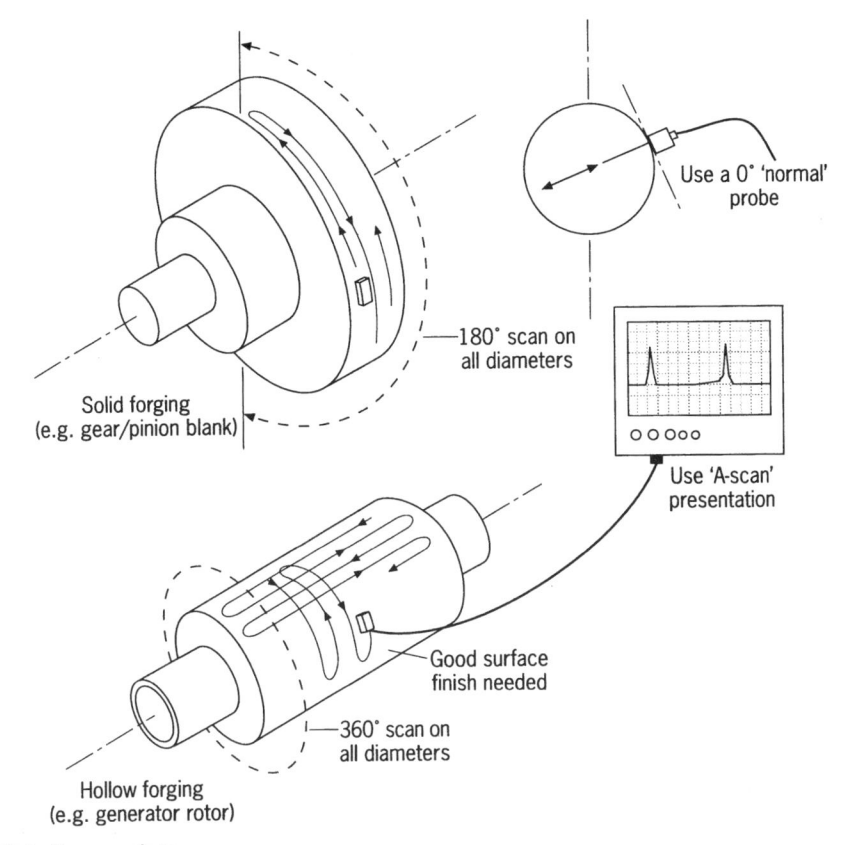

Note these points
- There is no common 'grading system' for discontinuities
- Acceptance criteria are difficult to define – much relies on specific agreements between manufacturer and purchaser
- BS 4124 is a good general standard. ASTM A418 specifically relates to rotor forgings

Fig 5.19   Ultrasonic testing of forgings

(10 MHz) to minimize the thickness of the near-surface 'dead zone' that is a feature of ultrasonic examination.
- The scanning method is usually quite simple on items with constant cross-section. It may be more complicated for components such as valves, which have more complex geometry.
- A normal (0 degree) probe is used for all preliminary scans. It may be necessary to use angle probes to accurately locate and size discontinuities.
- There are four basic methods of sizing discontinuities. The 20 dB drop and 6 dB drop methods outlined previously cannot be used in *all* applications and so their use is limited. The 'distance-amplitude' and 'distance-gain size' techniques have greater application. You can find descriptions of these techniques in BS 4124, or in NDT textbooks. Check which one is being used when you witness a test.
- It is important for the size of a discontinuity to be accompanied by a description of the technique that has been used to *determine* that size – it is a fact that sizing techniques are difficult, so it is wise to expect (and perhaps accept) some variability in the results.

Ultrasonic examination of welds

Ultrasonic examination is being increasingly accepted in industry as a technically robust method of verifying the integrity of welds in ferritic steel. Many manufacturers use ultrasonic examination as the exclusive volumetric NDT method for pressure vessel welds. The ultrasonic techniques used are more involved and varied than those used for standard forgings, castings, or plate. Compression and transverse 'shear' waves are used, the ultrasonic beam being transmitted using a variety of normal and angle probes from several locations on and around the weld. Results are displayed using the A-scope presentation. The technique cannot always be used effectively on austenitic steels unless special high-frequency probes are used, as the grain structure distorts the transverse wave-form.

A large number of weld types such as butt, nozzle, fillet and branch welds, can be examined using ultrasonics. Each weld-type has its own 'best technique' for finding defects. From a works inspector's viewpoint this means that ultrasonic weld examinations present a challenge of *comprehension* – the need to understand a range of techniques if you are to be able to witness such tests competently. Fortunately, help is available in the form of some particularly informative technical standards that address this subject.

Useful technical standards

The most useful standard is BS 3923 (Part 1). This covers ultrasonic examination of all the common types of welds that you will meet in vessels and other general engineering components. It is a very practical standard, showing the best techniques to use. There are explanations on how to identify defects but it does not give acceptance criteria – you are simply referred to the relevant 'application standard', which will often be a pressure vessel code. The equivalent American standard is ASTM E164: 1992 – this follows broadly the same principles as BS 3923 but is not quite so easy to use in a practical works situation.

An important point introduced by the technical standards is that of *levels of examination*; BS 3923 describes three levels of examination, levels 1, 2, and 3, for each type of welded joint thereby allowing three different 'degrees of verification' of weld integrity. The requirements for surface condition, number and extent of scans, probe type and weld surface conditions vary between the levels. Level 1 is the highest level of examination.

From an inspection viewpoint, level 2 examination is the *minimum required* on any welded joint to allow weld integrity to be verified for fitness for purpose. You should consider level 3 examination as 'advisory only' and not capable of finding all the possible weld defects that can affect the integrity of a welded joint. For this reason Fig. 5.20 is based on level 2 examination.

The weld assessment

There is a well-defined routine to follow when performing an ultrasonic test on a weld. The basis is the same for butt, nozzle, and fillet welds. When you witness these tests during a works inspection, make sure that five basic steps have been performed carefully – they need a little time to do correctly. The five steps are:

- Check the *parent material*. This is based on the BS 5996 method shown in Fig. 5.16. The purpose is to check the areas of parent material that the beam will pass through when the weld itself is being examined and to confirm the material thickness so that the beam paths for the angle probe 'views' can be determined. The examples in Fig. 5.20 show how the angle beam paths need to be reflected from the inside surface of the parent material in order to assess the weld itself.
- Check the *weld root*. This is most relevant to butt welds that have a separate root run (often MIG). For single-sided welds using a

Set-on nozzle with both bores accessible

- Scan A-B with two different angle probes
- Scan C (or D and E if C is inaccessible)
- Check BS 3923 for the right technique for a particular joint configuration

Note: these techniques are all for level 2 examination. This is the minimum level to ensure FFP.

15–50 mm

15–50 mm

Remember
- The most likely location of defects is in the weld ROOT
- If in doubt, look at BS 3923

Double-sided butt weld – both sides are accessible

Dress cap to 'near flat'

- Scan A-B with two different angle probes (for root and longitudinal defects)
- Normal scan C
- Longitudinal scan D with angle probe (for transverse defects)

Set-through nozzle – one bore only is accessible

15–100 mm

Inaccessible

- Scan A with angle probe
- Scan B-C with two different angle probes

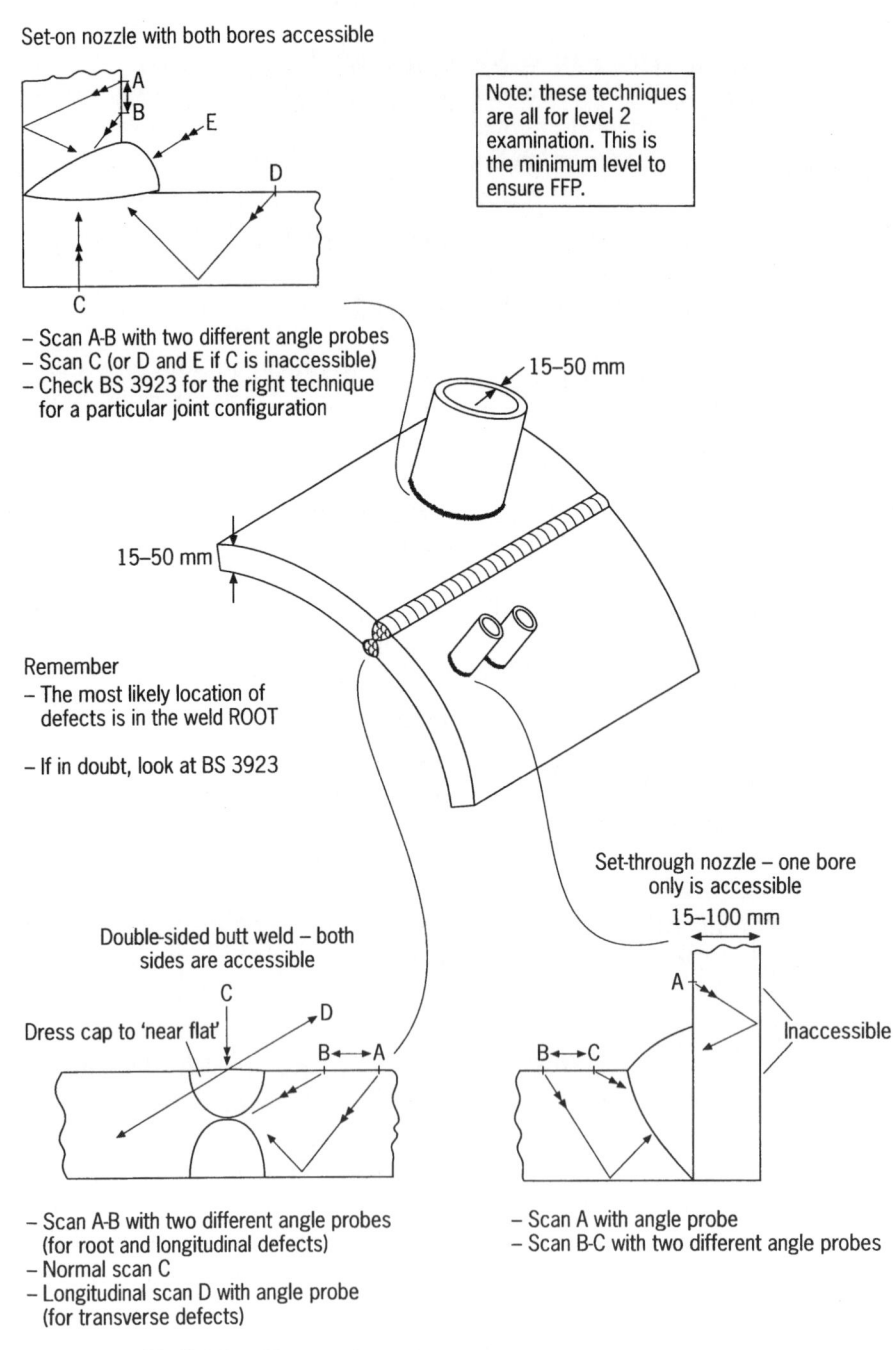

Fig 5.20   Ultrasonic testing of butt and nozzle welds

backing strip, the root will be undressed and must be examined for root cracks and incomplete root penetration. Note that the examination should be done twice, once from either side of the weld. The technique for double-sided welds is shown in Fig. 5.20.
- Check using a *normal* (0 degree) probe. All butt welds should be scanned over the surface of the weld with a 0 degree probe to detect any root concavity, lack of fusion into the backing strip, or poor inter-run fusion. Make sure that:
    - the weld cap is ground smooth (BS 3923 categories SP3 or SP4) to allow a proper examination.
    - probe frequency is a minimum of 4 MHz.
    - for thick material (say greater than 50 mm), separate scans are done from the upper and lower surfaces of the weld.
- Scan for longitudinal defects. Here, the beam angles and scanning direction are transverse (across the weld) in order to detect any defects that lie longitudinally along the weld axis. It is common to use several different angle probes – the best chance of detecting a defect is when the beam hits the surface of the defect at 90 degrees. Watch for cracks or lack of fusion lying along the fusion faces.
- Look for transverse defects. These are found by scanning along the axis of the weld. Note the following guidelines:
    - the weld cap should be ground smooth (BS 3923 grade SP3)
    - probe angle should be within 20 degrees of the normal. Two probes are needed on welds more than 15 mm thick.
    - thick welds (greater than 50 mm) should be examined from both sides.
    - corner and T-joints need special scanning techniques: refer to BS 3923 if you encounter these.

Evaluating defects

The evaluation of defects is a tricky operation. Although the principles are straightforward, it is the large variety of beam paths and different defect indications that makes it difficult. For this reason it is essential that you work closely with the ultrasonic operator when a defect is found. This is not an area where the inspector can afford to 'stand-back', relying on a later, more considered, analysis of the results – you have to gather your information whilst the test is being done. The best advice I can give is that you use the detailed information and checklists given in BS 3923 Part 1 whilst bearing in mind the *principles* under which defect assessment operates. These are:

- The assessment is related to the examination level that is being used. For level 1 examination, any indication above background noise (also called 'grass') needs to be investigated. For levels 2 and 3, only indications that cause a significant echo are investigated. The specific echo 'sizes' (related to screen height) are shown in BS 3923: note that they are very technique-specific.
- As a rule of thumb, any echo of more than 50 percent screen height (or a corresponding 50 percent reduction in back wall echo) merits investigation.
- Defects must be classified (by type), described and measured if they are to be reported with any real meaning. This means that the operator should complete a detailed report sheet when a defect is found. Be careful also of the way in which *you* report defects found – you must include full details of the type and size. It is poor practice just to refer to 'defects' without further qualification.

Radiography

Despite the fact that the *mandatory* requirement for radiographic NDT is weakening, you will still find it used extensively in general engineering manufacture. It has the clear advantage of producing a permanent visual record of the results of the examination, enabling the results to be reviewed and endorsed by all the necessary parties. A key technical advantage of radiography is its ability to identify important types of volumetric defects (porosity, inclusions, lack of fusion, and lack of penetration among others) and categorize them in terms of size. This makes radiography very suitable for finding those common types of defects found in multi-layer welds.

A lot of the technology and standards of radiographic (RG) testing come from pressure vessel practice. Hence you will find that the technical aspects of RG examination are very well covered by the various technical standards shown in Fig. 5.6 and in Chapter 6. You can expect definitive guidelines. Defect types are also well described. With acceptance criteria – as we discussed earlier – some interpretation and judgement is inevitably required.

In this section we will look at a typical radiographic test application, a full penetration butt weld and nozzle welds in steel pipework. Such

Welding and NDT

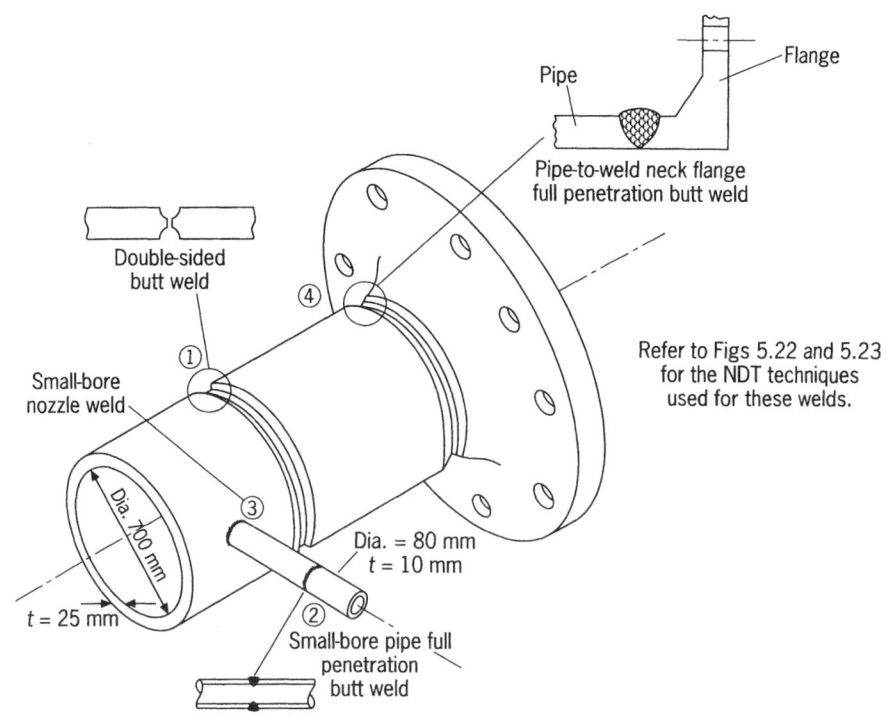

Fig 5.21 High pressure steam pipework – welds for volumetric NDT

pipework would be used for high temperature application under the ASME B31.1 code or one of the equivalent European standards.

An effective inspection consists of a well defined series of steps. We will follow these through in order.

Check the technique

Figure 5.21 shows the position of the four welds to be examined. Weld No. 1, the large-bore pipe butt weld, can be accessed from both sides. It can be easily radiographed using a single-wall technique. Weld No. 2 is a small-bore full penetration pipe butt weld. It can only be easily accessed from the outside, so it needs to be examined using a 'double-wall' method. In this case a double-wall, double-image technique is shown. Weld No. 3 is a nozzle weld of the set-through type, as the small tube projects fully through the pipe wall. This weld could be radiographed but the technique would be difficult, mainly due to the problem of finding a good location for the film. Attempts to bend the film around the outside of the small diameter nozzle weld would cause the film to crease, giving a poor image. As a general principle, nozzle-to-shell welds are more suited to being tested by ultrasonic means – radiography is difficult and impractical. Weld No. 4, the pipe-to-flange joint is of a common type often termed a 'weld neck' flange joint. Note that this is a single-sided weld with a root 'land' – it qualifies therefore as a full penetration weld and would require full NDT under most specifications and codes. This joint is capable of being radiographed using a single-wall technique similar to that shown for weld No. 1, as long as there is sufficient room for the film. That is, the width of the weld that can be examined should not be restricted by the position of the flange. In most cases you will find that there will be sufficient clearance. It would also be possible to use an ultrasonic technique on this weld if required, scanning from the pipe outer surface and flange face.

We will look in detail at welds 1 and 2. Before reviewing the results of any radiographic examination it is wise to look at the details of how the examination was performed – the examination *procedure*. This may be a separate document, or included as part of the results sheet. I have shown an example of 'technique information' for each of welds Nos 1 and 2 in Figs 5.22 and 5.23. Check the following points:

- *Single or double-wall technique?* Weld No. 1 is a single-wall technique – the film shows a single weld image shot through a single weld 'thickness'. Ten films are required to view this weld around its entire circumference. In contrast, weld No. 2 is a double-wall technique, suitable for smaller pipes. The offset double-image technique shown is used to avoid the two weld images being superimposed on the same area of film, and hence being difficult to interpret. Note the difference in *class* between the single- and double-wall techniques shown in the figures – single-wall is class A (from BS 2910).
- *Position of the films.* Note for weld No. 1 how the ten films, each approximately 230 mm in length, locate around the circumference. Starting from a datum point their position is indicated by lead numbers, so the position of any defects seen in the radiograph can be identified on the weld itself.
- *The type of source used.* You only need do a broad check on the suitability of the source *type*. For all practical purposes this will be either X-ray or gamma ray.
- X-rays produced by an X-ray tube are only effective on steel up to a material thickness of approximately 150 mm. Check the voltage level

A single-wall X-ray technique is used:
- the large bore allows access
- the source-outside technique is commonly used for large pipe

The technique looks like this:

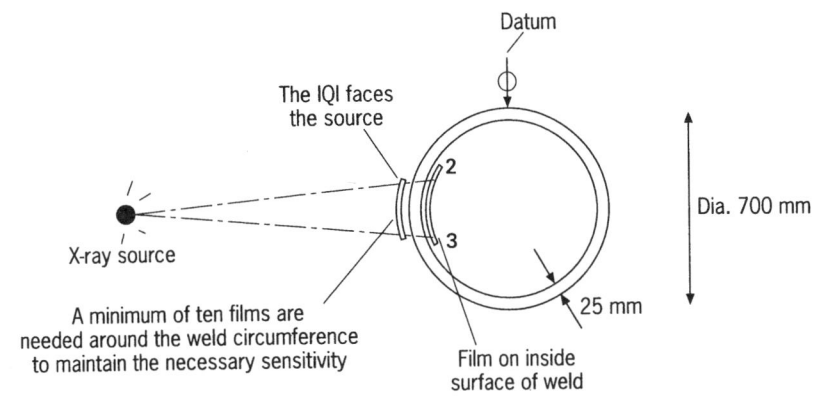

[Diagram: X-ray source outside a 700mm diameter pipe, with IQI facing the source, film on inside surface of weld, 25mm weld thickness, datum at 12 o'clock position, positions 2 and 3 marked. A minimum of ten films are needed around the weld circumference to maintain the necessary sensitivity.]

and is specified like this:

Specification	Explanation
Single wall	Only one weld 'thickness' shows on the film
Technique no. 1	A reference to BS 2910 which lists techniques nos 1–16
Class A	X-ray and single wall techniques give the best (class A) results
Fine film	BS 2910 mentions the use of fine or medium film grades
X-220kV	The X-ray voltage depends on the weld thickness

Notes
- The IQI is BS 3971 type I (BS EN 462–1)
- For 25mm weld thickness the maximum acceptable sensitivity given in BS 2910 is 1.3%. This means that the 0.32mm wire should be visible
- The IQI should be located in the area of <u>worst</u> expected sensitivity

Film marking (using lead symbols)

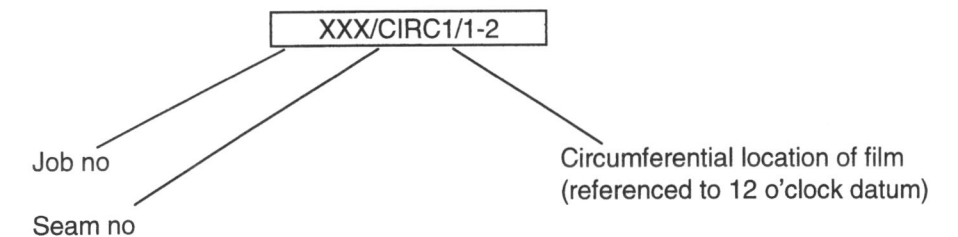

XXX/CIRC1/1-2

- Job no
- Seam no
- Circumferential location of film (referenced to 12 o'clock datum)

Fig 5.22 RG technique for large bore butt weld (No. 1)

A double-wall, double image technique is used.
The technique looks like this:

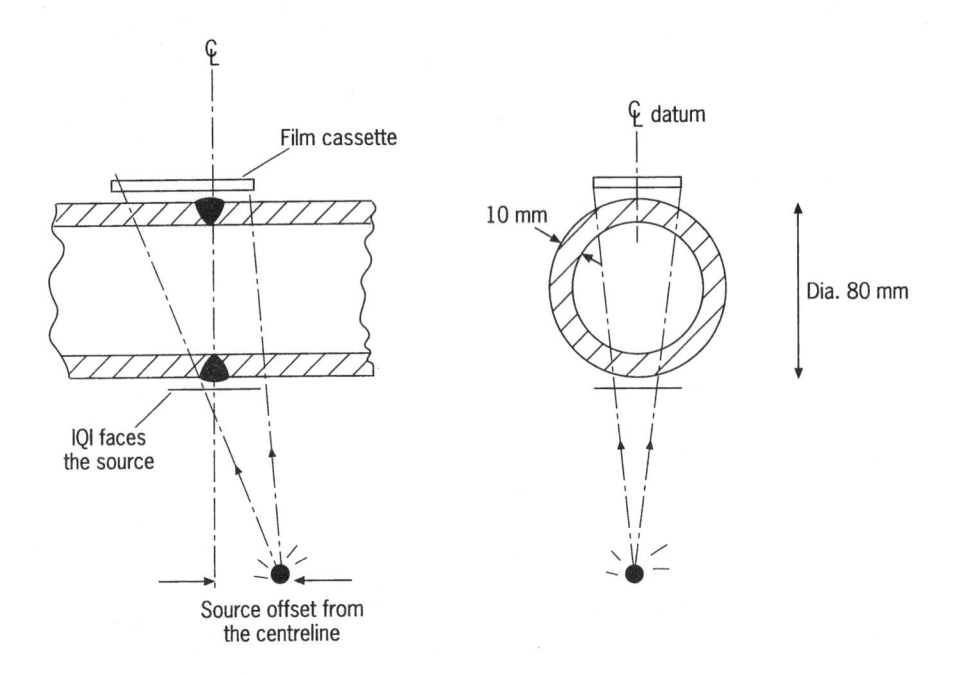

(Diagram: Film cassette above pipe with source offset from centreline below; IQI faces the source. Right side shows cross-section with 10 mm wall, Dia. 80 mm, ℄ datum.)

and is specified like this:

Specification	Explanation
Double-wall/image	Two weld 'thicknesses' show
Technique no. 13	A reference to BS 2910 which lists techniques 1–16
Class B	Double-wall techniques are inferior to single-wall methods
Fine film	BS 2910 mentions the use of fine or medium grades
Density 3.5–4.5	The 'degree of blackness' of the image

Notes
- The IQI is BS 3971 type I (BS EN 462–1)
- For 2 × 10mm thickness, the maximum acceptable sensitivity given in BS 2910 is 1.6%. This means that the 0.2mm wire should be visible
- A minimum of three films are required, spaced at 0°, 120°, and 240° from the datum
- The maximum OD for this technique is 90mm
- X-ray gives better results than gamma in techniques like this
- The source is offset to prevent superimposed images

Fig 5.23 RG technique for small bore butt weld (No. 2)

of the source unit: a weld thickness of up to 10 mm requires about 140 kV. Thicknesses in excess of 50 mm need about 400–500 kV. A practical maximum is 1200 kV.
- Gamma rays, produced by a radioactive isotope, can be used on similar thicknesses but definition is reduced. If a Cobalt 60 gamma-ray source is used, the best results are only obtained for material thickness between 50 and 150 mm. It does not give good results on thin-walled tube welds.
- It is not possible to compare accurately results obtained by X-ray and gamma ray methods.

Check the films

Although 'real-time' viewing methods are technically feasible, most works techniques use a photographic film which is developed after exposure. A visual record is therefore provided for review by all parties concerned. To check radiograph films properly consists of six essential steps. It is best to work through these in a structured way to ensure that the films are reviewed thoroughly.

- Check the film *location*. Make sure you relate each film to its physical location on the examined weld or component. Use the procedure sketch and position markers but *also* look for recognizable features such as weld-tees, flanges and surface marks to perform a double-check.
- Check *sensitivity*. This is a check to determine whether the radiographic technique used is 'sensitive enough' to enable defects to be identified if they exist. It is expressed as a *percentage* – a lower percentage sensitivity indicating a better, more sensitive technique. A typical quoted sensitivity is 2 percent, indicating that, in principle, the technique will show defects with a minimum size of 2 percent of the thickness of material being examined.
- The actual value of sensitivity is determined using a penetrameter, also called an image quality indicator (IQI). The general principle is that one of the IQI 'wire' diameters is a given percentage of the material thickness being examined. If this wire is visible, it shows that the RG technique is being properly applied. There are several commonly used types that you will see, they are described by BS EN 462-1 and ASME/ASTM E142 standards. Figures 5.24 and 5.25 show two types and how each is used to calculate the sensitivity. Note that you cannot make an accurate comparison between different

The European standard for IQIs is BS EN 462

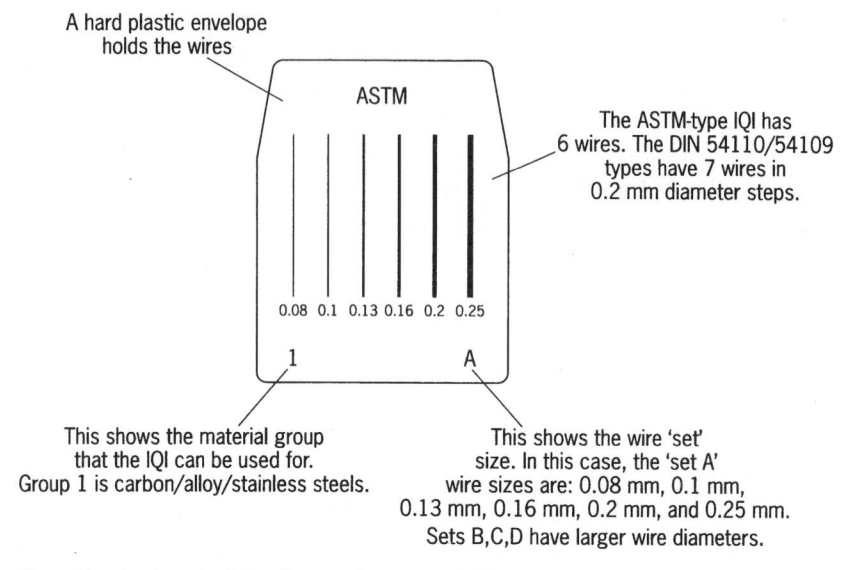

A hard plastic envelope holds the wires

ASTM

The ASTM-type IQI has 6 wires. The DIN 54110/54109 types have 7 wires in 0.2 mm diameter steps.

0.08 0.1 0.13 0.16 0.2 0.25

1   A

This shows the material group that the IQI can be used for. Group 1 is carbon/alloy/stainless steels.

This shows the wire 'set' size. In this case, the 'set A' wire sizes are: 0.08 mm, 0.1 mm, 0.13 mm, 0.16 mm, 0.2 mm, and 0.25 mm. Sets B,C,D have larger wire diameters.

- The objective is to look for the smallest wire visible
- Sensitivity = diameter of smallest wire visible/maximum thickness of weld
- If the above IQI is used on 10mm material and the 0.16mm wire is visible, then sensitivity = 0.16/10 = 1.6%
- Check the RG standard for the maximum allowable sensitivity for the technique/application being used

Fig 5.24 Using the wire-type penetrameter (IQI)

types – it is essential therefore to quote a sensitivity value against a specific IQI type for it to be capable of proper verification. If sensitivity is worse than that specified then the technique is unacceptable – it is not capable of finding relevant defects. From an inspector's viewpoint, there is normally little point in proceeding further. Issue an NCR at this point. Don't try and conclude *why* the sensitivity is poor – this is a complicated issue involving all of the technique parameters. Leave it to the specialists.

- Check *unsharpness* – properly termed 'geometric unsharpness'. This is a measure of the contrast of the image – the difference between the dark and light areas. Although methods have been devised to measure the degree of contrast is still involves subjectivity. Check that the image is 'sharp and defined', without any obvious blurring – you can get a reasonable impression by looking at the IQI. If you do find

Welding and NDT

The IQI number is shown here. This represents the thickness (t) in 0.001 inches. e.g. no. 20 is 0.020" thick

Nos. 10 to 180 are in common use

If in doubt, look at ASME V SE–1025 (identical to ASTM E1025)

The IQI has three holes, of diameter t, 2t, and 4t as shown

Dia. 4t
Dia. t
Dia. 2t

Note: the *thinner* the IQI (as a percentage of joint thickness) the *better* the sensitivity.

IQIs for use on non-ferrous material are designated by a series of notches. Steel ones have no notches.

Image quality designation is expressed as (X)-(Y)t:
(X) is the IQI thickness (t) expressed as a percentage of the joint thickness
(Y) (t) is the hole that must be visible

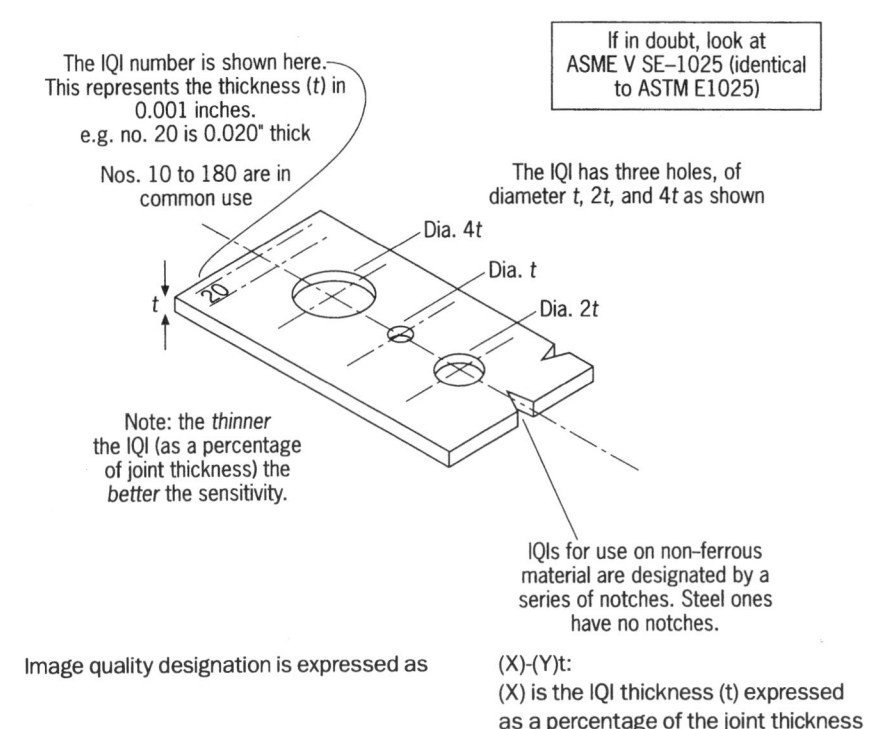

IQI designation	Sensitivity	Visible hole*
1–2t	1	2t
2–1t	1.4	1t
2–2t	2.0	2t
2–4t	2.8	4t
4–2t	4.0	2t

* The hole that must be visible in order to ensure the sensitivity level shown

Fig 5.25 How to read the ASTM penetrameter (IQI)

problems with unsharpness, check the technique sheet for details of the source-to-film and object-to-film distances used – incorrect distances are one of the common causes of unsharpness. Radiographic standards such as BS 2910 give the minimum distances required.

- Check the film *density*. You can think of density as the 'degree of blackness' of the image. It is determined by an instrument known as a densitometer. Typical values should be between 3.5 and 4.5. There is a separate standard, BS 1384 **(11)** if you need further details.

- *Now* view each film for *indications*. The principle is that indications (loosely termed 'defects') are identified and then classified according to type and size. Luckily, defect *classifications* are well accepted and defined – there is much less room for interpretation than with, for instance, the acceptance criteria for defects, once classified. Figure 5.6 included reference to standards containing reference radiographs. These show what the various indications look like, for both X-ray and gamma ray techniques. In practical works inspection situations, such reference radiographs are used less often than you might expect – a simpler list of generic defect types is normally used. You can expect to see a list like that shown in Fig. 5.26 appearing on the radiographic test results sheet. Defects are broadly classified as either major or minor. Major defects are those which impinge directly on the *integrity* of the weld – expect these to form the core content of the acceptance criteria. Major defects often need repairs. Defects classified as 'minor' mainly appear on the surface of the weld and have generally less effect on the weld integrity. You can see them visually. Note the abbreviations used for the various defects shown in Fig. 5.26; there is no universally accepted list of abbreviations but I have shown some common ones. Check that the radiograph test report sheet includes a key for the abbreviations it uses. All test reports tend to use abbreviations and present the results in tabular form.
- Marking up the films. This is a key role for the inspector. Work through the test report sheet viewing each radiograph film in turn, checking the reported indications carefully against what you see on the films. Mark important indications on the film with a chinagraph pencil, using the correct abbreviations. If a repair is required, mark the film 'R' in the top right hand corner, making a corresponding annotation on the report sheet. Keep each packet of films and its report sheet together and check they are correctly cross referenced in case they become separated.

These six steps are common to all radiograph review activities. Although routine, reviewing radiographs is a key aspect of checking for fitness for purpose. There are a few procedural guidelines which you may find useful:

- Before you start checking the films *ask* the manufacturer to summarize the results. Ask whether they are all acceptable and which acceptance criteria have been applied. *Wait* for the answer before you pick up the first film.

- Check the films showing indications *first*. Start with the worst ones, those exhibiting the major indications shown in Fig. 5.26. Don't waste time discussing minor indications or surface marks.
- Be aware of the acceptance criteria specified in the common pressure vessel standards (I have summarized these in Chapter 6). Surprising amounts of defects such as porosity *are* allowed – the amount depends on the particular criteria being applied. You must take these criteria into account. You cannot normally demand absolutely clear radiographs.
- It is best if any defects have already been repaired – but ask to look at both 'before' and 'after' radiographs to satisfy yourself what has been done.
- Remember the principle (see Chapter 2) that it is your job to guide a consensus on FFP. This is an important point in technical areas like this where interpretation and judgement are *always* part of the equation. Don't neglect the experience of the Third-Party inspector if you are reviewing weld radiographs relating to a statutory equipment item.

Destructive testing of welds

In contrast to NDT methods, which are essentially predictive techniques, the only *real* way to verify the mechanical integrity of a weld is to test a piece of it to destruction. There are two occasions on which you may be called to witness a weld destructive test; for the purposes of the welding PQR (described earlier in this chapter), or for the testing of a 'test plate' – properly called a production control test plate. The test plate often represents a pressure vessel longitudinal seam weld. It is tack-welded to the end of the seam preparation, welded at the same time, and then cut off and tested. Figure 5.27 shows the general arrangement and some points to check.

When are test plates required?

This is not an easy question to answer. The tendency of the main pressure vessel standards is to move away from specifying mandatory test plates for steel vessels. This has influenced the classification societies and independent Third-Party organizations with the result that they now take a similar view. With such loosening of prescriptive guidance the situation often reverts to the principle of whether test plates are specified in the 'agreement' between

Defect classification: points to look for

'MAJOR DEFECTS'

Defect	Appearance	Abbreviation (as DIN 8524)
Crack (longitudinal)	Fine dark wavy line along the weld axis	Ea
Crack (transverse)	Fine line across the weld axis	Eb
Isolated pores	Rounded indication darker than the surrounding image	Aa
Linear porosity	Joined-up lines of pores	Ab
Worm holes (piping)	Sharply defined, elongated indications	Ab
Slag inclusions	Sharp-edged dark shadows with irregular contour.	Ba
Poor root penetration	Continuous or intermittent dark band along the centre of the weld.	D
Poor root, side or inter-run fusion	Dark lines or bands along the weld.	C

'MINOR DEFECTS'

Defect	Appearance	Abbreviation (as DIN 8524)
Grinding mark	Light curved mark on weld surface.	–
Hammer mark	Light, rounded mark on weld surface.	–
Spatter	Easily visible on the weld surface.	–
Undercut	Shadows at edge of weld cap.	F
Root concavity	Shallow groove in the root of a butt weld.	–
Surface slag	Visible on the surface as a dark line.	–

Fig 5.26 Defect indications in weld radiographs

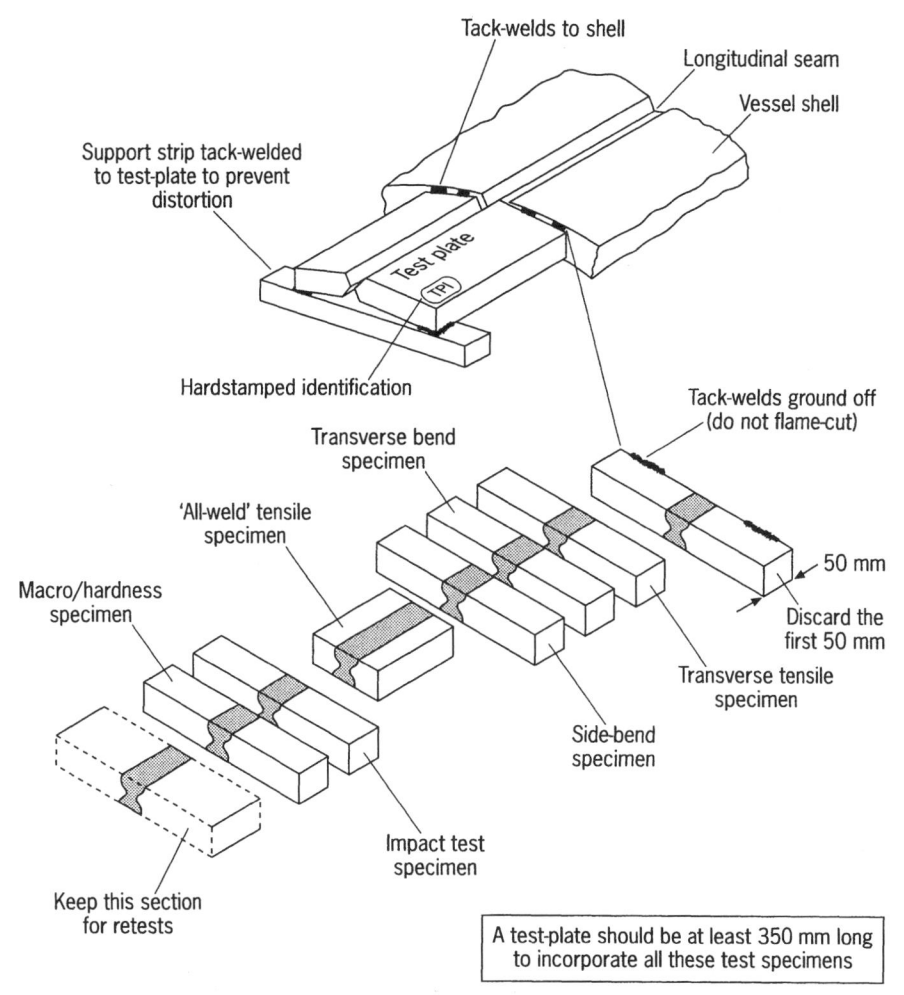

Fig 5.27 Test plate arrangement – vessel longitudinal seam

manufacturer and purchaser. This means that you have to check the contract specification. In the absence of a specific clause relating to test plates for steel pressure vessels, you can be *reasonably* safe in the assumption that they are not required.

You *will* see test plates specified for vessels or fabricated structures that use very thick material sections, complex alloys or unusual (and fundamentally unproven) weld procedures, or where the essential variables are particularly difficult to control. Here are some examples:

- steam turbine valve chest-to-casing fillet welds

- seamed boiler headers
- structural welds on crane bridges
- longitudinal welds on boiler drums (using material greater than 25 mm thick)
- welds between dissimilar materials.

Test acceptance criteria?

Once again, this is not an easy question to answer. Good guidance is hard to find. The main standard you will meet is BS 709 'Destructive testing of fusion welded joints and weld metal in steel' **(1)**. This standard explains the methodology of the tests but does not contain any advice on acceptance criteria. A document that does address acceptance criteria is PD 6493 **(12)** which is a British Standards 'Published document' so its status is more advisory than definitive. This is another of those areas where you have to use your judgement. To help you in your task I have provided, under each test heading, an outline of general practice for assessing the results of the tests. Treat these as guidance only – during witnessing such tests you will have to make an individual decision based on consideration of the material, type of welding and the application of the component.

The tests

The tests themselves are based on the mechanical tests covered in Chapter 4, but with the addition of special bend tests. Tensile, impact, and hardness tests follow the same methodologies as described, but with some relaxation on the requirements for specimen size and shape. This takes into account the physical difficulties of obtaining standard sized test specimens from a welded joint, particularly in thin steel plate. Refer to BS 709 if you need more details. The five main tests are shown in Fig. 5.28.

Transverse tensile test

This is a straightforward tensile test *across* the weld. Special reduced-section test pieces are acceptable and gauge lengths may differ, but otherwise the test follows the general principles explained in Chapter 4. Note the following guidelines on acceptance criteria:

- *The location of breakage is important.* A good weld should have the same structure and physical properties across the weldment, HAZ, and parent metal so the failure point could, in theory, be anywhere

1. Transverse tensile test
Weld cap ground off flat

2. 'All-weld' tensile test

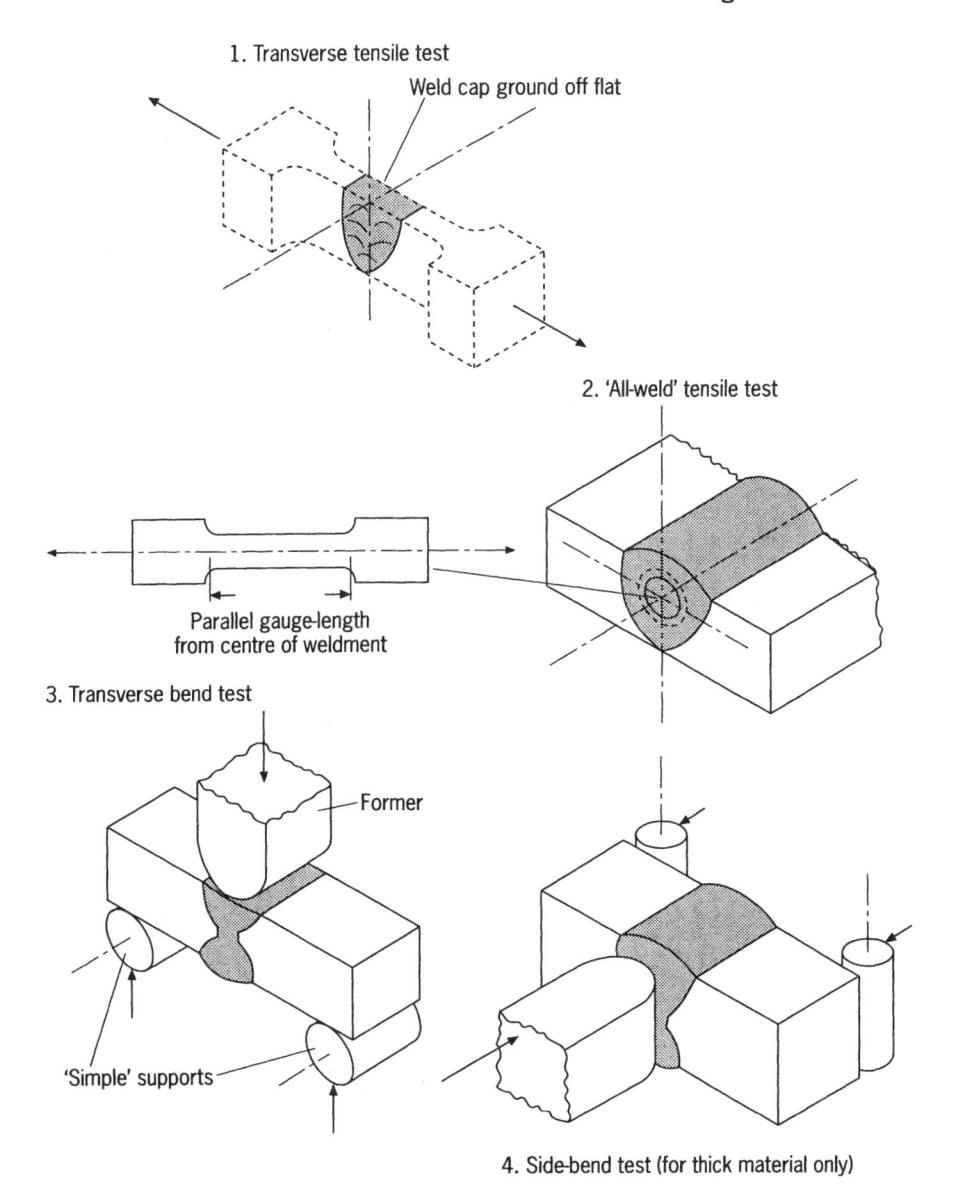

Parallel gauge-length from centre of weldment

3. Transverse bend test

Former

'Simple' supports

4. Side-bend test (for thick material only)

Fig 5.28 Tensile and bend tests

within the gauge length of the specimen. Be wary if the breakage occurs on clearly-defined planes *along the edge* of the weldment/HAZ interface – this could indicate problems of poor fusion or incorrect heat treatment.

- *Tensile strength*. There is no real reason why the yield/UTS values should be any different (less) than those of an unwelded specimen of parent metal. This means that you can use the parent material specification as your acceptance criteria. You may wish to allow an extra tolerance of, say, 10 percent on the values (perhaps to allow for inherent uncertainties in the welding process) but there is no stipulated reason why you *have* to. The same applies to elongation and reduction of area.
- *Fracture surface appearance*. You should see a quite dull 'cup and cone type' fracture surface, as you would expect on a normal tensile test specimen. Look for any evidence of marks on the fracture surface that would show evidence of metallic or non-metallic (slag) inclusions.

All-weld tensile test

'All-weld' tests are not very common. A tensile test is performed on a specimen machined out longitudinally from the weld. The gauge length diameter should be accurately located along the middle of the weld (see Fig. 5.28) so that it is all weldment. The main application of this test is for thick welds which have only single-side access or use a welding technique which is difficult to control. Guidelines on acceptance criteria are the same as I have given for the transverse tensile test.

Transverse bend test

Bend tests in which the welded specimen is bent round a roller or former are used to verify the soundness of the weldment and HAZ and to check that the HAZ is sufficiently ductile. The transverse test is used for both single- and double-sided welds. General acceptance criteria are:

- There should be no visible tearing of the HAZ, or its junction with the weldment.
- Cracking or 'crazing' of the material viewed on a macro level ($\times 200$ magnification) are unacceptable – it indicates weaknesses in the metallurgical structure caused by the welding.
- Check the *inside* radius of the bend, this is the area that is under compressive stress. This may show rippling or 'kinking' around the bend areas, but look for any radial cracks (perpendicular, or at an acute angle, to the specimen edge). These are initiated on slip planes in the metal structure and are early indicators of microstructural weakness.

Side bend test

This tests the soundness of the specimen *across* the weld using a similar former/roller set-up to the transverse bend test. It is generally only carried out on thicker materials (more than 10 mm). Although not normally specified directly, it is useful to take several side-bend specimens, at least one from each end of the test plate, and compare the results. This gives some indication of the consistency of the welding technique in a longitudinal direction (along the weld length). It is particularly useful for manual welds which are more likely to vary along the weld run. General guidelines on acceptance criteria are similar to those for the transverse bend test. It is worth paying particular attention to the edges of the weldment region – you will occasionally find problems here with multi-run welds on thicker material.

Macro and hardness test

This is quite a common test, see Fig. 5.29. It consists of etching and polishing a machined transverse slice through the weld then subjecting it to a close visual examination and hardness check. The specimen is polished using silicon carbide paper (down to P400 or P500 grade) on a rotating wheel. The metal surface is then etched using a mixture of nitric acid and alcohol (for ferritic steels) or hydrochloric acid, nitric acid, and water (for austenitic steels). You should check that the correct mixture has been used.

The size of test specimen is important. Make sure that the specimen is at least 10 mm thick, as shown in Fig. 5.29, and that there is enough material either side of the weld (in a transverse direction) to allow a

For HV 5 load, use 0.7 mm spacing in the HAZ
For HV 10 load, use 1 mm spacing in the HAZ
Use 2–3 times these spacings in the weldment and parent metal zones

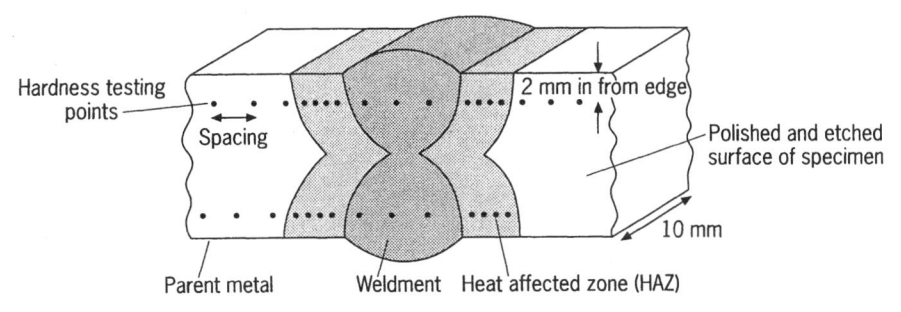

Fig 5.29 Macro/hardness checks across a weld

representative examination of the HAZ and parent material. About 10 mm of parent metal either side of the HAZ is ideal. Check that the weld cap has been left on.

Once polished and etched, a hardness 'gradient' test is carried out. This consists of testing and recording hardness values (any of the methods we discussed in Chapter 4 can be used) at closely spaced intervals across the weldment, HAZ, and parent metal regions. The *objective* is to look for regions which are too hard or areas in which the hardness gradient is steep, i.e. where hardness varies a lot across a small area. Either of these are potential weaknesses – they can cause fractures to be initiated. Note also the following points when you witness or review this test:

- Check the etched cross-section first by eye or using a hand-held magnifying glass. Look for obvious internal flaws, lack of fusion or inclusions (you are unlikely to see any surface defects on a transverse section like this). Identify and classify any defect you find. Do a location sketch.
- Examine the section using a microscope. Observe and compare the microstructure of the weldment, HAZ, and parent metal regions. Pay close attention to the *interfaces* between these regions – this is where you are most likely to find problems. Watch out for:
    - microcracks or voids
    - dramatic variations in the appearance of the microstructure across the weld
    - microscopic inclusions
    - an undesirable 'needle-like' appearance to the microstructure.

  If you see any of these, issue an NCR.

  The appearance of microstructure varies significantly between steels but as general guidance, for a low carbon pressure vessel steel, you should see a quite even medium grain structure with only gradual changes across the weldment, HAZ and parent metal regions.
- Concentrate on the hardness test, as there are some special points that need to be followed. Check that the line of test points is located accurately 2 mm in from the surface of the specimen, as I have shown in Figure 5.29. This is to eliminate erratic readings that can sometimes be found near the surface. The critical area for hardness is the HAZ – BS 709 specifies a spacing of 0.7 to 1.0 mm between test points in this region. This is close enough to allow a fair estimate to be made of the hardness gradient. Spacing in the body of the weldment and the parent metal region can be greater (perhaps 2–3 mm), as the expected hardness gradient will be lower in these areas.

Note how in Fig. 5.29 I have shown additional test points along the weldment/HAZ boundary. This is a good general practice to follow – it provides additional comfort that there are no areas of excessive hardness in this critical region.

Acceptance criteria for hardness levels require judgement. As a guide, you could require the hardest reading to still lie within the specified limits for the parent material: look at the material specification for this. It would not be unacceptable to add, say, +10 percent (but no more) to this as an additional tolerance. For maximum acceptable hardness *variation* across the three regions, a figure of 10 percent is sometimes used. I think this is a little restrictive, being based on tolerances that are allowed for hardness readings on unwelded materials. Perhaps 15–20 percent is better. These are only general guidelines. You have to use engineering judgement.

Other destructive tests which are sometimes done are; impact (Charpy V-notch), intercrystalline corrosion test (for austenitic stainless steel and pipe welds), fillet weld fracture test, and the fracture toughness (KIc/COD) test. All these tests are less common than the ones I have described. Look at BS709 if you need details.

Test reporting

Be careful how you report the results of destructive weld tests. Because acceptance criteria are not very well defined you have to present sound engineering descriptions to give foundation to the decisions that you make – particularly if you consider the test results unacceptable and issue (correctly) a non-conformance report. Close, concise technical description is what is required. Look forward to Chapter 15 on inspection reporting and the points I have made about 'logical progression' when you are writing technical descriptions in your reports.

Describe the results of destructive weld tests in more detail than you would for standard mechanical tests which have better-defined acceptance criteria. Here is a short checklist of points that you should cover:

- Describe the test specimen, its size (or gauge length) and where it came from in the test plate. Make explicit reference to whether or not it complies fully with BS709.
- For tensile and impact tests, describe *where* the fracture occurred, and the *appearance* of the fracture surface. Classify any flaws using the correct terminology (don't just refer to 'defects').

- For bend tests, record the angle of bend, the cross-sectional area of the specimen and the final radius of the inside face (the one under compression as the specimen is bent).
- For macro examination results, use accurate metallurgical terms to describe the microstructure. Try to avoid statements such as 'the microstructure was good' that don't actually mean anything. Better to say 'the HAZ did not show any evidence of undesirable martensite'. Refer to a text-book to clarify the terminology – but stick to basic descriptions.

KEY POINT SUMMARY: WELDING AND NDT

1. Welding and NDT are separate disciplines, but for inspection purposes you can consider them as being closely *linked*.

2. The key FFP criterion is material and weld integrity. The role of a works inspector is to *verify* this integrity – not to participate in detailed discussions about techniques.

3. Be prepared to use your judgement when considering acceptance criteria – the situation will *not* always be absolutely clear.

4. Weld defects have predictable causes and effects.

5. Full penetration welds need the most stringent NDT – this is due to the high probability of any weld defects being located in the weld root, which cannot be ground (back gouged) if access is restricted.

6. Welding 'controlling' documentation (WPSs, PQRs and welder approvals) is important – but it is not an end in itself. Don't concentrate too much on documentation instead of inspecting the weld itself.

7. Surface crack detection: dye penetrant (DP) and magnetic particle inspection (MPI) techniques really only provide an enhanced visual assessment.

8. Volumetric NDT: many equipment standards accept *either* radiographic or ultrasonic techniques – they do not find it easy to express a preference.

9. Try to learn a little about the various ultrasonic techniques (particularly the examination of welds) so you can understand what is happening when you witness a test.

10. Normally an inspector *cannot* demand absolutely 'clear' radiographs – you have to use the agreed acceptance criteria.

11. Destructive tests consist of tensile, bend and macro examination/hardness tests across the weld section. Learn to describe accurately the results of these tests.

References

1. BS 709: 1983. *Methods of destructive testing fusion welded joints and weld metal in steel.*
2. BS 2633: 1987. *Specification for Class I arc welding of ferritic steel pipework for carrying fluids.*
3. BS 2971: 1991. *Specification for Class II arc welding of carbon steel pipework for carrying fluids.*
4. BS 4570: 1985. *Specification for fusion welding of steel castings.*
5. BS EN 288 Parts 1 to 8: 1992. *Specification and approval of welding procuedures for metallic materials.*
6. BS EN 287-1: 1992. *Approval testing of welders for fusion welding.*
7. BS EN 26520: 1992. *Classification of imperfections in metallic fusion welds, with explanations.* This is an identical standard to ISO 6520.
8. BS EN 25817: 1992. Arc welded joints in steel – guidance on quality levels for imperfections.
9. BS 5289: 1983. Code of practice. *Visual inspection of fusion welded joints.*
10. BS 4124: 1991. *Methods for ultrasonic detection of imperfections in steel forgings.*
11. BS 1384 Parts 1 (1985) and 2 (1993). *Photographic density measurements.*
12. PD 6493: 1991. *Guidance on methods for assessing the acceptability of flaws in fusion welded structures.*

Chapter 6

Boilers and pressure vessels

Introduction

The inspection of boilers and pressure vessels forms a major part of a works inspector's role. Any power or process plant will have a large number of vessels for different applications. Some are complex, such as those forming the component parts of steam-raising plant or large condensers; others such as air receivers and low pressure or atmospheric vessels are of relatively simple design and construction. The scope is wide, but all present the inspector with a similar task.

In this chapter we will look at general principles and technical issues, drawing together relevant areas from previous chapters dealing with standards, materials of construction and NDT. Vessels provide a useful vehicle for showing how different inspection disciplines mesh together in a practical works inspection context. We will also look at the implications of the fact that many pressure vessels are subject to statutory requirements. I will attempt to explain the nature of *statutory certification* of vessels, why it is required and how it affects your inspection activities in the works. My objective is to try and dispel some of the confusion and misunderstandings that can easily arise in this area. This chapter contains figures giving 'technical summaries' of the most common examples of applications of pressure vessel practices and standards. These should cover most of the vessel types you will meet. Each summary contains technical information which you should find of direct use whilst you are doing works inspections and also when you prepare your inspection reports. Reports come later – the first step is to find some focus. We have to look at FFP.

Fitness-for-purpose-criteria

Pressure vessels can be dangerous. In use, they contain large amounts of stored energy. For this reason the main (and overriding) FFP criterion is *integrity* – we want to know that the vessel is safe and is not going to fail. Clearly there are many engineering aspects which contribute to this integrity; design, creep, fatigue, corrosion-resistance, and skill in manufacture, all have an effect. It would not be wrong to say that these all *contribute* to safety – but I think the term integrity is better, it is a more global definition.

How do you verify this integrity? Fortunately, over the past 100 years or so, industry has developed a set of 'norms'; commonly accepted activities (call them a system if you like) which together have the objective of assuring the integrity of manufactured pressure vessels. These norms contain some hard rules, and some softer rules and conventions – I am sure they are imperfect, but I ask you to accept that they are about the best you are going to find. I cannot recommend any better ones. Figure 6.1 shows the four norms. They are: independent design appraisal, traceability of materials, NDT, and pressure testing. I have shown them as being constrained within a closely controlled manufacturing process. This is an important control mechanism – look how it pushes and holds the four elements together. Your role as an inspector is to verify that all these elements are in place, have been completed, and that the control mechanism has indeed worked. By doing this you become, albeit in a slightly detached way, a part of the control mechanism.

In a real works inspection situation there are often several parties involved. The manufacturer, main contractor and statutory inspector all have an *interest* in FFP, separated perhaps by some slight differences in focus. I introduced this scenario of duplication of inspectors and inspections in Chapter 2 (make a quick review of this section now if necessary). Pressure vessel works inspections are the most common places for such duplication to occur – it is best to expect it. Try to rationalize things by accepting that this is an inherent part of the system of norms that has developed over time. If you are involved in the inspection of pressure vessels, you will very soon meet the issue of statutory certification. We will look now at how this forms a part of the concepts of integrity and FFP.

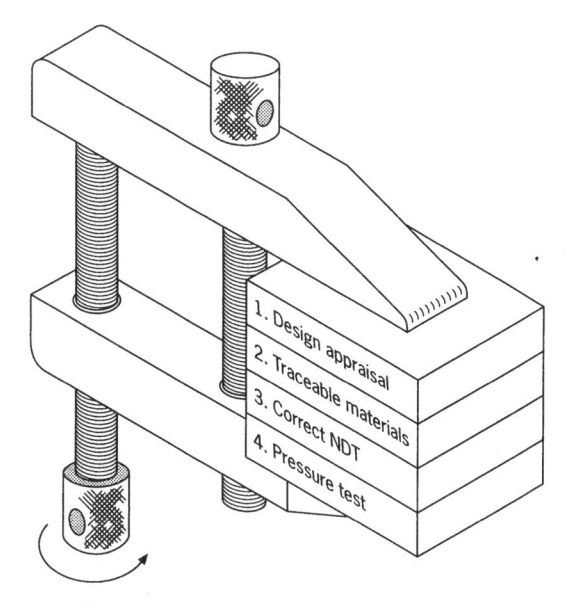

Manufacturing 'control'

Vessel integrity (and safety) is obtained by:
- Arranging for an independent design appraisal
- Using traceable materials
- Applying proven NDT techniques
- Doing a hydrostatic (pressure) test

and then

- Exerting proper (*meaning enough*) control over the manufacturing process

Fig 6.1 The pressure vessel 'norms'

'Statutory' certification

There are a number of common misunderstandings surrounding the statutory certification of pressure vessels. Don't be surprised if you occasionally find some confusion, even amongst experienced companies and their staff, as to what certification means and implies. It is useful therefore to have an understanding of the key issues so you can act effectively during a works inspection.

Why is certification needed?

There are four possible reasons why a pressure vessel needs to be certificated. They are:

- The need for certification is imposed or inferred by statutory legislation in the country where the vessel will be *installed and used*.
- The need for certification is imposed or inferred by statutory legislation in the country where the vessel is *manufactured*.
- The need for certification is imposed or inferred by the company that will provide an insurance policy for the vessel itself, and second and third party liabilities, when it is in use.
- The manufacturer, contractor or end-user *chooses* to obtain certification because he feels that:
  - it helps maintain a good standard of design and workmanship
  - it provides evidence to help show that legal requirements for 'due-diligence', and 'duty of care' have been met.

You will note that three of these reasons are as a result of certification requirement being imposed (or at least *inferred*) by an external player, whilst the other is a voluntary decision by one or more of the directly involved parties. Perhaps surprisingly, this voluntary route accounts for more than 30 percent of vessels that are certified. Of the other 70 percent, which are subject to certification because of the requirements of other parties, it is safe to say that some of these perceived requirements are undoubtably imaginary rather than real. The main reason for this is that there are some countries in the world where statutory requirements are unclear, sometimes contradictory, or non-existent. The more risk-averse vessel manufacturers and contractors often assume that certification *will* be necessary, even if evidence of this requirement is difficult to find.

What is certification?

Probably not *precisely* what you think. Certification is an *attempt* to assure the integrity (my term) in a way that is generally accepted by external parties. It uses accepted vessel standards or codes such as BS 5500 **(1)** and TRD **(2)** as benchmarks of acceptability and good practice. Certification does address issues of vessel design, manufacture and testing, but only insofar as these aspects are imposed explicitly by the relevant standard – not more. Certification is evidence therefore of *code-compliance*. Note that compliance with the ASME code **(3)** is a special case – it can be a statutory requirement in the USA.

So if a pressure vessel is fully certified then there is hard evidence that the vessel complies with the requirements of that design code or standard stated on its certificate.

Boilers and pressure vessels

In order to obtain full certification, the organization intending to issue the certificate will have to comply with the activities raised by the relevant code. These differ slightly between codes but the basic requirements are the need to:

- perform a quite detailed design appraisal
- ensure the traceability of the materials of construction
- witness NDT activities and review the results
- witness the pressure test
- monitor the manufacturing process
- issue a certificate (BS 5500 'form X' or its equivalent).

Notice the similarities between these activities and the norms in Fig. 6.1. Look also at the one main difference: the inspection presence of the certification body is a valuable contribution to the 'control mechanism' of the manufacturing process *but it is not all of it*. It is only a part. To avoid misconceptions it is useful to look at what certification is not, and to examine a few of its limitations. Vessel certification is normally:

- Not a *guarantee* of integrity (there can be no such thing).
- Not a statement of fitness for purpose.
- Nothing to do with project-specific engineering aspects of the vessel such as the position of nozzles, dimensional accuracy, corrosion resistance, mounting arrangements, instrumentation, or external painting and internal protection (including shotblasting and preparation).
- Not a facility for the manufacturer to off-load contractual or product liability responsibilities on to the certifying organization; (who will rarely actually *approve* drawings or documents – they will simply be stamped 'reviewed'). Similarly, detailed technical aspects such as concessions are more likely to be 'noted' rather than 'agreed' as part of the certification process.

Often it is these limitations of the certification exercise which are misunderstood, rather than the objectives of the inspections that are carried out by the certifying organization.

Who can certificate vessels?

There are two aspects to this, *independence* and *competence*.

Independence

The main pressure vessel codes, including BS 5500 and TRD, require that vessels, if they are to comply fully with the standard, are certificated by an organization which is independent of the manufacturer. This is generally taken to mean that there should be no direct links in the organizations that would cause commercial (or other) pressures to be imposed on the objectivity of the certificating organization's actions.

Competence

In many European countries (this can be where the vessel is to be either manufactured or used) there exists a restricted list of organizations that are deemed competent to perform the certification of vessels. By definition these organizations have been assessed, usually by governmental authorities, for both independence and competence. ASME vessels are, once again, a special case.

In the UK, you will simply meet the statement that vessel certification has to be performed by a 'competent' person or organization. Don't expect to see detailed guidance as to what qualifies a person or organization as being competent. Competent status will normally *only* be challenged if there is an accident, failure incident, or negligence claim – then the onus will be on the person or organization claiming the status to prove it. There are various registration schemes in existence by which organizations may try to improve their perceived status as a 'competent body'. In the UK, for conventional (non-nuclear) plant, it is a fact that *none* of these are imposed directly by any statutory instrument or regulation. They are all voluntary. The majority of such schemes involve the organization submitting to audits based on BS EN ISO 9001 **(4)**, sometimes with additional requirements added. Of those organizations that do provide certification services for pressure vessels, some subscribe to such schemes because they see some benefit. Others do not. Please draw your own conclusions.

In the eyes of the technical standards (excluding ASME) that *require* certification (or 'inspection and survey' is the term sometimes used), all organizations that can meet the independence criteria have the same status. This means that any of the following organizations can do it:

- a classification society
- an insurance company

Boilers and pressure vessels 155

- an independent inspection company
- a government department
- (in theory) an uninvolved manufacture or contractor.

Remember that this rather loose situation is often tightened by the commercial preferences of insurance companies, plant purchasers and end-users.

There are no hard and fast rules which state that all parts of the certification process have to be performed by the same organization. You may find in practice that the design appraisal part is separate. In some industries it is traditionally done by a 'design institute' or indigenous classification society – the works inspection part is then sub-contracted to a separate inspection company after competitive tender. This is perfectly acceptable, as long as the certifying organization is able to accept the validity of the design appraisal performed by the other party. They must also be able to issue a full and unqualified certificate for the vessel. You may find Fig. 6.2 useful, it illustrates particularly the independence aspect of the certifying organization. Also, look again at Fig. 3.3 in the 'Specifications, standards and plans' chapter (Chapter 3). The best way of understanding certification is to consider these figures together – both have something to contribute.

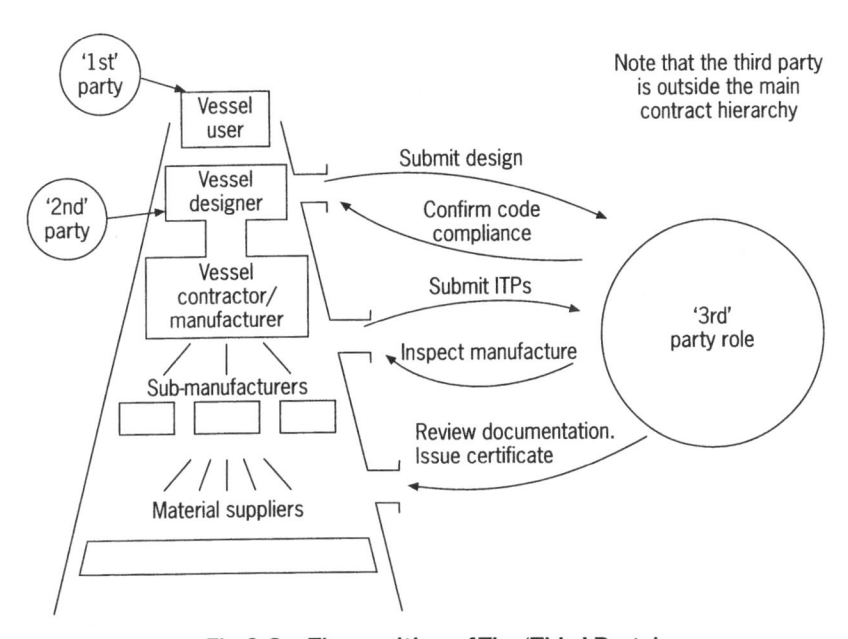

Fig 6.2　The position of The 'Third Party'

Working to pressure vessel codes

It is difficult to find a contract specification for an industrial plant that does not refer to one of the series of internationally recognized technical codes and standards for pressure vessels. These are more generally referred to as pressure vessel 'codes'. One of the core tasks of works inspection is to work within the requirements of these codes – or more precisely, you need to learn the techniques of *inspecting to* a specific code. This is not easy to do well without a little forethought.

The main problem is the size and sheer complexity of the code documents. Each has many hundreds of pages making it quite impractical to carry the codes with you to a works inspection. Even if you were able to carry a microfiched or computer-disk version (these are available) you would still find difficulty. This is due mainly to the extensive list of associated technical standards to which the codes refer. The answer is *selectivity*. Remember that you need only be concerned with those parts of a vessel code that have an impact on the activities of works inspection. Look carefully at the vessel codes and try to get an overall insight into the structure of the documents. Although there are some differences between the American (ASME) and European-based (BS, TRD, etc.) codes, both groups follow similar principles. For the design of the pressure vessel, most of the detailed information is incorporated in the code document itself, whereas for the inspection aspects (and to a lesser degree the testing requirements) then the main purpose of the code document is to cross-reference applicable subsidiary technical standards. The result is that vessel codes have a strong tendency to be design-orientated (rather, perhaps, than *inspection-orientated*).

This means that to *inspect* a vessel to a pressure vessel code you only need a limited, and therefore manageable, amount of information from that code. You then also need a selective amount of information from subsidiary technical standards. The best place to start is with Fig. 6.3. Look carefully at the content. It lists *only* those pieces of information that an inspector needs from a vessel code. The subjects coincide broadly with the entries that you will find in the index of a vessel code. You can use this figure as a general guide, before considering one of the specific codes. For more specific details on *where* to find this information in a code, see Figs 6.4 to 6.6, which summarize the situation for the BS 5500, ASME VIII, and TRD codes. Use the information in these figures as a place to start. If you are inspecting to one of the vessel codes that I have not listed (there are many, but most are not commonly used for international projects), then start with the index reference of the subject and progress from there. Be

1. Responsibilities

First, you need information on the way that the code allocates *responsibilities* including:
- Relative responsibilities of the manufacturer and the purchaser.
- The recognized role of the independent inspection organization.
- The technical requirements and options that can be agreed between manufacturer and purchaser.
- The way in which the certificate of *code compliance* is issued, and who takes responsibility for it.

2. Vessel design

Important points are:
- Details of the different construction *categories* addressed by the vessel code.
- Knowledge of the different classifications of welded joints.
- Knowledge of prohibited design features (mainly welded joints).

3. Materials of construction

You need the following information about materials for pressurized components of the vessel:
- Materials that are *referenced* directly by the code – these are subdivided into plate, forged parts, bar sections, and tubes.
- The code's requirement for other (non-specified) materials, in order that they may be suitable for use as *permissible* materials.
- Any generic code requirements on material properties such as carbon content, UTS or impact value.
- Specific requirements for low temperature applications.

4. Manufacture, inspection and testing

The relevant areas (in approximate order of use) are:
- Requirements for material identification and traceability
- NDT of parent material
- Assembly tolerances (misalignment and circularity)
- General requirements for welded joints
- Welder approvals
- Production test plates
- The extent of NDT on welded joints
- Acceptable NDT techniques
- Defect acceptance criteria
- Pressure testing
- Content of the vessel's documentation package

Fig 6.3   Information you need from a pressure vessel code

Information you need on:	How to find it in BS 5500:
1. Responsibilities	Section 1.4 places the responsibility for code compliance firmly on the manufacturer.
• Certification	Look at section 1.1 – it confirms that BS 5500 applies only to vessels built under independent survey. Section 1.4 specifies that code compliance is documented by the use of 'Form X' – this is issued by the manufacturer and countersigned by The Inspecting Authority (the 'third party').
• Manufacturer/Purchaser agreement	Table 1.5 lists approximately 50 points for separate agreement.
2. Vessel design	Vessel design is covered by section 3 (the longest section) of BS 5500
• Construction categories	Section 3.4 defines the three construction categories (1, 2, and 3) – they have different material and NDT requirements.
• Joint types	Section 5.6.4 defines types A and B welded joints, which have different NDT requirements (see Fig. 6.11 in this book).
3. Materials	Materials selection is covered by section 2.
• Specified materials	Section 2.1.2(a) references the British Standard materials that are specified. The main ones are BS 1501 (plates), BS 1503 (forgings), and BS 3601 to 3606 (tubes).
• Other permissible materials	Section 2.1.2.1(b) explains that other materials are allowed as long as they meet the criteria specified in this section, and in section 2.3.2.
• Material properties	Tables 2.3 give the specified material properties for all pressurized materials.

	See BS5500	
4. Manufacture, Inspection and Testing	section	**Summary**
• Material identification	4.1.2	Material for pressure parts must be 'positively identified' (but note there is no reference to material certificate 'levels' (as in EN 10204).
• NDT of parent material	5.6.2	This is not mandatory (see 4.2.1.2 for NDT of cut edges).
• Assembly tolerances	4.2.3	Alignment and circularity (out-of roundness) are closely specified. Tolerances depend on the material thickness.
• General requirements: welded joints	4.3	The manufacturer must provide evidence of 'suitable' welding.
• Welder approvals	5.2/5.3	WPS and PQRs must be approved. Welders must be qualified.
• Production test plates	5.4	Not mandatory for carbon steel vessels. Purchaser option.
• Extent of weld NDT	5.6.4	Different extent for Cat 1, 2, 3 vessels and type A or B welds.
• NDT techniques	5.6.4	US and RG testing are both acceptable volumetric NDT techniques.
• Defect acceptance criteria	5.7	Use tables 5.7(1) to 5.7(4) depending on the technique used.
• Pressure testing	5.8	A witnessed test is required. Section 5.8.5.1 gives the formula for calculating test pressure.
• Documentation package	1.4.4	Form 'X' is mandatory. Section 1.5.2.2 lists the minimum documentation required for code compliance.

Fig 6.4   BS 5500 – a summary

Boilers and pressure vessels 159

Information you need on:	How to find it in ASME Section VIII
1. Responsibilities:	Part UG-90 lists 19 responsibilities of the manufacturer and 14 responsibilities of the Authorized Inspector (A.I).
• Certification	For vessels to carry the ASME 'stamp' they must be constructed and inspected fully in compliance with the code. The manufacturer must have a 'Certificate of Authorization'. Parts UG 115-120 specify marking and certification requirements.
• Manufacturer/Purchaser agreement	Although such agreements are inferred, there is no explicit list of 'agreement items'.
2. Vessel design	There are two 'divisions of vessel', Div 1 and Div 2.
• Construction categories	Most vessels will qualify under Div 1. Div 2 is for specialized vessels with restricted materials and onerous testing requirements.
• Joint types	Part UW-3 defines category A, B, C, D welded joints, which have different NDT requirements (see Fig. 6.15 in this book).
• Design features	Figures UW-12, 13.1, 13.2, and 16.1 show typical acceptable (and some unacceptable) welded joints. Most of the information is for guidance, rather than being mandatory.
3. Materials	
• Permissible materials	Part UG-4 references materials specified in ASME Section II. Part UG-10 allows alternative, or incompletely specified, material to be recertified. Typical specified materials are: ASTM SA-202 (plates), SA-266 (forgings), SA-217 (castings) and SA-192 (tubes).
4. Manufacture, inspection, and testing	See ASME VIII part Summary
• Material identification	UG-94 Material for pressure parts must be certificated and marked. This infers traceability.
• NDT of parent material	UG-93(d) Visual examination is inferred. DP/MPI is only mandatory for certain material thickness and applications.
• Assembly tolerances	UG-80 and UW-33 Circularity (out-of roundness) and alignment tolerances are closely specified.
• Welder approvals	UW-28, 29, 48 WPS, PQR, and welder qualifications are required.
• Production test plates	UG-84 Test plates are required under the conditions set out in UG-84.
• Extent of weld NDT	UW-11 UW-11 gives priority to RG techniques. US is permitted in certain exceptional cases.
• NDT techniques	UW-51, 52, 53 ASME Section V techniques are specified.
• Defect acceptance criteria	UW-51, 52 UW-51 gives acceptance criteria for 100% RG testing. UW-52 gives acceptance criteria for 'spot' RG testing.
• Pressure testing	UG-99 A witnessed test is required. UG-99 (b) specifies a test pressure of 150% working (design) pressure.
• Documentation package	UG-115 to 120 Marking and documentation requirements are mandatory for full code compliance.

Fig 6.5   ASME VIII – a summary

Information you need on:	How to find it in TRD
1. Responsibilities:	TRD 503 specifies that an Authorized Inspector shall issue final vessel certification. There are legislative restrictions under the relevant Dampfkessel (steam boiler)
• Certification	regulations controlling which organizations can act as Authorized Inspector for TRD boilers.
• Manufacturer/Purchaser agreement	Such agreements are inferred and can override some parts of the standard – but there is no explicit list of 'agreement items'.
2. Vessel design	There are four main groups, I, II, III, IV, defined in TRD 600 onwards.
• Construction categories	Joints are classified by material thickness and type rather than by location in the vessel.
• Joint types	The TRD 300 series of documents covers design aspects.
3. Materials	TRD 100 gives general principles for materials. In
• Permissible materials	general, DIN materials are specified throughout. Other materials can be used (TRD 100 clause 3.4) but they need full certification by the Authorized Inspector. Typical specified materials are DIN 17102/17155 (plates), DIN 17175/17177 (tubes), DIN 17100/St 37-2/DIN 17243 (forgings), and DIN 17245/17445 (castings). The standards refer to DIN 50049 (EN 10204) material certification levels in all cases.

4. Manufacture, inspection, and testing	See TRD Section	Summary
• Material identification	TRD 100 (3.4)	Material for pressure parts must be certificated and marked. EN 10204 3.1 A and 3.1 B certificates are required, depending on the material and its application.
• NDT of parent material	HP 5/1 TRD 110/200,	Visual examination and US testing is generally required.
• Assembly tolerances	TRD 301, TRD 104/201	Circularity (out-of-roundness) and alignment tolerances are closely specified.
• Weld procedures	TRD 201	WPS, PQRs are required for all important welds. Tests are specified.
• Welder approvals	TRD 201	Annex 2 specifies welder approval tests to DIN 8560.
• Production test plates	TRD 201	Test plates are required. The number and type of tests to be carried out are closely defined.
• Extent of weld NDT	HP/0 HP5/3	Tables are provided defining NDT requirements for various materials.
• NDT techniques	HP5/3	DIN standards are referenced for all NDT techniques.
• Defect acceptance criteria	HP5/3	HP5/3 gives acceptance levels for RG and US examinations. Variations can be agreed (clause 6) by discussions between all parties.
• Pressure testing	TRD 503	A witnessed test is required at 120–150% design pressure, depending on application (see TRD 503 clause 5).
• Documentation package	TRD 503	A detailed list of documents is not explicitly provided, however TRD 503 specifies that evidence be provided of material type, heat treatment and NDT. This infers a full documentation package (similar to BS 5500 and ASME VIII) is required.

Fig 6.6   TRD boiler code – a summary

deliberate. Go slowly and carefully until you find the particular clause that you need.

Effective inspection of vessels means letting the manufacturer do some of the work. It is perfectly acceptable to ask the manufacture to identify a relevant code section or clause and then to demonstrate his vessel's compliance with it. It is not wise, however, to take this to extremes – you should know the basic material, NDT, acceptance criteria, and inspection requirements. This will enable you to ask the right questions. To help you with this I have provided Figs 6.7 to 6.9 which are 'quick reference charts' to the key material, NDT and defect acceptance criteria aspects of three main vessel codes. You can use these before and during vessel inspections. Figure 6.10 is a general reference chart identifying the main subsidiary standards – if you are doing the inspection correctly you will need to consult some of these to check more detailed technical information.

Applications of pressure vessel codes

Most of the international pressure vessel codes have been developed to the point where there is a high degree of cross-recognition, between countries' statutory authorities, of each other's codes. Core areas such as vessel classes, design criteria and requirements for independent inspection and certification are based on similar guiding principles. Increasingly, vessel codes are also being adopted for use on other types of engineering components and equipment. This is most evident with BS 5500 and ASME, which are fundamentally *design-based* codes. Their parameters for allowable stresses and factors of safety are used for guidance in the design of other equipment. This generally involves equipment items having thick cast sections, such as steam turbine casings and large cast valve chests, or those which have similar construction to pressure vessels, such as high pressure tubed heaters and condensers. In applications where pressure loading is not such a major issue you will still find the inspection and testing parts of vessel codes being used. The welding and NDT requirements of both BS 5500 and ASME VIII/ASME V are frequently specified for use in gas dampers and ductwork, structural steelwork, tanks, and similar fabricated equipment.

Vessel codes as 'an intent'

The use of some of the content of vessel codes for other equipment can make the activities of a works inspector a little more difficult. You can

Component	BS 5500	ASME VIII	TRD
Plates (shell and heads)	BS 1501 : Part 1 Gr 164 : Carbon steel Gr 223/224 : C–Mn steel	ASTM SA-20 (General requirements)	DIN 17155 and
	BS 1501 : Part 2 Gr 620/621 : low alloy steel	ASTM SA-202 : Cr–Mn–Si alloy steel	DIN 17102 ferritic steels
	BS 1502 : Part 3 Gr 304/321 : high alloy steel	ASTM SA-240 : Cr–Ni stainless steel (see ASME II)	(see TRD 101)
Forged parts (nozzles and flanges)	BS 1503 C–Mn steel	ASTM SA-266: Carbon steel	DIN 17100: grades St 37.2, St 37.3, St 44.2 and St 44.3
	BS 1503 stainless steel (austenitic or martensitic)	ASTM SA-336 : Alloy steel	
		ASTM SA-705 : Stainless and heat resistant steel	DIN 17243 : High temperature steels (see TRD 107)
Castings (where used as a pressure-part)	Special agreement required (see section 2.1.2.3) BS 1504: Carbon, low-alloy, or high-alloy steel	ASTM SA–217: stainless and alloy steels ASTM SA–351: austenitic and duplex steels	DIN 17245: grades GS-18 Cr Mo 9–10 and G-X8 Cr Ni 12. DIN 17445 austenitic cast steel (see TRD 103)
Pressure tubes	BS 3604: Parts 1 and 2 ferritic alloy steel BS 3605: Part 1 austenitic stainless steel	ASTM SA–192: Carbon steel ASTM SA–213: Ferritic and austenitic alloy tubes (seamless) ASTM SA–249: austenitic tubes (welded)	DIN 17175 and 17177: Carbon steel (see TRD 102)
Boiler/superheater tubes	BS 3059: Part 1: C–Mn steel BS 3059: Part 2: austenitic steel	ASTM SA-250: ferritic alloy ASTM SA-209 C–Mn alloy	DIN 17066: grades 10 Cr Mo 9–10 and 14 MoV 6–3 for elevated temperature applications (see TRD 102)
Sections and bars	BS 1502 Carbon or C–Mn steel	ASTM SA–29: Carbon and alloy steels	General standard steel DIN 17100–various grades (see TRD 107)
	BS 1502 low-alloy or austenitic stainless steel	ASTM SA–479: Stainless and heat-resistant steels.	

Fig 6.7 Pressure vessel codes – material requirements (quick reference)

BS 5500	**Cat 1 vessels** • 100% RG or US on type A joints (mandatory). See figure 6.11 in this book. • 100% RG or US on type B joints above a minimum thickness (see table 5.6.4.1.1) • 100% crack detection (DP or MPI) on all type B joints and attachment welds • Crack detection on type A joints is optional (by agreement only) **Cat 2 vessels** • Minimum 10% RG or US of the aggregate length of type A longitudinal and circumferential seams to include all intersections and areas of seam on or near nozzle openings. • Full RG or US on the nozzle weld of one nozzle from every ten. • 100% crack detection (DP or MPI) on all nozzle welds and compensation plate welds. • 10% crack detection (DP or MPI) on all other attachment welds to pressure parts. **Cat 3 vessels** • Visual examination only: no mandatory NDT • Weld root grind-back must be witnessed by the Inspecting Authority.
ASME VIII	**Div 1 vessels** • The extent of NDT us not straightforward. Look first at part UW-11 then follow the cross-references. Generally, the following welds require 100% RG. • Butt welds in material $\geq$ 38 mm thick • Butt welds in unfired boiler vessels operating at $\geq$ 3.4 bar G • Other welds are typically subject to spot RG testing prescribed by UW-52. **Div 2 vessels** • All pressure shell welds require 100% RG
TRD	• Longitudinal welds (LN): Generally 100% RG or US is required. • Circumferential welds (RN): Between 10% and 100% RG or US depending on material and thickness • T-Intersections (St) in butt welds: Generally 100% RG or US • Nozzle and fillet welds: Generally 100% DP or MPI • The complete details of NDT extent are given in AD-Merkblatt HP 0 and HP 5/3 Tables 1, 2, 3.

Fig 6.8 Pressure vessel codes – NDT requirements (quick reference)

164 Handbook of Mechanical Works Inspection

	Visual examination	Radiographic examination
BS 5500 (Cat 1)	Planar defects : not permitted Weld fit-up 'gap' : max 2mm Undercut : max 1mm Throat thickness error : +5/-1mm Excess penetration : max 3mm Weld 'sag' : max 1.5mm	Cracks, lamellar tears, lack of fusion and lack of root penetration: not permitted Porosity : max size of isolated pores is 25% of material thickness, up to a maximum diameter, which depends on the material thickness.
	Root concavity : max 1.5mm	Wormholes : max 6mm length x 1.5mm width.
	Weld cap overlap : not permitted	Inclusions : allowable dimensions of solid inclusions depend on the position of the inclusion and the type of weld (refer to BS 5500 Table 5.7(1).
	Refer to Tables 5.7(1) and 5.7(3) of BS 5500 for full details. Ultrasonic defect acceptance criteria are given in Table 5.7(2)	

	Radiographic examination (100%)	'Spot' radiographic examination
ASME VIII	Cracks Lack of fusion, any type } not permitted Lack of penetration	Cracks Lack of fusion } not permitted Lack of penetration
	Rounded indications : see ASME VIII appendix 4	Slag or cavities : max length 2/3t to maximum of 19 mm
	'Elongated' indications (e.g. slag or inclusions) : max length is 6mm to 18mm, depending on the material thickness	Porosity : no acceptance criteria See UW-52 for full details
	See UW-51 for full details	

	Radiographic examination	Ultrasonic examination
TRD	Cracks Lack of side wall fusion } not permitted	Longitudinal flaws : allowable up to a length *approximately* equal to material thickness. See HP 5/3 Table 5.
	Incomplete root fusion : not permitted on single-sided welds	Transverse flaws : max of three per metre length. See HP 5/3 clause 4.4
	Solid and gaseous 'inclusions' : max length 7mm (t < 10mm) or 2/3 t for 10 mm < t < 75 mm	
	Tungsten inclusions: max length 3 to 5 mm depending on metal thickness	
	See AD – Merkblatt HP 5/3 for full details	

Fig 6.9 Pressure vessel codes – defect acceptance criteria (quick reference)

Boilers and pressure vessels 165

SUBJECT	BS 5500	ASME VIII	TRD
Materials			
Plates	BS 1501, 1502 (BS 5996 for NDT)	ASTM SA-20, SA-202 (ASTM SA-435 for NDT)	DIN 17155, 17102
Forgings	BS 1503 (BS 4124 for NDT)	ASTM SA-266, SA-366	DIN 17100, 17243
Castings	BS 1504 (BS 4080 for NDT)	ASTM SA-217, SA-351	DIN 17245, 17445
Tubes	BS 3604, 3605, 3059, 3603	ASTM SA-192, SA-213	DIN 17175, 17177
Materials testing			
Tensile tests	BS EN 10002 (BS 18) BS 3668	ASTM SA-370	DIN 17245, 17445, 50145
Impact tests	BS 131 (BS EN 10045-1)	ASTM E-812	DIN 50115
Hardness tests	BS 240, (BS EN 10003) BS 860	ASTM SA-370, E340	DIN 50103
Chemical analysis	Use individual material standards	ASTM SA-751, E354	DIN EN 10036 DIN EN 10188
Welding			
Welding techniques	BS 5135	ASME IX	AD – Merkblatt HP- 5/1, HP-5/3
WPS/PQR/approvals	BS 4870 (BS EN 288) and BS 4871 (BS EN 287)	ASME IX, ASTM SA-488	AD HP-2/1 DIN 8563 TRD 201 DIN 8560
NDT			
Radiographic techniques	BS 2600, 2910	ASTM E94, E1032	AD HP-5/3, DIN 54111
Image quality	BS 3971 (BS EN 462-1)	ASTM E142,	DIN 54109
Ultrasonic techniques	BS 3923, 4080	ASTM SA-609, SA-745, E273	DIN 54125, 54126
Dye penetrant tests	BS 6443	ASTM E165, E433	DIN 54152
Magnetic particle tests	BS 6072	ASTM E709, E1444, SA 275	DIN 54130
Destructive tests	BS 4870 (BS EN 288), BS 709	ASTM SA-370	HP-5/2, HP-2/1
Pressure testing	BS 3636 (gas tightness only)		TRD 503

Fig 6.10 Pressure vessel codes – some referenced standards

foresee the problem – an item such as a large steam condenser may be specified as being 'to BS 5500 intent'. From a design stress point of view this is fine – the problem comes when you try to apply the inspection and testing requirements of the standard. Items such as material traceability, the amount of NDT and defect acceptance criteria often just do not fit. This leaves you with the scenario of *partial* code compliance – or, more correctly, with the task of inspecting against partial code compliance. You will meet this situation quite often in works inspections. Try to formulate a thought pattern to deal with it. Your main task is to rationalize the way *and the extent* to which the elements of the vessel codes contribute to fitness for purpose of the equipment. You have to see them in the right context – try the following guidelines:

- Look for those component *parts* of the piece of equipment that are the same as those used in the vessel code (see Fig. 6.16 for a typical example). Treat these parts as you would a full code pressure component, applying the code requirement just as rigidly.
- Where a component does not match the type of component mentioned in the code (i.e. the component for which the code was originally intended), you need to apply a little judgement. Be prepared to make a decision based on your engineering knowledge. Look closely at the FFP criteria for the component in question then try to apply as many of the code requirements as you can *without* there being a direct technical contradiction. For example, there is no reason why the ASME code requirement for hydrostatic test pressure cannot be applied to a fabricated intake chest for a sea-water pumping system. Subsidiary standards, particularly those relating to materials testing and NDT, can similarly be applied without any technical contradictions to many equipment types. If in doubt, think again about the FFP criteria for the equipment. Then use your judgement.
- *A golden rule*. Treat the technical requirements of the vessel codes as good and proven engineering practice. Don't be persuaded that they are 'too specialized' to be applied to other equipment. Be careful of over-bearing material traceability requirements though, most non-statutory equipment will not exhibit full material traceability. Partial traceability using predominantly EN 10 204 type 3.1B or 2.2 certificates is common (and accepted) in most industries.

The technical summary examples in Figs 6.11 to 6.22 show applications of the common pressure vessel codes. One example has the status of

Boilers and pressure vessels 167

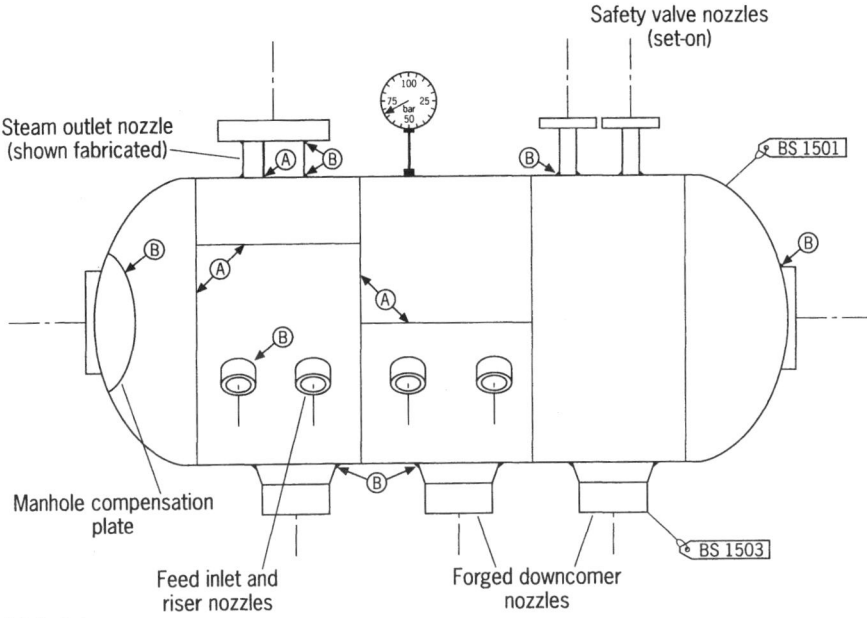

[Figure: Boiler drum diagram showing Steam outlet nozzle (shown fabricated), Safety valve nozzles (set-on), BS 1501, BS 1503, Manhole compensation plate, Feed inlet and riser nozzles, Forged downcomer nozzles, with Type A and Type B weld locations marked]

Materials
- Specified materials are given in BS 5500 section 2. Other materials can be permissible if specification and testing requirements are met
- No maximum plate thickness is specified
- There are temperature limitations for ferritic steels. Austenitic materials are permissible down to approx $-200°C$

Welded joints
- Types A and B joints are as shown

NDT requirements

Location	NDT extent and technique	Acceptance criteria
Plate material	US examination to BS 5996 (optional)	BS 5996
Weld preparations	Visual examination is mandatory	By agreement
Type A welds	100% RG or US is mandatory. Surface crack detection (DP or MPI) is optional	RG: BS 5500 Table 5.7 (1) US: BS 5500 Table 5.7 (2)
Type B welds	100% RG or US (some exceptions based on thickness) 100% DP or MPI, including attachment welds	As above BS 5500 Table 5.7 (3)
Test plate	Not mandatory	By agreement

Fig 6.11 Boiler drum to BS 5500 Cat 1

BOILER DRUM TO BS 5500 Cat 1

KEY POINTS

1. Code compliance
   - It is unusual for boiler drums not to be specified as fully compliant with Cat 1 requirements. Remember that independent survey and certification is mandatory for full BS 5500 compliance.

2. Materials of construction
   - If 'non-specified' materials are used, do a full check on the material properties. BS 5500 Section 2 specifies the checks that are required for materials to be permissible.
   - Check that all pressure-retaining material is positively identified and is traceable to its source. As a guide, high alloy steels should have EN 10 204 3.1A certificates and other pressure parts should have 3.1B traceability – but these EN certificates are not mandatory.

3. Manufacturing inspection
   - Check the weld preparations (after tack – welding) for the correct root gap and accuracy of alignment. Look carefully for distortion.
   - Tolerances and alignments are important. Check the following:

     Circumference:
     for O.D $\leqslant$ 650 mm, tolerance is $\pm$ 5mm
     for O.D > 650 mm, tolerance is $\pm$ 0.25% of circumference

     Straightness:
     Max deviation is 0.3% of total cylindrical length.

     Circularity:
     Circularity is represented by I.D max – I.D min at any one cross-section. The maximum allowable is:
     ID max – ID min $\leqslant$ [0.5 + 625/O.D] to a maximum of 1%

     Surface alignment:
     For material thickness (*e*)

     Longitudinal joints, $e \leqslant$ 12mm, max misalignment is *e*/4
     12 mm < *e* < 50 mm, max misalignment is 3mm

     Circumferential joints, $e \leqslant$ 20 mm, max misalignment is *e*/4
     20 mm < *e* < 40 mm, max misalignment is 5 mm

4. Documentation
   Final certification should use the 'Form X' shown in BS 5500 Section 1. Check that the correct wording has been used.

Fig 6.12 BS 5500 Cat 1 – key points

Boilers and pressure vessels

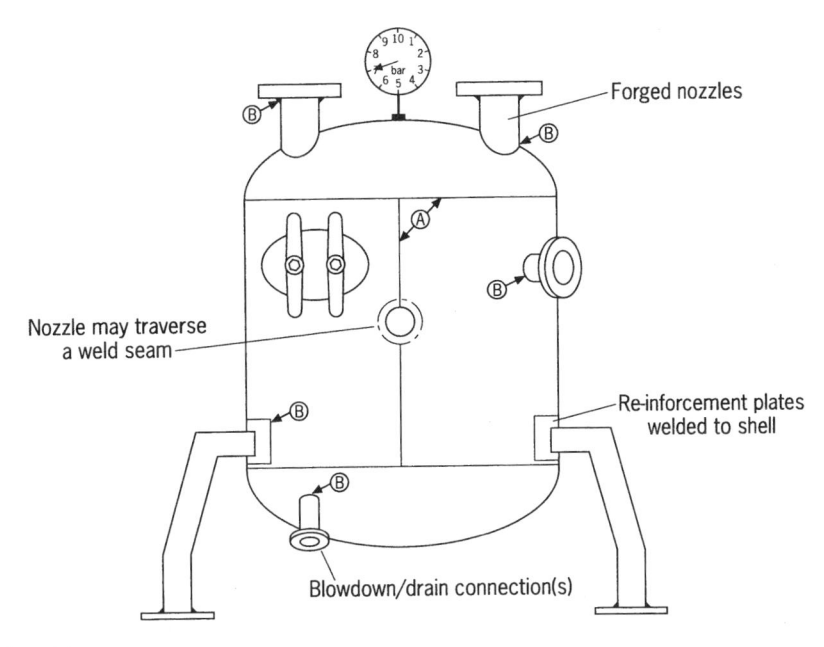

Materials
- Plate thickness is limited to: 40mm for carbon steel (groups M0 and M1)
  30mm for C-Mn steel (group M2)
  40mm for austenitic steel
- Specified materials are the same as for Cat 1 vessels-see BS 5500 Tables 2.3

NDT requirements

Location	NDT extent and technique	Acceptance criteria
Plate material	US examination to BS 5996 (optional)	BS 5996
Weld preparations	Visual examination is mandatory	By agreement
Seam welds (type A)	Minimum 10% RG or US of the aggregate length but including all intersections and seam areas near nozzles (as shown)	Use BS 5500 Tables 5.7(1)–(4). There is a special re-assessment technique if defects are found - see BS 5500 – Fig. 5.7
Nozzle welds (type B)	Full US or RG on one nozzle per ten. 100% DP or MPI on all nozzle welds	BS 5500 Table 5.7 (1) or (2) Table 5.7 (3)
Compensation plate welds	100% DP or MPI	BS 5500 Table 5.7 (3)
Attachment welds to pressure parts	10% DP or MPI	BS 5500 Table 5.7 (3)
Test plate	Not mandatory	By agreement

Fig 6.13 Pressure vessel to BS 5500 Cat 2

PRESSURE VESSEL TO BS 5500 Cat 2

KEY POINTS

1. Code compliance
   - As with Cat 1 vessels it is unusual for Cat 2 pressure vessels not to be specified as fully compliant with code requirements. Independent survey and certification is mandatory.

2. Materials of construction
   - BS 5500 Cat 2 vessels have limits placed on maximum plate thickness – this limits the design pressure.
   - Material traceability requirements are the same as for Cat 1 vessels.

3. Manufacturing inspection
   - Note that permanent weld backing-strips are allowed for Cat 2 vessels. Check that they are of the correct 'compatible' material.
   - Alignment tolerances are the same as for Cat 1 vessels (see Fig. 6.12 in this book).
   - Because only a 'percentage' volumetric NDT is required it is doubly important to make sure that critical areas are investigated. These are:
     - Intersections of longitudinal and circumferential joints.
     - Areas of weld seam which are within 12 mm of nozzle openings.
     - The ends of weld seams, particularly the 'beginning' of the run.
   - For radiographic examination make sure that the '10% of aggregate length' requirement includes at least one radiograph from each weld seam. This is good practice.
   - Hydrostatic test pressure is the same as for Cat 1 vessels.

4. Documentation
   Final certification should use the 'Form X' shown in BS 5500 Section 1. Check that the correct wording has been used and that the 'Category 2' designation is clearly indicated.

Fig 6.14   BS 5500 Cat 2 – key points

Boilers and pressure vessels 171

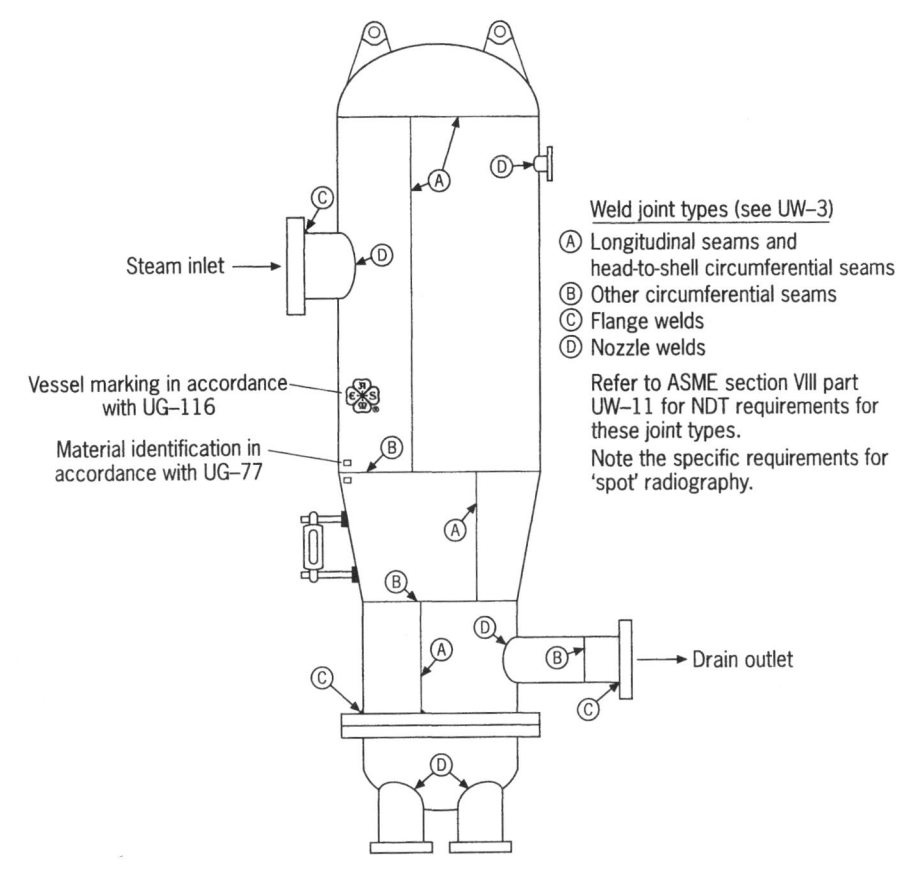

Steam inlet →

Vessel marking in accordance with UG–116

Material identification in accordance with UG–77

→ Drain outlet

Weld joint types (see UW–3)
Ⓐ Longitudinal seams and head-to-shell circumferential seams
Ⓑ Other circumferential seams
Ⓒ Flange welds
Ⓓ Nozzle welds

Refer to ASME section VIII part UW–11 for NDT requirements for these joint types.
Note the specific requirements for 'spot' radiography.

Key points
- A vessel is only fully 'ASME-compliant' if it has the ASME stamp.
This requires:
    – the manufacturer to be assessed and certificated to produce ASME vessels
    – construction to be monitored by an ASME-authorized inspector (only)
    – full compliance with ASME sections VIII, II, V and IX
- Manufacturing tolerances are:
    – permissible out-of-roundness (IDmax - IDmin) ⩽ 1% D
    – permissible butt-weld misalignments are:

Thickness	Joint types	
(t)	A	B, C, D
⩽12.7mm	t/4	t/4
12.7mm – 19mm	3.2mm	t/4
19mm – 38mm	3.2mm	4.75mm
38mm – 51mm	3.2mm	t/8

Fig 6.15   HP feed heater to ASME VIII Div 1

172 Handbook of Mechanical Works Inspection

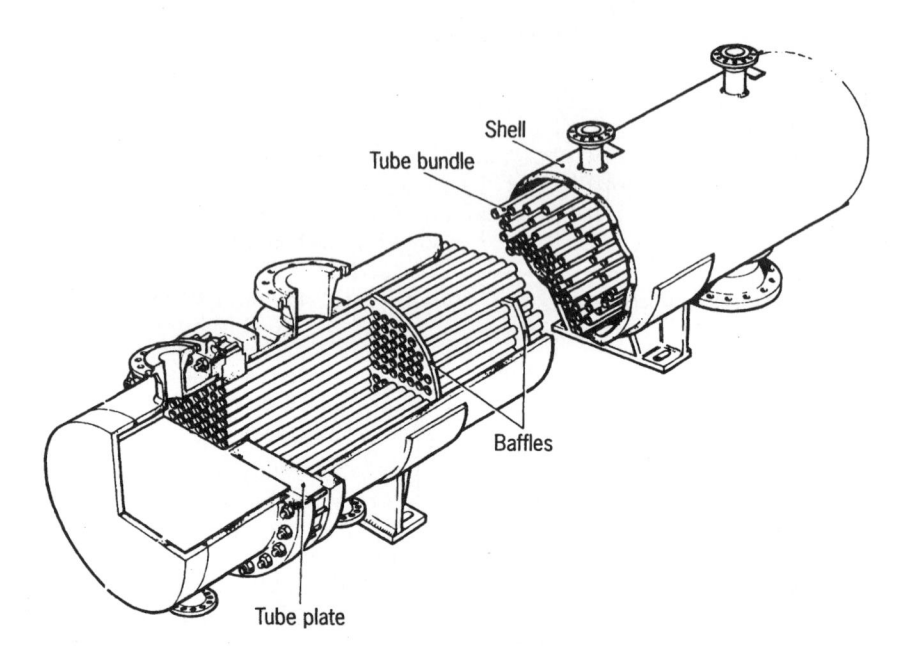

Materials (particularly the internals) are unlikely to be those specified by ASME II

Stress calculations do use the assumptions of ASME VIII (but often only for the outer shell – not the tubes)

Tube-tube plate components may be based on TEMA (rather than ASME) requirements

The vessel is not ASME-stamped

NDT extent is not governed by the ASME joint types ABCD. A simpler system is often used.

NDT techniques and defect acceptance criteria may be ASME-specified or modified by purchaser/manufacturer agreement.

This figure shows 'typical' characteristics of an 'ASME-intent' vessel. There are many possible variations from the ASME code – the figure shows the common ones.

Fig 6.16   Heat exchange to 'ASME VIII intent'

HEAT EXCHANGER BUILT TO 'ASME VIII INTENT'

KEY POINTS

1. **Code compliance**
   - A vessel built to 'ASME intent' will not, by definition, be ASME 'stamped'. This means that there is no verification of full code compliance and the vessel will not be recognized (either by statutory authorities or purchasers) as a vessel that complies fully with the ASME code. Hence the reference to ASME becomes a guideline to engineering practices only. For most vessels built in this way the adaption of ASME VIII practices is often limited to the outer pressure shell only – as in the heat exchanger shown in Fig. 6.16.

2. **Materials**
   - You will probably find that vessels built to 'ASME intent' will not use ASME II specified materials, nor will materials have been recertified as required by ASME VIII part UG–10. 'Equivalent' materials should be checked to make sure that elevated and low temperature properties are equivalent to the ASTM-referenced materials in ASME II. The most common differences you will find relate to material impact properties.
   - Material traceability requirements of ASME are not too difficult to reproduce. A system using EN 10204 certificates level 3.1B for pressure parts is broadly compliant with the requirements of ASME VIII part UG–94.

3. **Manufacturing inspection**
   - The ASME Authorized Inspector (AI) will normally be replaced with a different third party inspectorate. Although the inspection role will probably be similar to that defined in UG–90 an unauthorized inspectorate cannot authorize 'ASME-stamping' of the vessel.
   - NDT is an important area. Most 'ASME-intent' vessels will not use the ABCD joint types specified in ASME VIII and will probably use a simplified extent of NDT (such as RG or US on seam welds and DP/MPI on nozzle welds). ASME defect-acceptance criteria (UW-51 and 52) can be used but are sometimes replaced by different agreed levels. Try to clarify which practices for NDT extent and defect acceptance criteria are in force, before manufacturing progresses too far – as this is one of the most common areas for confusion.

4. **Documentation**
   - It it rare that documentation content causes significant problems with 'ASME-intent' vessels. Most competent manufacturers' documentation practices can comply.

Fig 6.17   ASME VIII 'intent' – key points

Class definitions

Class	Definition	Design points
I	Unlimited size and design pressure (P)	Flat end-plates unacceptable. Test plates required for longitudinal seams.
II	P ⩽ 35 barG maximum and P (barG) × i.d (mm) ⩽ 37 000	Test plates required for longitudinal seams - but only transverse tensile and bend tests.
III	P ⩽ 17.5 barG maximum and P(barG) × i.d (mm) ⩽ 8 800	Test plates not required.

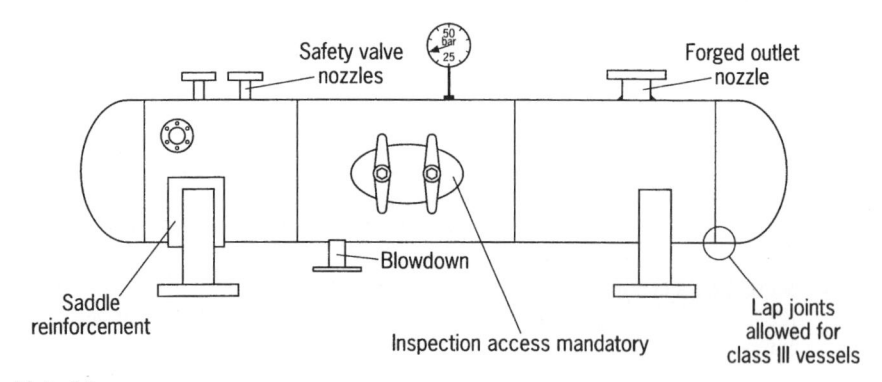

Materials
- Class I receivers use grades of steel from: BS EN 10 207
  BS 1501 (plate)
  BS 1503 (forgings)
  BS 1502/970 (bar).
- Class III receivers can also use plate grades BS 1449 and BS 4360.
- Impact tests are required if design temperature is <0°C (as in BS 5500 App. D)

NDT requirements

Location	NDT required Class I	Class II	Class III	Acceptance criteria
Parent plate	Not specified	None	None	Unacceptable defects are: • cracks or l.o.f
Circumferential and longitudinal welds	100% RG	None	None	• elongated slag inclusions (see BS 5169)
Nozzle welds and other welds	Not specified	None	None	• total porosity >6mm² per 25mm wall thickness in any 645mm² of weld area

Fig 6.18   Air receivers to BS 5169 (classes I, II, III)

AIR RECEIVERS TO BS 5169

KEY POINTS

1. Code compliance
   - BS 5169 requires that 'competent supervision' is required for class I receivers but does not specify full independent inspection. In *practice* it is common for air receivers to be considered as having equivalent 'statutory status' to BS 5500 vessels – hence they are normally subject to independent survey and certification.
   - It is rare to see any type of 'partial code compliance' specified for these air receivers.

2. Materials of construction
   - For class I receivers, BS 1449 and BS 4360 plate materials are not permitted.
   - Material tensile and bend tests are required.
   - Material certificates are mandatory for class I receivers. They are not specified for classes II and III unless the design temperature is outside the range -10°C to +120°C.

3. Manufacturing inspection
   - WPS/PQR/welder approvals are required for classes I and II vessels.
   - Most seams are made using double-sided butt welds. If a single-sided weld is used, a backing strip is permitted, subject to satisfactory PQR tests.
   - Longitudinal weld test-plates are required for class I and II receivers only.
   - For lap welds (allowed for class III head-to-shell joints), the pre-welding fit-up is important. Check that the plates have a tight 'sliding' fit.
   - Hydrostatic test pressure is 125% design pressure (P) for class I and 150% P for class II.

4. Documentation
   Although a formal 'Form X' type certificate is not mandatory, BS 5169 does require the manufacturer to issue a 'Certificate of Construction and Test'.

Fig 6.19 BS 5169 – key points

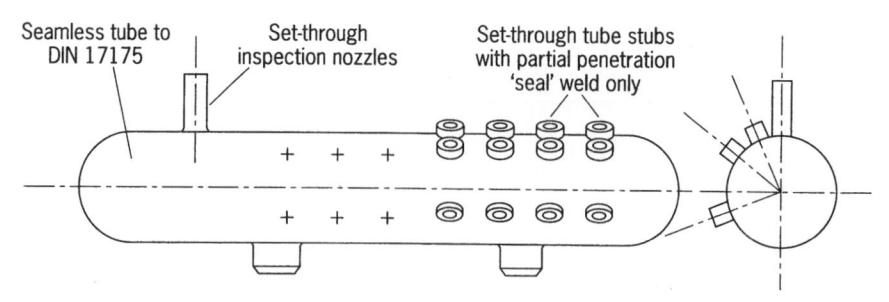

176 Handbook of Mechanical Works Inspection

Materials
TRD 104 specifies overall mechanical property requirements for seamless tubes. They are:

- minimum impact value (ISO V-notch) of 31J–41J (depending on the rated tensile strength)
- minimum elongation at fracture of 16% (using a tangential specimen)
- seamless tubes require a DIN 50 049 (EN 10 204) type 3.1 (B) material certificate

Tolerances are:

- ±1% on bore diameter
- straightness deviation 0.3% of cylindrical length
- maximum out-of-roundness (U) <2% of nominal diameter

Fig 6.20 Superheater header to TRD boiler code

being 'code intent' – the scenario I have just described. Each application shows the way that the requirements of a specific vessel code have been ascribed to the total item, or to parts of it (this would be stated in the contract specification). I have summarized the material and NDT requirements – and explained some common interpretations and key points on code compliance, manufacturing and documentation. Use these figures carefully – this is core information that you will need when inspecting these common types of vessels.

Inspection and test plans (ITPs)

The use of ITPs is well accepted throughout the pressure vessel manufacturing industry. The ITP is one of the most useful *working* documents; used in a key monitoring and control role it summarizes the activities of manufacturer, contractor, and statutory certification organization. Owing partly to the statutory nature of pressure vessels,

Boilers and pressure vessels 177

[Diagram of a vertical boiler with labels: Fabricated uptake, Fabricated shell, Fabricated furnace, Firehole, Furnace, 'Ogee' ring, End plate]

Class definitions

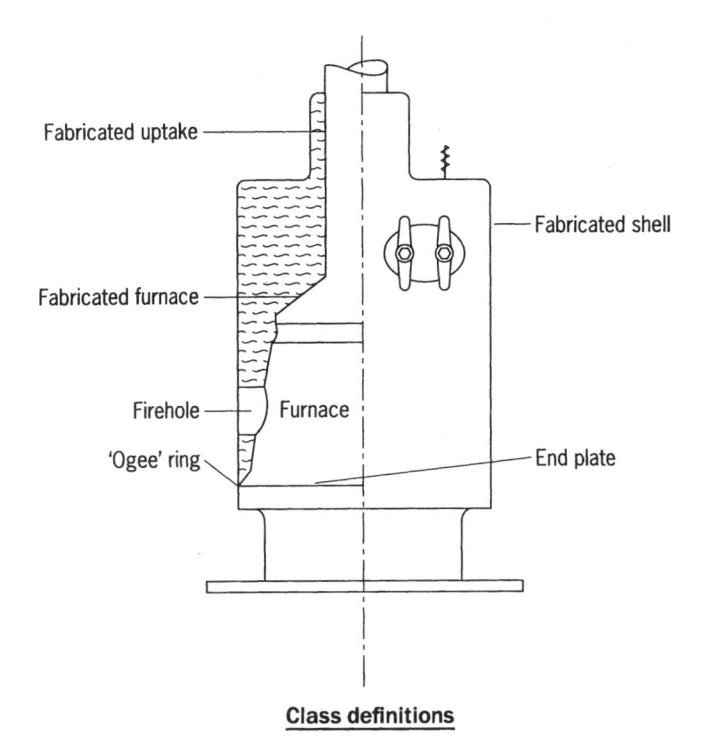

Class	Definition	
I	Design pressure (P) > 7.2 bar and P/10 × diameter (d) mm > 920	There are not a large number of construction differences between class I, II and III vessels. The main difference is in the NDT requirements
II	Neither of the class I definitions apply	
III	P ≤ 3.8 bar and P (bar)/10 × (d) ≤ 480	

NDT requirements

Location	NDT requirements			Acceptance criteria
	Class I	Class II	Class III	
Parent end plate	100% US	100% US	Not required	BS5996: Q-grade B4 and E3
Other parent plate	Edges only	Edges only	Visual exam only	BS 5996: Q-grade E3 (edges)
Longitudinal butt welds	100% RG or US	10% RG or US	Visual exam only	BS 2790 gives 3 sets of

Fig. 6.21 continued over page

Fig. 6.21 continued

Circ^L welds shell to end-plates	100% RG or US	100% RG or US	Visual exam only	acceptance criteria for:
Other circ^L welds	10% RG or US	10% RG or US	Visual exam only	• Visual examination (see below)
T-butt furnace welds	25% RG or US	25% RG or US	Visual exam only	• Radiographic examination
Attachment welds	25% DP or MPI	10% DP or MPI	Visual exam only	• Ultrasonic examination

BS2790 has its own visual acceptance criteria for welds. They are:

- undercut $\leqslant$ 0.5mm
- shrinkage $\leqslant$ 1.5mm
- overlap not permitted
- excess penetration $\leqslant$ 3mm
- weld reinforcement must be smoothly 'blended in'.

Fig 6.21 Shell boiler to BS 2790 (classes I, II, III)

you can expect ITPs to have a well-defined set of technical steps. This tends not to be the case with witnessing responsibilities, however – the number and extent of witness points can vary significantly between contracts. In general, it is the contractual agreement between the parties that defines the activities to be witnessed during manufacture. Remember that the extent of mandatory inspection required for statutory purposes is not always explicitly defined – relevant legislation quotes the main vessel codes as examples of what is considered good technical practice, and some of these are clearer than others.

For these reasons the typical vessel ITP shown in Fig. 6.23 concentrates on the *technical* steps included in a good ITP rather than trying to define responsibilities for witnessing these activities. Also included is the inventory of typical documentation items relevant to each step of the ITP. The general topic of ITPs is covered in Chapter 3 of this book. Note the following particular guidelines when dealing with pressure vessel ITPs.

- *Code compliance.* A good pressure vessel ITP will refer to the relevant sections of the applicable code for major topics such as welding procedures, production test plates, and NDT.
- *Acceptance criteria.* A reference to the defect acceptance criteria to be used should be shown explicitly in the ITP. If it is not, make a point of holding early discussions with the vessel manufacturer about this.

SHELL BOILER TO BS 2790

KEY POINTS

1. Code compliance
   - BS 2790 states that independent survey and certification is required.
   - Recent additions to the standard require that manufacturers must have an operating quality system in place.
   - It is rare to see any kind of 'partial code compliance' specified – although class III vessels have a very low level of verification of integrity.

2. Materials of construction
   - These shell boilers are made predominately from Carbon or Carbon-Manganese steels such as BS EN 10028/BS 1501 (plate) and BS 1503 for forged components. Other similar steels with < 0.25% Carbon can be used under the provisions of the standard. Check the elevated temperature properties if this is the case.
   - Test plates are required. Those representing single sided full penetration butt joints require extra root-bend tests.
   - Pressure-part material must be identified, but the standard does not explicitly require 'full traceability'.

3. Manufacturing inspection
   - WPS/PQR and welder approvals (EN 287, 288) are required.
   - Full penetration welds must be used for longitudinal and circumferential seams. Remember the importance of eliminating root defects in this type of weld.
   - Longitudinal joints have an alignment tolerance of t/10, to a maximum of 3mm.
   - Weld repairs are allowed – the same acceptance criteria are used.
   - There are 3 sets of defect acceptance criteria quoted in the standard. They are similar to those in BS 5500.
   - Make sure that attachment welds do not cross seam or nozzle welds.
   - Hydrostatic tests are performed at 150% design pressure for 30 minutes.

4. Documentation
   - There is no mandatory certificate format. The manufacturer must issue a certificate to be authorized by the independent inspection organization.
   - A full document package (similar to that for BS 5500) is normally provided.

Fig 6.22 BS 2790 – key points

ACTIVITIES	RELEVANT DOCUMENTATION
1. Design appraisal	Certificate from independent organization that the vessel design complies with the relevant code.
2. Material inspection at works (forgings, casting, plate and tubes)	Identification record/mill certificate (includes chemical analysis)
2.1 Identification/traceability	Witness identification stamps
2.2 Visual / dimensional inspection	Test report (and sketch)
2.3 UT testing of plate	Test certificate
2.4 Mechanical tests	Mechanical test results (including impact)
3. Marking off and transfer of marks	Material cutting record (usually sketches of shell/head plate and forged components).
4. Examination of cut edges	MPI/DP record for weld-prepared plate edges.
5. Welding procedures	
5.1 Approve weld procedures	WPS/PQR records
5.2 Check welder approvals	Welder qualification certificates
5.3 Verify consumables	Consumable certificate of conformity
5.4 Production test plates	Location sketch of test plate
6. Welding	
6.1 Check weld preparations	Check against the WPS
6.2 Check tack welds and alignment of seams/nozzle fit-ups	Record sheet
6.3 Back chip of first side root weld, DP test for cracks	DP results sheet
6.4 Visual inspection of seam welds	WPS and visual inspection sheet
7. Non-destructive testing before heat treatment	
7.1 RG or US of longitudinal and circumferential seams	RG or US test procedure
7.2 RG or US of nozzle welds	Defect results sheet
7.3 DP or MPI of seam welds	NDT location sketches
7.4 DP or MPI of nozzle welds	
7.5 Defect excavation and repair	Repair record (and location sketch)
8. Heat treatment/stress relieve	
8.1 Visual/dimensional check before HT	Visual/dimensional check sheet
8.2 Heat treatment check (inc test plates where applicable)	HT time/temperature charts
9. Non-destructive testing after HT	
9.1 RG or US of longitudinal and circumferential seams	RG or US test procedures
9.2 RG or US of nozzle welds	NDT results sheets
9.3 DP or MPI of seam welds	NDT location sheets
9.5 DP or MPI of attachment welds, lifting lugs and jig fixture locations	
10. Final inspection	
10.1 Hydrostatic test	Test certificate
10.2 Visual and dimensional examination	Record sheet
10.3 Check of internals	Record sheet
10.4 Shotblasting/surface preparation	Record sheet
10.5 Painting	Record of paint thickness and adhesion test results
10.6 Internal preservation	Record of oil type used
10.7 Vessel markings	Copy of vessel nameplate
10.8 Packing	Packing list
10.9 Documentation package	Full package, including index
10.10 Vessel certification	Form X or equivalent
10.11 Concession details	Record of all concessions granted (with technical justification)

Fig 6.23   Pressure vessels – typical ITP content

Remember that interpretation and judgement is still required for defect acceptance criteria written in vessel codes.
- *Hold points*. The use of 'hold points', where a manufacturer must stop manufacture until an inspector completes an interim works inspection, should be treated carefully. Expect manufacturers to paint a gloomy picture of how hold points can delay the manufacturing programme. In practice this *is* often the case. Try to see the manufacturer's viewpoint and limit *formal* hold points to the major manufacturing steps, and the final inspection/hydrostatic test. This does not mean that you have to give up your right to *witness* important manufacturing steps. You can just as easily do it informally, and without disruption. This means that interim visits *are* required. Don't expect all manufacturers to do your job for you.
- *Documentation review*. Some manufacturers will only start to compile the documentation package for a vessel after that vessel has successfully passed its visual/dimensional inspection and hydrostatic test. Typically, the package is compiled over a period of some two or three weeks after the hydrostatic test, during which time the vessel is shot blasted, painted, and packed ready for shipping. Such a system is not very helpful – any problems of missing or incorrect documentation are invariably discovered too late and it is not unknown for inspectors to be placed under pressure to release the vessel before all documentation has been properly reviewed. This whole scenario is, quite frankly, bad practice. If, as an inspector, you understand the contractual equipment release mechanism (I have explained this subject in Chapter 3) then you should never be caught like this. Make sure that you review all the key documentation during the manufacturing programme. Your review should be substantially complete before you visit the works to witness the hydrostatic test and visual/dimensional check. Expect to have to provide several reminders about early compilation of the documentation package. *Ask* to see it – don't just remind everyone of its importance.

Pressure testing

Virtually all vessels designed to operate above atmospheric pressure will be subject to a pressure test – most often in the manufacturers' works. Witnessing pressure tests is therefore a common inspection task. It should always be shown as a witness point for all parties concerned on a vessel's ITP and is an integral part of the role of the Third-Party organization or authorized inspector when the vessel is subject to statutory certification.

The point of a pressure test

The *objective* of pressure tests is sometimes misunderstood. It is part of the system of verifying the integrity of a vessel – remember that it is one of the norms shown in Fig. 6.1 at the beginning of this chapter. It also has limitations. The stresses imposed on a vessel during a pressure test are effectively static; they impose principal stresses and their resultant principal strains on the vessel. This means that what they test is the resistance of the vessel only to the principal stress and strain fields, not its resistance to cyclic loadings (that cause fatigue), creep, or the other mechanisms that have been shown to cause vessels to fail. This is an important point. The pressure test is *not* a full test of whether the vessel will fail as a result of being exposed to its working environment. Frankly, the incidence of steel vessels actually failing catastrophically under a works pressure test is small, almost negligible. I can only reinforce the point I have made in several chapters of this book, that a pressure test is not a 'proving test' for vessels that have not been properly checked for defects (particularly weld defects). It is also not a proving test for vessels where unacceptable defects have been found – so that the vessel can be somehow shown to be fit for its purpose, in spite of the defects. So:

- a pressure test is a test for leakage under pressure and
- that is about *all* it is.

The standard hydrostatic test

This is the most common pressure test performed on steel vessels. It is also commonly known as a hydraulic test. It is a routine test used when the vessel material thickness and allowable stresses are well defined and there are no significant unknown factors in the mechanical aspects of the design. For single-enclosure vessels (such as drums, headers, and air receivers) a single hydrostatic test is all that is required. For heat-exchange vessels such as heaters, coolers, and condensers, a separate test is performed on each 'fluid side' of the vessel. Figure 6.24 shows the guidelines to follow when witnessing a hydrostatic test. All the main pressure vessel codes provide a formula for calculating the test pressure as a multiple of design

USE THESE GUIDELINES WHEN WITNESSING A STANDARD HYDROSTATIC TEST ON ANY VESSEL

Vessel configuration
- The test should be done after any stress relief.
- Vessel components such as flexible pipes, diaphragms and joints that will not stand the pressure test must be removed.
- The ambient temperature must be above 0°C (preferably 15–20°C) and above the brittle fracture transition temperature for the vessel material (check the mechanical test data for this).

The test procedure
- Blank off all openings with solid flanges.
- Use the correct nuts and bolts, not G-clamps.
- Two pressure gauges, preferably on independent tapping points, should be used.
- It is essential for safety purposes to bleed all the air out. Check that the bleed nozzle is really at the highest point and that the bleed valve is closed off progressively during pumping, until all the air has gone. It is best to witness this if you can.
- Pumping should be done slowly (using a low capacity reciprocating pump) so as not to impose dynamic pressure stresses on the vessel.
- Test pressure is stated in BS 5500, ASME VIII or the relevant standard. This will not overstress the vessel (unless it is a very special design case). If in doubt use 150% design pressure.
- Isolate the pump and hold the pressure for a minimum of 30 minutes.

What to look for
- Leaks. These can take time to develop. Check particularly around seams and nozzle welds. Dry off any condensation with a compressed air-line; it is possible to miss small leaks if you do not do this. Leaks normally occur from cracks or areas of porosity.
- Watch the gauges for pressure drop. Any visible drop is unacceptable.
- Check for distortion of flange-faces etc. by taking careful measurements. You are unlikely to be able to measure any general strain of the vessel – it is too small.
- If in any doubt, ask for the test to be repeated. It will not do any harm.

Fig 6.24   The standard hydrostatic test – guidelines

pressure. For inspection purposes you can assume that design pressure is the same as working pressure – you may also find it referred to as rated pressure in some standards. As a general rule, if the manufacturer has not calculated the test pressure precisely it is common to use 150 percent design pressure for a minimum period of 30 minutes.

Pneumatic testing

Pneumatic testing of pressure vessels is a 'special case' testing procedure, used when there is a good reason for preferring it to the standard hydraulic test. Common reasons are:

- Refrigeration system vessels are frequently constructed to ASME VIII and pneumatically tested with nitrogen.
- Special gas vessels may have an unsupported structure and so be unable to withstand the weight of being filled with water.
- The vessels are used in critical process applications where the process of even minute quantities of water cannot be tolerated. You are unlikely to meet such specialized vessels in power generation or general industrial plant contracts.

Pneumatic tests are dangerous. Compressed air or gas contains a large amount of stored energy, so in the unlikely event that the vessel does fail this energy will be released catastrophically. The vessel will effectively explode, with potentially disastrous consequences. For this reason there are a number of well-defined precautionary measures to be taken before carrying out a pneumatic test on a vessel – and some safety aspects to be considered during the testing activity itself. These are shown in Fig. 6.25 along with more general guidelines on witnessing a pneumatic test.

Vacuum leak testing

Vacuum tests (more correctly termed 'vacuum leak' tests) are different to the standard hydrostatic and pneumatic tests previously described. The main applications you will see are for condensers and their associated air ejection plant. This is known as 'coarse vacuum' equipment, designed to operate only down to a pressure of about 1 mm Hg absolute. Most general power and process engineering vacuum plant falls into this category. There are many other industrial and laboratory applications where a much higher 'fine' vacuum is

Precautionary measures before a pneumatic test

- BS 5500 requires that a design review be carried out to quantify the factors of safety inherent in the vessel design. NDT requirements are those specified for the relevant Cat 1 or Cat 2 application plus 100% surface crack detection (MPI or DP) on all other welds.
- ASME VIII (part UW-50) specifies that all welds near openings and all attachment welds should be subject to 100% surface crack detection (MPI or DP).
- It is *good practice* to carry out 100% volumetric NDT and surface crack detection of all welding prior to a pneumatic test – even if the vessel code does not specifically require it.

The test procedure

- The vessel should be in a pit, or surrounded by concrete blast walls.
- Ambient temperature should be well above the brittle fracture transition temperature.
- Air can be used but inert gas (such as nitrogen) is better.
- Pressure should be increased very slowly in steps of 5–10% – allow stablization between each step.
- BS 5500 specifies a maximum test pressure of 150% design pressure.
- ASME specifies a maximum test pressure of 125% design pressure, but consult the code carefully – there are conditions attached.
- When test pressure is reached, isolate the vessel and watch for pressure drops. Remember that the temperature rise caused by the compression can affect the pressure reading (the gas laws).

Fig 6.25 The pneumatic test – guidelines

186 Handbook of Mechanical Works Inspection

specified but this is a highly specialized area, outside the common works inspection field.

The objective of a coarse vacuum test is normally as a proving test on the vacuum system rather than just a vessel itself. A typical air ejection system will consist of several tubed condenser vessels and a system of interconnecting ejectors, pipework, valves, and instrumentation. The whole unit is often skid-mounted and subject to a vacuum test in its assembled condition. A vacuum test is much more 'searching' than a hydrostatic test. It will register even the smallest of leaks that would not show during a hydrostatic test, even if a higher test pressure was used. Because of this the purpose of a vacuum leak test is not to try and verify whether leakage exists – rather it is to determine the *leak rate* from the system and compare it with a specified acceptance level.

Leak rate and its units

The most common test used is the 'isolation and pressure-drop method'. The vessel system is evacuated to the specified coarse vacuum level using a rotary or vapour-type vacuum pump and then isolated. Note the following salient points:

- The acceptable leak rate will probably be expressed in the form of an allowable pressure rise (p). This has been obtained by the designer from consideration of the leak rate in torr litres per second.

$$\text{Leakrate} = \frac{\mathrm{d}p \times \text{volume of the vessel system}}{\text{time } (t) \text{ in seconds}}$$

Note the units are torr litres per second (torr l s^{-1}). For inspection purposes you can consider 1 torr as being effectively equal to 1 mm Hg. Note also that when discussing vacuum, pressures are traditionally expressed in absolute terms – so a vacuum of 759 mm Hg below atmospheric is shown as +1 mm Hg.

- It is also acceptable to express leak rate in other units (such as 1 μm Hg s^{-1} – colloquially known as a 'lusec') and other combinations. These are mainly used for fine vacuum systems. If you do meet a unit that is not immediately recognizable, it is easiest to convert it back to torr l s^{-1} via the SI system.
- Because leak rate is a function of volume, the volumes of all the system components: vessels, pipes, traps, bypasses, and valves need

> **PREPARATION IS IMPORTANT**
>
> 1. Do a standard hydrostatic test on the installation before doing the vacuum test – to identify any major leaks.
> 2. Leak-test small components (pipes, valve assemblies, instrument branches, etc.) before assembly. Submerging the pressurized component in a water bath is the best way.
> 3. The inside surfaces of all components must be totally clean and dry. Even small amounts of moisture, porous material or grease will absorb air and release it during the test giving erroneous readings ('virtual leaks').
> 4. Visually check and clean all flange faces before assembly. Polish out radial scratches. Use new pipe joints. Strictly, liquid or paste jointing compound should not be used – it can mask leaks.
> 5. Do the test under dry, preferably warm (10°C minimum) ambient conditions.
> 6. The configuration should be such that various sections of the system can be isolated from each other. This helps leaks locate the position of leaks.

Fig 6.26   Vacuum leak test – guidelines

to be calculated accurately. It is not sufficient just to use the approximate 'design' volume of the vessels.

The test procedure

The procedure for the 'isolation and pressure drop' test is simple. Evacuate the system, close the valve and then monitor pressure rise over time. The main effort, however, needs to be directed towards the preparation for the test – it is surprisingly easy to waste time obtaining meaningless results if the preparation is not done properly. This happens quite frequently. Figure 6.26 shows points to check before carrying out a vacuum leak test. Use this as an action list if you witness an aborted test because of major leaks, or when you feel (by asking questions – look *again* at the section in Chapter 2 of this book under the heading 'Asking and Listening') that the preparation may not have been done properly. Then repeat the test. It is wise not to be *too* hasty in issuing an NCR for excessive leakage until you have implemented these points. This is more to save you time and effort by avoiding abortive repeat works visits to witness poor tests, rather than to give the equipment manufacturer the benefit of the technical doubt.

Leaks are often difficult to locate. If the observed leak rate is above the specified levels but still relatively small make a double-check on the tightness of the pressure gauge and instrumentation fittings – they are a common source of air ingress, particularly if they are well used and have slightly worn union connections. The next step is to isolate the various parts of the system from each other to identify the leaking area. The system can then be pressurized using low pressure air and a soap/water mixture brushed onto suspect areas. Concentrate on joints and connections – leaks will show up as bubbles on the surface.

Visual and dimensional examination

The visual and dimensional examination is part of the final inspection activities carried out on a pressure vessel. Final inspection is mandatory for vessels which are subject to statutory certification – as well as being a normal contractual witness point. It is not a difficult exercise but it does benefit from a structured approach and the use of checklists to aid reporting. The visual examination and dimensional check can be done before or after the standard hydrostatic test. It is perhaps more common to do it afterwards – this enables internal and external examinations to be done during the same visit. At this stage the vessel is awaiting shot-blasting and painting. Although the visual examination and dimensional check are normally carried out together, we can look for clarity at each activity separately.

The visual examination

The purpose of the visual examination is to look for problems that are likely to affect integrity. It is also an important way in which you can gain *clues* about any poor manufacturing practices that may have been used during those manufacturing activities that were not witnessed by an external inspector.

The vessel exterior

The basic examination principles are similar for all vessels. Check these points:

- *Plate courses.* Check the layout of the plate courses against the original approved *design* drawings (not the ones that you may find on the shop floor next to the vessel). It is not unknown for manufacturers to change the layout of the plate courses to make more economical use of their stock plate. If this has occurred, make sure that the new layout has not caused design changes such as placing nozzle openings across, or too near to, welded seams.
- *Plate condition.* Check for dents and physical damage. Look for any deep grinding marks – obvious grooves, deeper than 10 percent of plate thickness, caused by the edge of a grinding disc, are cause for concern.
- *Surface finish.* General mill-scale on the surface of the plate is acceptable before shotblasting. Check for any obvious surface 'rippling' caused by errors during plate rolling.
- *Reduced thickness.* Excessive grinding is unacceptable as it reduces the effective wall thickness. Pay particular attention to the areas around the head-to-shell joint, this area is sometimes heavily ground to try and blend in a poorly aligned seam.
- *Bulging.* Check the whole shell for any bulging. This is mainly caused by 'forcing' the shell or head during tack welding to compensate for a poor head-to-shell fit, or excessive out-of-roundness of the shell.
- *Nozzle flange orientation.* Check that the nozzles have not 'pulled' out of true during fabrication or heat treatment of the vessel. This can cause the nozzle flanges to change their alignment relative to the axis of the vessel. A simple check with a steel tape measure is adequate.
- *Welding.* Make a visual examination of all exterior welding (use the guidelines in Chapter 5, which are based on BS 5289 **(5)**). You can get an indication as to whether the correct weld preparations were used by looking at the width of the weld caps. Check that a double-sided weld has not been replaced with a single-sided one, perhaps because the manufacturer has found access to the inside of the vessel more difficult than anticipated. Watch for rough welding around nozzles, particularly small ones of less than 50 mm diameter. It can be difficult to get a good weld profile in these areas – look for undesirable features such as undercut, incomplete penetration, or a too-convex weld profile.

The vessel interior

It is important to make a thorough inspection of the inside of the vessel. This cannot be done properly by just looking through the manhole door

– you have to climb inside with a good light to be able to make an effective inspection. Check these points:

- *Head-to-shell alignment.* Most manufacturers take care to align carefully the inside edges of the head-to-shell circumferential joint. Check that this is the case and that there is a nice even weld-cap all the way around the seam.
- *Nozzle 'sets'.* Check the 'set-through' lengths of those nozzles protruding through into the vessel. Again, you should use the approved design drawing.
- *Weld seams.* Do the same type of visual inspection on the inside weld seams as you did on the outside. Make sure that any weld spatter has been removed from around the weld area.
- *Corrosion.* Check all inside surfaces for general corrosion. Light surface staining may be caused by the hydrostatic test water and is not a cause for concern. If the vessel specification does not call for internal shot blasting, such staining should be removed by wire brushing. In general there should be no evidence of mill-scale on the inside surfaces – if there is, it suggests the plates have not been properly shotblasted before fabrication.
- *Internal fittings.* Check that these are all correct and match the drawing. In many vessels, internal fittings such as steam separators, feed baffles, and surge plates are removable, with bolt threads being protected by blind nuts. The location of internal fittings is also important – make sure they are in the correct place with respect to the 'handing' of the vessel. It is also worth checking the fit of the manhole door and any inspection covers.

If you feel any uncertainty about the results of an internal inspection, it is best to address them immediately, before the manufacturer starts the process of preservation and packing of the vessel. Check any doubtful areas of welding for defects using a dye penetrant test (see Chapter 5). Small defects should be ground out. Make a note of all the areas you looked at and *describe* any defects that you find – remember to make a location sketch.

The dimensional check

It is normal to carry out a dimensional check of pressure vessels. Although a vessel is not a 'precision item', the positions of the fittings attached to the vessel shell are important – alignment of connecting pipework can be affected and pipe stress calculations are done assuming correct alignment to fixed points such as vessel flanges. Misalignment must therefore be avoided. Practically, the dimensional check can be done either before or after the hydrostatic test. Any strains or distortions that do occur will be small, and difficult to detect by simple measurement methods. It is normal for the vessel manufacturer to have completed a dimensional examination report (this is a simplified sketch of the vessel showing only the key dimensions) before the inspector arrives. This makes your task a little easier, eliminating the need to check against several different drawings. Dimensional checking can be done using a steel tape measure, with the use of a long steel straightedge and large inside or outside callipers for some dimensions.

Dimensional tolerances for pressure vessels tend to be quite wide. Follow the general tolerances shown on the manufacturing drawings. If any tolerances look particularly large, say more than ± 5–6 mm, double check them; note that there is a technical standard DIN 8570 **(6)** that gives general guidance for tolerances on fabricated equipment. Use the following guidelines when doing the dimensional check:

- *Datum lines*. First locate the datum lines from the drawings. Each vessel should have two, a longitudinal datum (normally the vessel centreline) and a transverse datum. The transverse datum is normally *not* the circumferential weld line – it will be located 50–100 mm inwards from the seam towards the dished head (see Fig. 6.27). It should be indicated on the vessel by deep centre-punch marks.
- *Manway location*. Check the location of the manway with respect to the longitudinal datum line.
- *Manway flange face*. Check that this flange face is parallel to its indicated plane. A tolerance of ± 1 degree is acceptable.
- *Nozzle location*. One of the more important sets of dimensions is the location of the nozzles in relation to the datum lines. It is easier to measure from the datum line to the edge of each nozzle flange rather than to try and estimate the position of each nozzle centreline.
- *Nozzle flange faces*. Check these by laying a long straightedge on the flange face and then measuring the distance between each end of the straightedge and the vessel shell. It may also be possible to use a graduated spirit level in some cases. Nozzle flange faces should be accurate to within 0.5 degrees from their indicated plane. Check also the dimension from each nozzle flange face to the vessel centreline – a tolerance of ± 3 mm is acceptable.
- *Flange bolt holes*. Check the size and pitch circle diameter of bolt holes in the flanges. It is universal practice for bolt holes to straddle

the horizontal and vertical centrelines, unless specifically stated otherwise on the drawing.
- *Vessel 'bow' measurements.* Both vertical and horizontal vessels sometimes 'bow' about their axial centreline. This is normally the result of uneven stresses set up during fabrication and heat treatment. The amount of acceptable bow depends on the length (or height) and diameter of the vessel. A small vessel of 3–5 m long and a diameter of up to 1.5 m should have a bow of less than about 4 mm. A typical steam drum of approximately 10 m long and 2.5 m diameter could have a bow of perhaps 6–7 mm and still be acceptable. Larger amounts of bow than these approximate levels are generally undesirable. You can detect bow by sighting along the external surface of the vessel by eye. The extent can be measured using a taut wire. Check at three or four positions around the circumference to get a full picture of the extent of any bowing.
- *Vessel supports.* Horizontal vessels usually have simple saddle-type supports. Check that these are accurately made so that the vessel sits level. A tolerance of ± 3 mm is good enough – perhaps a little less for longer storage vessels which will have a greater tendency to bend if inaccurately mounted. Vertical vessels may have tripod-type saddle supports, or a tubular mounting plinth which fits over the lower end of the vessel shell (see Fig. 6.27). Check this type carefully for accuracy – the vessel should stand vertically to within 1 degree. The best time to check a tubular plinth is before it is welded over the vessel shell.

It is not uncommon for a dimensional check on a pressure vessel to show a few dimensions that are marginally out-of-tolerance. These are not *necessarily* a case for rejection of the vessel. What is needed is a good sketch in your inspection report indicating where the deviations are. You should say whether such deviations affect the fitness for purpose of the vessel (as would be the case, for instance, with excessive bow or major misalignment of nozzles) or whether they are merely cosmetic. Make specific mention of any dimensional inaccuracies that will have an effect on the amount of site work required to connect the vessel to its piping systems.

Vessel markings

The marking and nameplate details of a pressure vessel are an important source of information when a vessel is received at its construction site. A

Boilers and pressure vessels 193

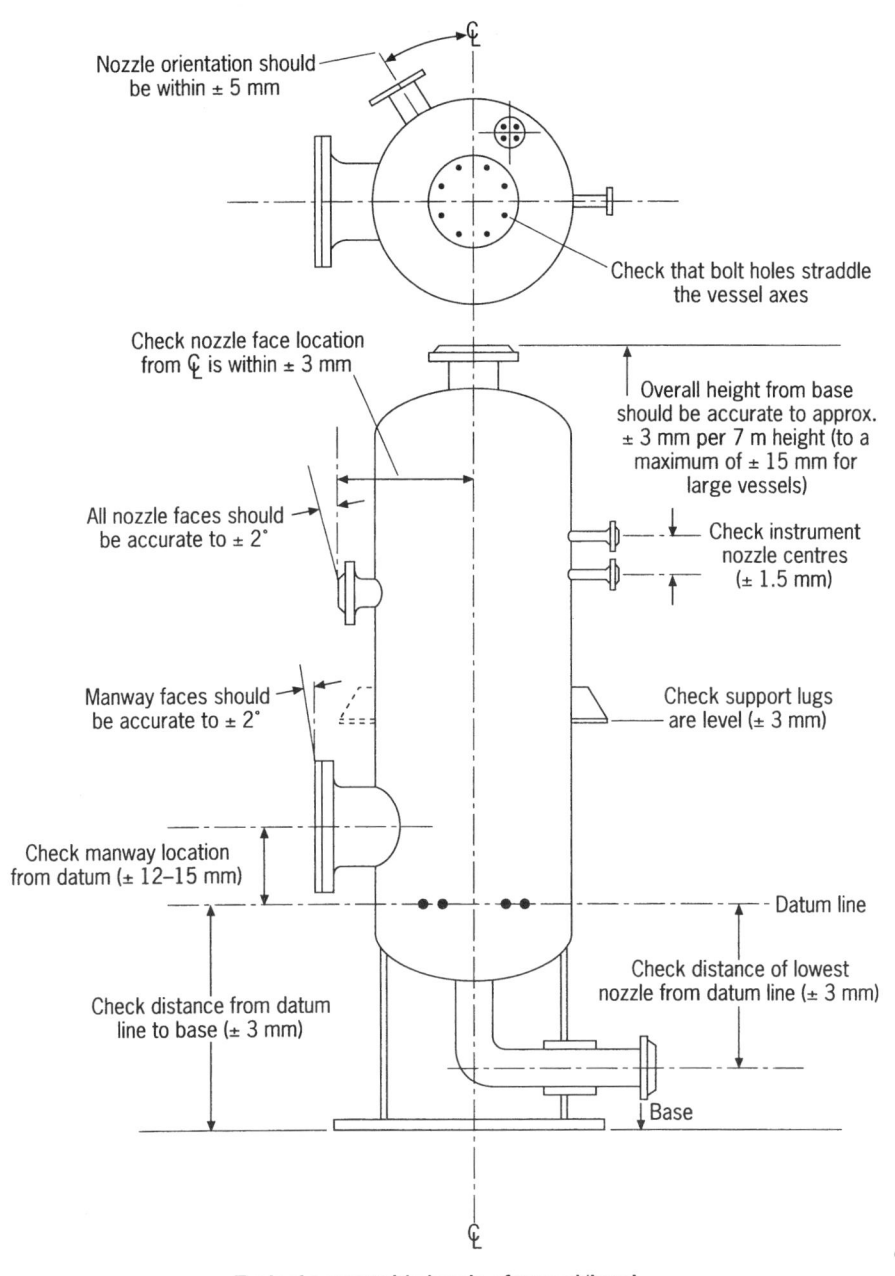

- Nozzle orientation should be within ± 5 mm
- Check that bolt holes straddle the vessel axes
- Check nozzle face location from ℄ is within ± 3 mm
- Overall height from base should be accurate to approx. ± 3 mm per 7 m height (to a maximum of ± 15 mm for large vessels)
- All nozzle faces should be accurate to ± 2°
- Check instrument nozzle centres (± 1.5 mm)
- Manway faces should be accurate to ± 2°
- Check support lugs are level (± 3 mm)
- Check manway location from datum (± 12–15 mm)
- Datum line
- Check distance of lowest nozzle from datum line (± 3 mm)
- Check distance from datum line to base (± 3 mm)
- Base

Typical acceptable levels of vessel 'bow'

Height (mm)	Diameter (mm)		
	<1200	<1300	>1700
<3000mm	2.5	2	2
3000-9000mm	4	7.5	6
>9000mm	5	10	8

Fig 6.27 Checking vessel dimensions

large plant can have several hundred vessels, constructed to different codes and for varying applications, so positive identification is a distinct advantage. Many of the pressure vessels will be subject to a further

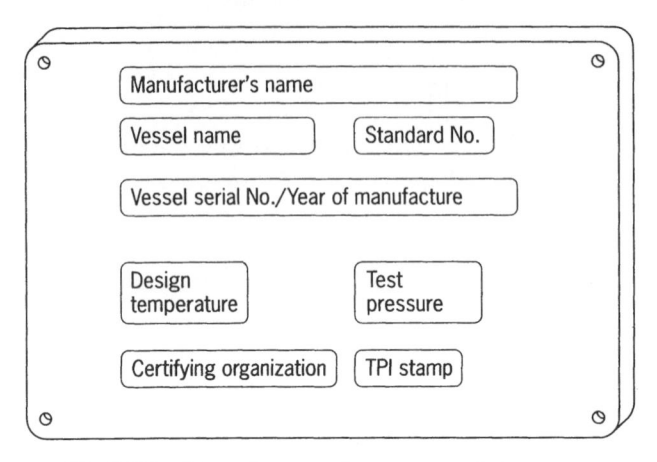

```
┌─────────────────────────────────────────────┐
│  ┌───────────────────────────────────┐      │
│  │ Manufacturer's name               │      │
│  ├──────────────────┐ ┌──────────────┤      │
│  │ Vessel name      │ │ Standard No. │      │
│  └──────────────────┘ └──────────────┘      │
│  ┌───────────────────────────────────┐      │
│  │ Vessel serial No./Year of manufacture │  │
│  └───────────────────────────────────┘      │
│  ┌──────────────┐    ┌──────────────┐       │
│  │ Design       │    │ Test         │       │
│  │ temperature  │    │ pressure     │       │
│  └──────────────┘    └──────────────┘       │
│  ┌──────────────────┐ ┌──────────┐          │
│  │Certifying organization│ │ TPI stamp │    │
│  └──────────────────┘ └──────────┘          │
└─────────────────────────────────────────────┘
```

Fig 6.28   Vessel nameplate–essential content

hydrostatic test after installation, so information about the design pressure and works test pressure needs to be clearly shown. The correct marking of a vessel also has statutory implications, it is inherent in the requirements of vessel codes and most safety legislation that the safe conditions of use are clearly indicated on the vessel. Vessel marking is carried out either by hard-stamping the shell or by using a separate nameplate. It is preferable to use a separate nameplate on vessels with plate thickness of less than about 7 mm, or if the vessel is designed to operate at low temperatures. Figure 6.28 shows the content and layout of a typical vessel nameplate. Check the nameplate during the final inspection visit, using the following guidelines:

- The nameplate details should be completed *before* the vessel leaves the manufacturers' works. This must include all the data shown in Fig. 6.28. The only common exception to this is if it has been clearly agreed that a standard hydrostatic test will not be performed in the works but will be performed later at the construction site. This occasionally happens and, surprisingly, does not contradict the requirements of most vessel codes.
- Check that the statutory inspector has hard-stamped the nameplate. This will act as a general assurance to construction site staff that the

statutory aspects of design and manufacture have been properly addressed.
- *Concessions*. A vessel manufactured with concessions from full code compliance must be identified as such. There are several ways to do this, the main criteria being that the concessions are brought to the full attention of the construction site and operation staff. Any resultant limitations on the use of the vessel can then be properly addressed. For BS 5500 vessels, the normal way is to add the suffix XX to the vessel serial number on the nameplate. This shows that concessions are in force. If you see this designation, check that the technical details, and justification, of the concessions are adequately explained in the vessel's documentation package. Make a special note in your inspection report.
- Make sure the nameplate is *firmly fixed* to the vessel. There should be a steel mounting plate welded to the shell and the nameplate should be bolted or riveted to it. Do not accept loose nameplates – they will inevitably get lost.

It is the practice of some statutory inspectors to take a 'pencil rubbing' of the completed vessel nameplate – this creates a permanent record which is included in the final documentation package. You can do this if you have the time.

Non-conformances and corrective actions

If non-conformances do exist in a pressure vessel they are normally not too difficult to find and identify. The main reason is that the underlying requirement for statutory inspection encourages the use of detailed ITPs and a high level of inspection scrutiny. The problem is more one of corrective action; non-conformances are often only discovered when the vessel manufacture is well advanced or complete, at which point the corrective action options can be limited. You will find two types of non-conformance, those where it is possible to initiate an active correct action to restore full FFP compliance and those where the only action possible is a retrospective one. This second category rarely gives a perfect solution and often has to result in a code concession being applied to the vessel.

An important part of the role of a works inspector involved in pressure vessel inspections is to agree *solutions* to non-conformances. These take many forms but you can expect a lot of pressure to be placed on you by manufacturers to accept *retrospective* solutions. Normally these will involve some measure of compromise of the FFP criteria. For this reason you will find vessels a good 'proving ground' for the strategic and tactical approaches I introduced in Chapters 2 and 3. You should also be aware of the influence of cost when dealing with pressure vessel issues. Try to follow the principle of agreeing solutions to vessel non-conformances directly with the manufacturer, as long as you feel technically comfortable with what you are doing. This is the most effective and economical way to do it. The alternative approach, that of referring even minor points to technical specialists, is of course acceptable – but the time and cost implications will rise accordingly. Ask yourself if this is effective inspection before deciding your own course of action.

Bearing these principles in mind we can look at eleven of the more common non-conformances that are found in pressure vessels. I have tried to present these broadly in the order of frequency with which you should find them, starting with the most common.

Missing documents

This is very common. Sometimes it is caused by missing ITP steps, or it may be the result of poor communication with sub-suppliers further down the manufacturing chain. You have to come to a quick decision as to whether a particular missing document really compromises FFP. In about 90 percent of cases it probably will not – the *activity* will have been performed correctly but the document record will have been mislaid. The remaining 10 percent of cases will impinge on FFP (these are generally NDT related) and lead to some uncertainty about weld integrity. The best steps to take are:

- Decide whether FFP may be in jeopardy.
- Ask questions. Look for evidence that indicates whether all the necessary activities have been done (in spite of the missing document).
- Make your best efforts to resolve the issues quickly. Contact sub-suppliers if necessary.
- Issue an NCR, making it clear exactly which document is missing. Give your decisions in your report – say whether or not you think the missing document affects FFP.

Incomplete statutory certification

There are many situations that can result in the conditions for statutory certification being incomplete. There may be outstanding design appraisal questions, missing documents, or observations made during the manufacturing process; all can cause the certifying organization to be hesitant. For all inspectors other than those representing the certifying organization this should be a rather 'black and white' issue. The objective is to obtain a certificate of *unqualified* code-compliance – so place the onus on the certifying organization to state clearly why it feels this is not possible. Ask for precise reasons, not just expressions of general discontent. Be careful to check that any vessel certificates that are subsequently issued are complete and unqualified – they should meet fully the wording of BS 5500 'form X' or the ASME equivalent. Learn to be wary of certificates that have evasive wording or additional exclusions. If in doubt, look at the certificate under the assumption that a serious failure incident has occurred and you are required to *prove why* you accepted the certificate as an assurance that the vessel was fully compliant with the relevant code. We discussed at the beginning of this chapter that the vessel codes are stated in statutory documents as accepted examples of good practice. This means it is poor practice to accept qualified vessel certificates.

Once again, the best solution is to find technical solutions quickly, before manufacture progresses to a point where the only solution involves a permanent non-compliance with the vessel code. Few responsible certification organizations will issue an unqualified vessel certificate on this basis.

Incomplete material traceability

Don't confuse this with incorrect material properties – which is a different (more important) issue. Unless you have very firm evidence that FFP *is* compromised (in which case the vessel should be rejected and remanufactured) this is normally a retrospective exercise. For pressure shell components, the best step is to specify a retest of the material specimens from the production test plate (see Chapter 5). This changes a documentation problem into an engineering activity. It allows a more objective solution than will an exercise of 'certificate-chasing' from material sub-suppliers – which takes a lot of time and will not reinstate any traceability chain that has been broken (see Fig. 4.6). Assess the retest results carefully and explain in your report what was done and why. For non-pressurized components such as vessel saddles and frames you may wish to take a more relaxed view, to reflect the lesser effect these items have on the fitness for purpose of the vessel.

Incorrect dimensions

With large vessels it is not uncommon to find a few dimensions that are marginally outside the drawing tolerances. As long as any inaccuracy is not caused by serious distortions or bowing, the effects are generally of a minor nature. It is wise to exercise a little restraint – an NCR would only be properly justified if the vessel was clearly 'the wrong size'. For completeness, you should record the out-of-tolerance dimensions (using a sketch) particularly if these involve the nozzle positions. This way any implications for the site connection of pipework can be assessed.

Head-to-shell misalignment

Misalignment falls into a different category to the other dimensions of a pressure vessel. Maximum allowable misalignment is carefully calculated to keep the discontinuity stress (caused by the different response of the shell and head to internal pressure) within defined limits. These discontinuity stresses can become very high if the allowable misalignment is exceeded. If you find such misalignment, issue an NCR. There is little that can be done to rectify the situation short of remanufacturing the vessel. It is not advisable to pursue concessions to the vessel code in such instances – the technical risks are too high.

Incorrect weld preparations

The most common faults you will find are:

- wrong weld preparation angles
- assymetrical weld preparations machined the wrong way round
- incorrect root gaps (after tack-welding).

It is difficult to make general statements about the acceptability of incorrect weld preparations. The weld preparation design is part of the weld procedure specification (WPS) for a particular welded joint which is then qualified by the use of an approved procedure qualification record (PQR), showing the results of tests carried out on that configuration. Once this WPS/PQR link is broken by the use of an incorrect weld preparation, it is likely that the strength and integrity of the weld joint will be affected. As a general rule it is best to request that the preparation be remachined to the correct configuration, if this can be done without removing so much parent material that it will cause 'mismatch' of other joints in the vessel.

If remachining is not possible, the correct action is to specify that the incorrect joint be qualified (i.e. a new PQR), by making a test piece and subjecting it to the necessary non-destructive and destructive tests. This is particularly important for nozzle-to-shell welds as their strength characteristics tend to be less predictable than for simple single or double-vee butt welds.

Incorrect weld procedure specifications (WPSs)

Treat this in a similar way to that for incorrect weld preparations – but anticipate more serious consequences. An incorrect WPS is often only discovered after the welding has taken place, hence there may be no real chance to correct it without cutting the vessel and rewelding. The only other solution is to try and qualify the actual WPS used by carrying out the PQR steps mentioned previously. Be careful not to be too enthusiastic about recommending this action, however – take a close look at the incorrect WPS first. If the modifications include any major changes to the weld root (for instance the lack of a MIG root-run where this has been specified previously), or to the filler (consumable) material, it is unlikely that any subsequent attempt to qualify the new WPS will be successful. It will probably fail the tests. For other errant weld variables, the technical risks involved are less. Make it clear in your report that you have taken the essential variables into consideration before making your decision.

Incorrect material properties

This refers to a situation where you find the material properties are marginally outside the specification tolerances, not a situation where a totally incorrect material has been used in error. On balance, the mechanical properties of a pressure vessel material are of greater importance than the chemical composition. This is because the mechanical properties are a *function* of the chemical analysis and because mechanical properties can be changed by heat treatment, which leaves the chemical analysis nominally unchanged. There are two main types of non-conforming mechanical properties that you may encounter. The simplest case is that of out-of-tolerance tensile properties – the most common being that the yield strength is too low. For most pressure vessel designs, marginal differences in tensile strength can frequently be compensated for by the factor of safety already built in to the design. A corrosion allowance will also be included, although under most vessel codes this must not be included in the material thickness value used for stress calculations. Marginal differences in tensile properties alone should rarely cause a vessel to be declared unfit for purpose. In the first instance your NCR should specify a recalculation exercise to demonstrate whether the revised factor of safety (using the actual material tensile strength) meets the code requirements. Assess the results carefully in conjunction with the certification organization.

If it is the impact or hardness test results that are out-of-specification, the effects are likely to be more serious. Low impact values or high hardness readings imply increased brittleness of the material. Brittleness is related to crack propagation, one of the mechanisms which contributes significantly to practical failure mechanisms. A sound strategy is to specify a programme of retests, even for marginal out-of-specification results. Remember that a minimum of three specimens is required, because of the inherent difficulties in obtaining reproducible results from impact tests (see Chapter 4). If the retest results confirm that the impact properties are genuinely too low, this casts a serious shadow of doubt on the fitness for purpose of the vessel. It is not easy to demonstrate *explicitly* the effects of poor impact strength – you will find that none of the vessel codes are very prescriptive in this area. It is difficult therefore for vessel designers to demonstrate acceptability in the way that is possible with reduced tensile properties.

You can minimize wasted time and effort by anticipating these outcomes in advance. I feel that it is well justified to reject a vessel that has out-of-specification impact properties. If you can demonstrate firm evidence of increased brittleness you are on safe technical ground – it is unlikely that the certifying organization will overrule this type of decision. Be wary of proposed solutions which advocate 'downgrading' of the vessel to operate at a lower pressure. There is often little technical coherence in such arguments.

Missing NDT

A missing NDT is an important omission, but not one which should cause too many problems. The solution is simply to specify that the NDT be repeated on the vessel in its current condition. Surface crack detection using DP or MPI is straightforward and ultrasonic testing can replace radiography – this is acceptable under most vessel codes. Specify the technique to be used, and the relevant acceptance standard, in your NCR. There are few valid reasons why ultrasonic testing cannot be done on a completed pressure vessel – any rough surface finish can be improved by grinding and the critical areas of butt and nozzle welds are easily accessible by a skilled operator (see Chapter 5). It is poor practice not to ensure that all the specified NDT tests are properly carried out. NDT results are important evidence that the integrity of welded components has been properly assessed. Their review is an important inspection responsibility.

Remember the common misconception (introduced in several chapters of this book) that a hydrostatic test at an increased pressure can be considered a substitute for missing NDT. It is worth repeating that although the 'surface arguments' for this approach can seem convincing, the underlying technical rationale is poor. Hydrostatic tests will identify leakage – but they will not necessarily reveal the type of defects that contribute to the onset of failure. The correct NDT is essential to demonstrate the integrity and FFP of the vessel.

Unrecorded repairs

Occasionally during visual examination of a vessel you may find evidence of unrecorded repairs. This is more common in welds than in the 'body' of components such as nozzles. Repairs are allowed, but should have been recorded, along with details of the repair procedure used. First, check whether the repair procedure and records really are unavailable – they may exist, but just not be included in the document package submitted for review. If you conclude that a repair is genuinely unrecorded then you are justified in taking further action to make sure that no *other* similar repairs exist, and to check the integrity of those unrecorded repairs that have been done. Start with 100 percent surface NDT, followed by a percentage volumetric examination in the most critical areas (butt weld tee-joints and those full penetration weld types shown in Fig. 5.2 of this book). Make sure that your NCR specifies clearly the type and extent of NDT you feel is necessary.

Hydrostatic test leaks

If there is any leakage at all, even from temporary flanges or pipe connections, issue an NCR. Leakage from welds is usually an indication of either extensive porosity or cracks. The question you must ask is: 'Why were these not discovered during the NDT activities?' Assess the answers carefully – you need to decide whether there are other unseen risks to the integrity of the vessel, in addition to those defects that are causing visible leaks. Areas of porosity and cracking can be excavated and then rewelded. Make sure that the repair procedures are properly reviewed and approved first – some materials will require further heat treatment of the whole vessel to eliminate stresses induced during the repair welding. Record all repairs using sketches if necessary.

# KEY POINT SUMMARY: PRESSURE VESSELS

Fitness for purpose

1. The overriding FFP criterion is the *integrity* of the vessel – particularly the welded joints. The industry has developed four 'norms' (commonly accepted activities) to try and provide an assurance of integrity. They are:

   - an independent design appraisal
   - material traceability
   - prescribed NDT activities
   - hydrostatic (pressure) testing.

Certification

2. The requirement for vessel certification may be imposed by an end-user, purchaser, insurance company or statutory authority.
3. Vessel certification *only* shows that a vessel complies with the design code (BS, ASME, TRD, etc.). It is not a guarantee of integrity or FFP. It is normally unrelated to project-specific requirements.

Pressure vessel codes

4. Vessel codes deal mainly with design, rather than inspection. Much of the inspection-related information you will need is contained in referenced 'subsidiary' technical standards, not in the code document itself.
5. Expect to see vessel codes used for other types of fabricated and cast components. In such cases a more flexible approach is required.
6. A hydrostatic test is a test for leakage, and a vessel's ability to withstand static principal stresses. When vessels do fail, they normally do so because of *different* failure mechanisms. So, a hydrostatic test is not a substitute for correct NDT.

Final inspection

7. The final visual/dimensional examination of a vessel benefits from a careful, structured approach. Use a checklist.
8. There are 11 main categories of non-conformance that you should be aware of. Some have only retrospective (and imperfect) solutions if discovered too late in the manufacturing process. If you are not careful (or are badly organized), the *costs* of deciding corrective actions will rise steeply.

References

1. BS 5500: 1994. *Specification for unfired fusion welded pressure vessels.*
2. Technische Regeln fur Dampfkessel (TRD) – *Boiler code of practice for pressure vessels.* Vereinigung der Technischen Uberwachungs-Vereine. E.V. Germany. Also the Arbeitsgemeinschaft Druck Behalter (AD Merkblatter) range of standards.
3. *The ASME boiler and pressure vessel code: 1995 edition.* An internationally recognized code, published by the American Society of Mechanical Engineers.
4. BS EN ISO 9001: 1994. *Model for quality assurance in design, development, production, installation and servicing.*
5. BS 5289: 1983. *Code of practice. Visual inspection of fusion welded joints.*
6. Deutscher Normenausschus (DIN) 8570. 1987. Part 1: *General tolerances for welded structures – linear and angular dimensions.* Part 2: *Geometrical tolerances.*

Chapter 7

Gas turbines

Gas turbines (GTs) are not known for their tolerance to poor engineering practices. Build a few basic manufacturing faults into a machine, send it to site, and very soon it just won't work. Gas turbines are particularly interesting from an inspection viewpoint – you are unlikely to meet other machines of such complexity utilizing such a large number of very precise manufacturing operations. Perhaps because of this you can expect the manufacturing businesses that make them to be amongst the most technically competent that you will meet.

Fitness-for-purpose criteria

It is a little unusual that machine thermal efficiency, which GT designers continually strive to improve, actually imposes very little (if at all) on the work of an inspector. Frankly, there is little point in trying to verify thermal efficiency during normal works inspection activities – efficiency is almost entirely dependent on design, and then adjustment (which is done on site to meet the ambient conditions and planned loading regime). Leave it to others. The FFP criteria which it is necessary to consider are:

- *Running integrity*. This is particularly important in GTs because of their high temperatures, high speeds, and tight tolerances. Be prepared to pay a lot of attention to balancing and vibration tests – these are pivotal parts of the manufacturing process.
- *Systems function*. Due to their complex technology GTs rely heavily on their support systems. Auxiliary systems to supply atomizing air, fuel, hydraulic and lubricating oil, and power take-off are generally installed on the GT skid and hence form part of the manufacturer's

scope of supply. Control and safety systems are also important. One of the prime FFP objectives is to ensure that these systems are all integrated correctly – functional tests are therefore more important and more involved than with other types of prime movers which have simpler support systems. So, a good inspector of gas turbines is one who pays *attention to detail*. This is definitely not the place for a 'broad brush' approach to FFP.

Basic technical information

GTs divide broadly into single-shaft machines and two-shaft machines, in which the gas generator and power turbine have separate shafts. Most designs used for power generation are of the single shaft type. Sizes range from small units of 1 MW up to approximately 230 MW. The main GT licensors classify their machines using 'frame sizes' or model numbers. Site installations can be either open cycle or combined cycle, in which the GT exhausts to a waste heat recovery boiler. The same GTs are used for either. Expect power generation GTs to be designed for continuous base load operation. The main fuel is most often natural gas, with the capability to burn distillate oil as a stand-by. This means that most GTs you will see will incorporate a dual fuel system on the skid. Auxiliary equipment mounted on the skid may vary – it depends on the way in which the GT will be incorporated into the site installation.

Acceptance guarantees

Acceptance guarantees for gas turbines are an uneasy hybrid of explicit and inferred requirements. In most contract specifications you will find that there are four explicit performance guarantee requirements: power output, net specific heat rate, auxiliary power consumption, and NO_x emission level. These are heavily qualified by a set of manufacturer's correction curves which relate to the various differences between reference conditions and those experienced at the installation site. You cannot check any of these (accurately) during a factory no-load running test. This does not mean that witnessing gas turbine works tests is a formality. The reality is that the opposite is the case. By being present at the works test you will be accepting responsibility for the *utility* of the installation. Remember that you are not there only to verify the explicit performance guarantees – what you are looking for is fitness for purpose. That is why, in a good GT purchase specification, you will find a significant number of more expansive engineering requirements – often listed under 'technical particulars' or enclosed in a technical schedule. These are inferred FFP requirements and will include:

- *Governing characteristics.* The range of speed adjustment and droop.
- *Overspeed settings.* Typically designed to operate at 110 percent full speed for the first electronic trip and 112 percent for the second electronic trip or mechanical *bolt* mechanism.
- *Vibration and critical speeds.* Expect to see acceptable vibration levels specified in two ways: bearing housing vibration measurement and shaft relative vibration measurement.
- *Noise levels.* These are also sometimes specified in two ways. A level may be specified at 1 metre from the acoustic enclosure and at the site boundary limit (this also incorporates the effect of other equipment, hence is not a function of the gas turbine only).

There is a further important aspect to gas turbine acceptance guarantees. Because of the complexity of the machines the technical standards used as a basis for procurement (e.g. ISO 3977/BS 3863) **(1)** make specific reference to 'agreements between purchaser and supplier'. This infers a technical agreement on specific engineering or test matters that are not well defined in applicable standards. It is important to make sure that you obtain details of such agreements, if they exist. Some purchasers *will* have specific requirements which you may find difficult to anticipate.

Specifications and standards

Gas turbine manufacture is one of the few areas of power plant technology in which international technical standards have not developed as quickly as manufacturers' own procedures. The main reason is that intense technical competition between a relatively small group of manufacturers, firmly linked to the three or four predominant licensors, has caused the technology to develop very quickly. Note the way in which aero-derivative gas turbines have rapidly increased in power output and how many mechanical aspects of these turbine designs have changed over a very few years.

I have said that published GT standards lag behind the best industrial practice. This decreases slightly the power of the external inspector when he is witnessing work tests. If you want to continue to be *effective* you have to be ready to modify slightly the way in which you view the role of standards when dealing with gas turbines. Expect to find them less prescriptive – remember the need for useful *qualifying information*, which I introduced in Chapter 2.

Standards that you may meet are listed below. Their content varies from the more general (at the beginning of the list) to more specific technical standards towards the end of the list.

ISO 3977 *Guide for gas turbine procurement*, is identical to BS 3863. Treat this as a guidance document, useful for information on definitions of cycle parameters and for explaining different open and closed cycle arrangements. There is a useful framework around which to structure the purchaser/manufacturer agreement. I do not think this standard is detailed enough to act as a stand-alone purchase specification.

ISO 2314 *Gas turbine acceptance tests* (identical to BS 3135) (**2**). This is not a step-by-step procedure for carrying out a no-load running test. It contains mainly technical information on parameter variations and measurement techniques for pressures, flows, powers, and so on. You may need this standard if you become involved in detailed performance calculations.

ANSI/ASME PTC 22 *Gas turbine power plants* (**3**). This is one of the performance test codes (PTC) family of standards. The introduction to this standard states its content as *procedures for testing of gas turbines*. Its content is quite limited, covering broadly the same area as ISO 2314 but in less detail (so it is perhaps not essential reading).

API 616 *Gas turbines for refinery service* (**4**). In the mould of most API standards, this provides good technical coverage. There is a bit of everything to do with gas turbines in here. The best way to apply it is first to check if it is an explicit requirement of the purchase specification. If it is not, be warned that you may find it does not fit exactly the characteristics of some proprietary gas turbines – particularly those of European licensors' design. Some manufacturers have significant experience of submitting concession applications listing an inventory of those technical parts of API 616 that they do not agree with. This does not mean it is not a good informative standard, just that it must be applied in the correct context.

ISO 1940/1 (identical to BS 6861 Part 1) (**5**) and VDI 2060 cover balancing of the rotors. These are useful standards which are surprisingly similar. You will need ISO 1940 if you want to understand balancing – it gives acceptable unbalance limits (which do tend to comply with manufacturers' own acceptance limits).

ISO 10494 (similar to BS 7721) (**6**) and ISO 1996 are standards relating to GT noise levels. You will see them used in most industries. They are good standards which are easy to understand.

Vibration standards used are as follows:

- Bearing housing vibration is covered by VDI 2056 (group T) (**7**). This is a commonly used standard for all rotating machines. Remember that it uses vibration velocity (mm/sec) as the deciding parameter.
- Shaft vibrations using direct-mounted probes are covered by API 616 or ISO 7919/1 (**8**) (also commonly used for other machines). The measured vibration parameter is amplitude. VDI 2059 (Part 4) (**9**) is sometimes used, but it is a more theoretical document which considers the concept of non-sinusoidal vibrations – it is less suitable for direct use during works inspections.

In addition to these standards there are also many engineering materials standards that apply to the steels used in gas turbine construction.

Inspection and test plans (ITPs)

Inspection and test activities are a cornerstone of gas turbine manufacturing. A brief investigation into the manufacturer's assembly process will reveal the presence of a large body of documents – you will find that together they form a comprehensive and coherent system of assurance. From an inspector's viewpoint, the number and complexity of manufacturing processes can prove difficult to handle – it is all too easy to become submerged in plans and schedules. The answer is *selectivity*. A good gas turbine ITP for use by external inspectors has been specifically designed for that purpose alone. This is in contrast to simpler equipment, where the ITP serves the dual purpose of being a manufacturing 'guide' and having a role in structuring external input. You should find that the ITP for a complete gas turbine and skid will occupy perhaps three or four pages. You may find a separate ITP for the rotor – this is because the rotor is often manufactured at a separate location, closely controlled by the technology licensors. There may be slight variations of ITP content for the starter and auxiliary components. These are frequently sub-contracted items and hence tend to be subject to a lower level of direct control.

The number of external inspectors' witness points on the consolidated ITP will be surprisingly small. Check that these essential steps are included:

Before assembly of the rotor into the casing

- Rotor runout measurement.
- Rotor dynamic balancing.
- Rotor overspeed testing.
- Interim documentation review.

After assembly

- Measurement of radial and axial blade clearances.
- No-load running test. Not all manufacturers will do this if it is not specifically requested in the purchase contract. Don't expect that any of the technical standards impose it as a requirement to help you.
- Full functional test of all on-skid systems (that does mean *full* functional tests).
- Final detailed documentation review.

Many gas turbine contracts have been successfully concluded to the satisfaction of all parties (more or less) using only these essential witnessed inspection steps. Further inspection costs more money. Consider carefully whether this is justified.

Rotor runout measurement

Prior to final assembly of the blades to the gas turbine rotor the rotor runout is measured. The objective is to identify excessive ovality in the rotating mass which will produce out-of-balance forces, and to ensure that there is no significant eccentricity or taper of the journal bearing surfaces. Although runout measurement is a relatively straightforward task there are several points that merit special attention.

- The correct value to measure is total indicated runout (TIR). This is the biggest recorded *difference* in the dial test indicator (dti) – reading as the rotor is turned through a complete revolution. Figure 7.1 shows a typical results format.
- Make sure the rotor is correctly mounted, otherwise the results may

be meaningless. Mounting is usually on a large centre-lathe. Check that:
— The headstock is rotated with the oil supply circulating for at least 30 minutes. This will ensure the headstock adopts a central position.
— The headstock should drive the rotor via two universal joints to maintain concentricity of the rotor centreline.
— The mounting stand for the dial test indicator should be securely mounted when each measurement is taken. Make sure, particularly when a tall floor-mounted stand is used, that it does not wobble.

- *Acceptance limits.* The maximum acceptable TIR is usually defined by the manufacturer rather than specified directly by a technical standard. Expect to see three levels of limits as shown in Fig. 7.1. The tightest limit is for the bearing journals (typically 10–15 µm). Radial surfaces of the turbine blade discs should have a limit of 40–50 µm. Axial faces of the discs often have a larger limit, perhaps 70–90 µm. The exact limits used depend on the design.

Rotor balancing

Gas turbine rotors are all subjected to dynamic balancing. This is a key part of the manufacturing witness programme. Balancing is carried out with all the blades installed. Procedures can differ slightly. Turbines that have separate compressor and turbine shafts may have these balanced separately (this is common on larger three-bearing designs), although some manufacturers prefer to balance the complete rotor assembly. Although the dynamics of balancing can be a little confusing, as we are dealing with vector quantities, the simple physical shape of most rotors means that the procedure and calculations are reasonably straightforward. The key fact to remember is that you are dealing with *two* correction planes.

Which limits to work to?

The figure in which we are mainly interested is the limit of acceptable unbalance. Note that this is frequently expressed per correction plane (as in ISO 1940), but I have occasionally seen it quoted in manufacturers' test procedures as the vector resultant of the allowable unbalance on the two correction planes. The correct unit of unbalance is gramme metres (gm).

You may often find that manufacturers impose their own limits without an explicit reference to one of the technical standards. It is

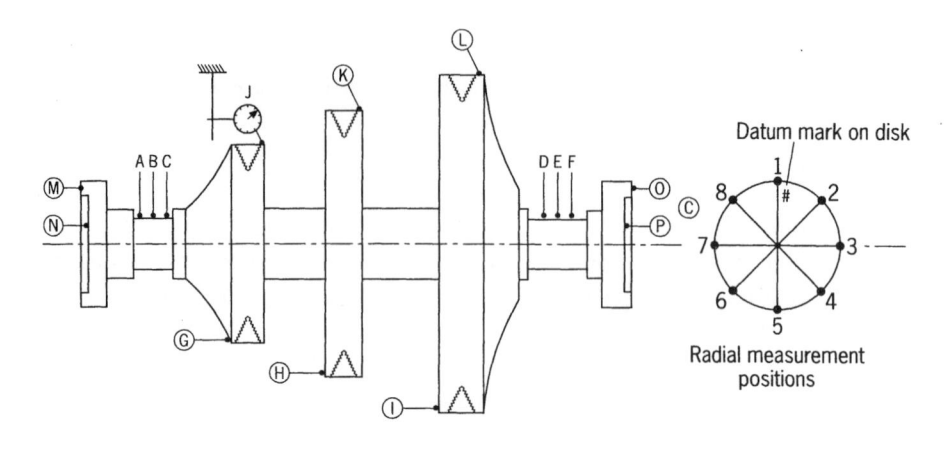

Measurement point	Typical limit (μm)	1	2	3	4	5	6	7	8	TIR
A, B, C	15									
D, E, F	15									
G, H, I	70									
J, K, L	45	13	15	47	41	18	21	31	35	34
M, N, O, P	15									

Specimen result:
measured TIR is 47μm−13μm=34μm
this is less than the limit of 45μm ∴ ACCEPTABLE

Fig 7.1 Turbine rotor runout measurement

worth doing a quick check to see how their limits comply with ISO 1940 Grade 2.5 (which is a generally accepted standard for turbine rotors) – use the chart in the standard. I have always found that manufacturers meet, or exceed this grade. I have shown in Fig. 7.3 the way in which the acceptable limits should be expressed to allow easy comparison with the test results.

The balancing test

As shown in Fig. 7.2, the witnessed test consists of a set of well-defined steps. Follow them through in this order:

- The rotor is supported by its journals in a balancing machine, and rotated at test speed (normally 10–15 percent of rated speed). The unbalance of the drive coupling is determined, then the coupling is disconnected and reconnected to the rotor with 180 degrees displacement. The process is then repeated to confirm the coupling

Gas turbines 213

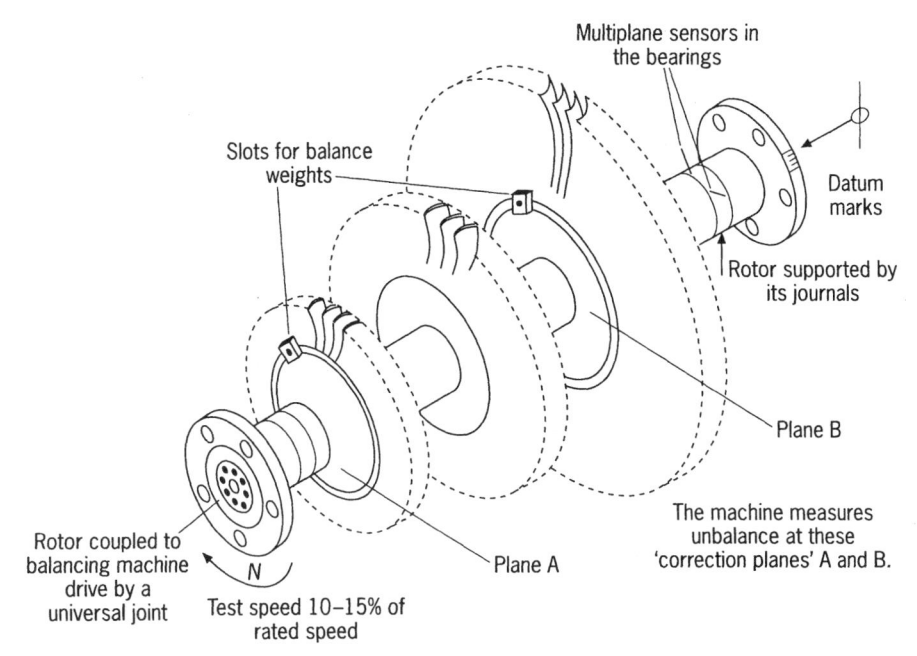

Fig 7.2 Two plane (dynamic) balancing of a GT rotor

unbalance, which is programmed into the balancing machine so it will compensate for it. As a result of this last run we will now have an approximate *initial unbalance* reading for the rotor.

- The next step is *stabilization*. The rotor is run near test speed to enable the blades to adopt their running position (they are loose, not fixed, remember). Make sure the stabilization period lasts until the readings are constant, showing the blades have stabilized (the correct way to do this is to raise and lower the rotational speed slightly and see if the unbalance readings change, ≤ 5 percent is acceptable). The time needed to stabilize can be anything from 1 to 24 hours. It is longer for large compressor-stage rotors.
- Once the rotor is stabilized it is returned to test speed. Test readings are now taken. Figure 7.3 shows the form that the balancing readout will take (most will also provide a print-out).
- *Interpretation*. For each correction plane (note that, strictly, these are also 'measurement planes' as well as correction planes) the unbalance mass is displayed along with its phase angle relative to a datum. The mass figure will have been calculated as acting at an effective radius equal to that of the balance weight slot on that particular rotor (Fig. 7.2 shows where these are for a turbine-end rotor). Now compare

these with the acceptable limits (in gm) per correction plane – or sum the two (vectorially) and then compare, if you are working to a 'resultant' limit. Figure 7.3 demonstrates this technique and shows a typical set of results. You will see that in this example the unbalance limits per plane are different, depending upon the angle between the vectors. This is because an angle of more than 90 degrees results in an additional 'couple' on the rotor – so the reduced limits are chosen to take this into account. Don't worry *too* much about this, ISO 1940 treats this as a second-order effect. However, a good manufacturer will take it into account.

- *Adding balance weights.* If the rotor is within its unbalance limits there is no need to add any further weights. If weights are needed, let the test staff decide their position – it is not always straightforward because there may be bolt holes or existing weights occupying the ideal position where weights need to be added. Note the location of weights for your report. Then retest.
- *Balancing machine accuracy check.* Be careful not to be confused by this. After balancing the rotor to within its unbalance limits it is common to perform an accuracy check on the balancing machine. The purpose of this is to eliminate cumulative errors due to measurement inaccuracies, etc. It is done by adding more known weights to the rotor, firstly to compensate for the measured unbalance and then, again, to increase the unbalance. By reviewing the machine readings it is possible to determine the accuracy level. You should witness this part of the test. Look for an acceptable accuracy level of better than ± 5 percent for the mass reading and ± 5 degrees for phase angle.

Gas turbine rotors do sometimes give problems after installation. If this happens, your balancing witness report may become an important reference. Have you included the following?

— A clear sketch of the position of the balance weights, separately annotated to differentiate between weights added at the shaft balancing stage and those at the assembled rotor balancing stage.
— A proper comparison of the acceptable limits and the actual test results.
— Records of the balance vector and resultant, so you can show that you did understand what was going on.
— Confirmation that you *really* witnessed stabilization of the rotor before taking residual unbalance readings.

Gas turbines 215

The balancing machine readout is:

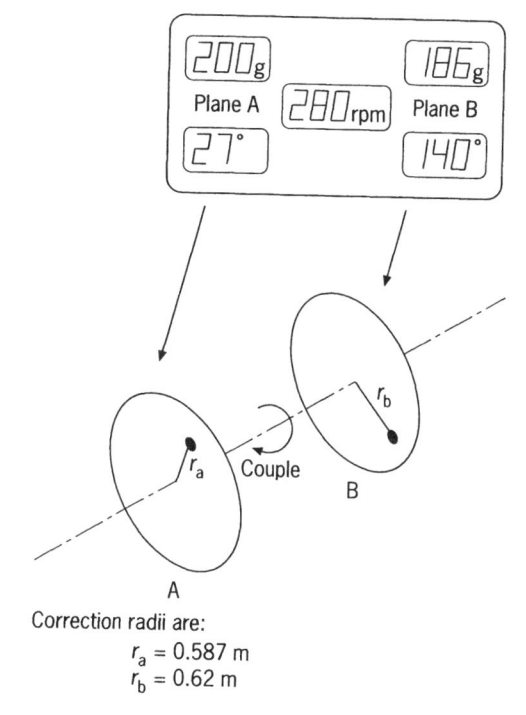

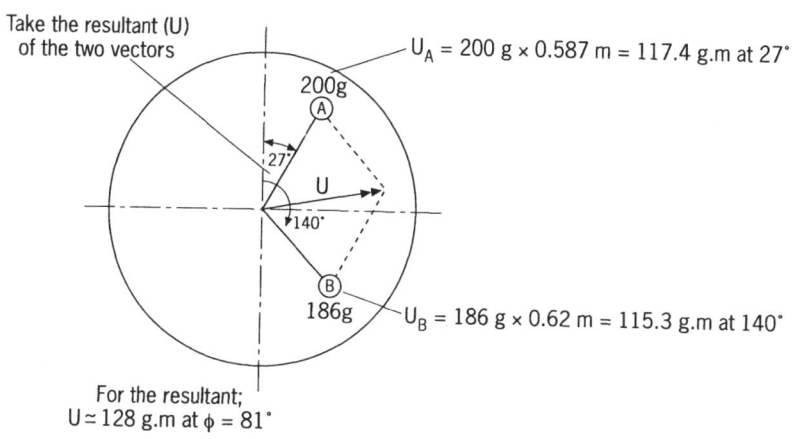

```
┌─────────────────────────────────┐
│  200g              186g         │
│  Plane A  280rpm   Plane B      │
│  27°               140°         │
└─────────────────────────────────┘
```

Correction radii are:
$r_a = 0.587$ m
$r_b = 0.62$ m

Take the resultant (U) of the two vectors

$U_A = 200$ g × 0.587 m = 117.4 g.m at 27°

$U_B = 186$ g × 0.62 m = 115.3 g.m at 140°

For the resultant;
$U \approx 128$ g.m at $\phi = 81°$

Show the results like this

Angle of resultant U_{Max}	U_{Actual}
$\phi < 90°$ 165g.m	128g.m
$\phi > 90°$ 123g.m	–

This rotor is within limits

Fig 7.3 Rotor balancing – interpreting the results

Rotor overspeed test

Gas turbine rotors are subjected to an overspeed test with all the compressor and turbine blades in position. The purpose is to verify the mechanical integrity of the stressed components without stresses reaching the elastic limit of the material. It also acts as a check on vibration characteristics at the rated and overspeed condition. The test consists of running the rotor at 120 percent rated speed for three minutes. Drive is by a large electric motor and the test is performed in a concrete vacuum chamber to eliminate windage. Full vibration monitoring to VDI 2056 or API 616 is performed as mentioned earlier.

Check the following points when witnessing this test:

- Watch the vibration levels during run-up. This will give you an indication of the first critical speed.
- Once the rotor reaches 120 percent speed, watch closely the vibration readings during the full three minutes overspeed run. You should be looking for any *change* in the pattern of readings. This would indicate yielding or movement of components, either of which are cause for serious concern.
- Following the overspeed test make a visual inspection of the rotor in the test chamber. Look for:
    — Loose or displaced balance weights (check the drawings to see where the weights should be).
    — Any visible damage or evidence of yielding. Be particularly vigilant around the blade roots. It is worthwhile to measure the radial length of some of the longer blades, if the specialized equipment is available, then compare your readings with the tolerances shown on the drawings.

An important note: do not, *under any circumstances* allow the speed to exceed 120 percent, even momentarily. If it does happen, the only solution is to fully strip the rotor and measure each blade individually for evidence of yielding and elongation.

Blade clearance checks

The measurement of radial blade clearances is an important witness point in the gas turbine assembly programme. The purpose is to verify running clearances between the ends of the rotor blades and the inside of the casing. It is necessary to guard against incorrect clearances which may be a result of rotor/bearing misalignment, incorrect blade dimensions, or inaccurate assembly of the casing sections. Clearances which are too large will result in reduced stage efficiency. If the clearances are too tight, the blades may touch the inside of the casing and cause breakage, particularly at the compressor end.

Preparation

The test is carried out after the rotor has been finally installed in the casing, and the casing bolts have been tightened to their design torque. The rotor is rotated very slowly (there is often no lubricating oil system installed yet) after manual lubrication of the bearings. Allow 15–30 minutes of this slow rotation to ensure that the rotor adopts a stable configuration.

You should make the following checks at this stage (see Fig. 7.4):

- Check the accuracy of the micrometer depth gauge against an accurate calibration block.
- Each measurement hole in the casing should have a spot-faced recess to form an accurate datum for the depth gauge stock. Check also that the casing wall thickness is hard-stamped next to each hole (they will not all be the same so the thickness measurement should be to an accuracy of 0.01 mm).
- Ensure that any axial play of the rotor position is 'taken up'. The rotor should be in its running position to simulate the position it will adopt when subject to gas forces in operation.
- *Acceptable tolerances.* The manufacturer should have a proforma results sheet on which to record the test reading. Make sure that it has the tolerance levels applicable to each compressor and turbine stage. A general tolerance is unacceptable.

Taking the measurements

Figure 7.4 shows the normal measurement locations. There are six radial locations – note that there is one immediately either side of the horizontal to identify any mismatch in the mating of the two casing halves.

There are some key points to observe during the measurement:

- Zero the depth gauge after each reading – otherwise errors will creep in.
- Be aware of the acceptable radial clearances. These are measured

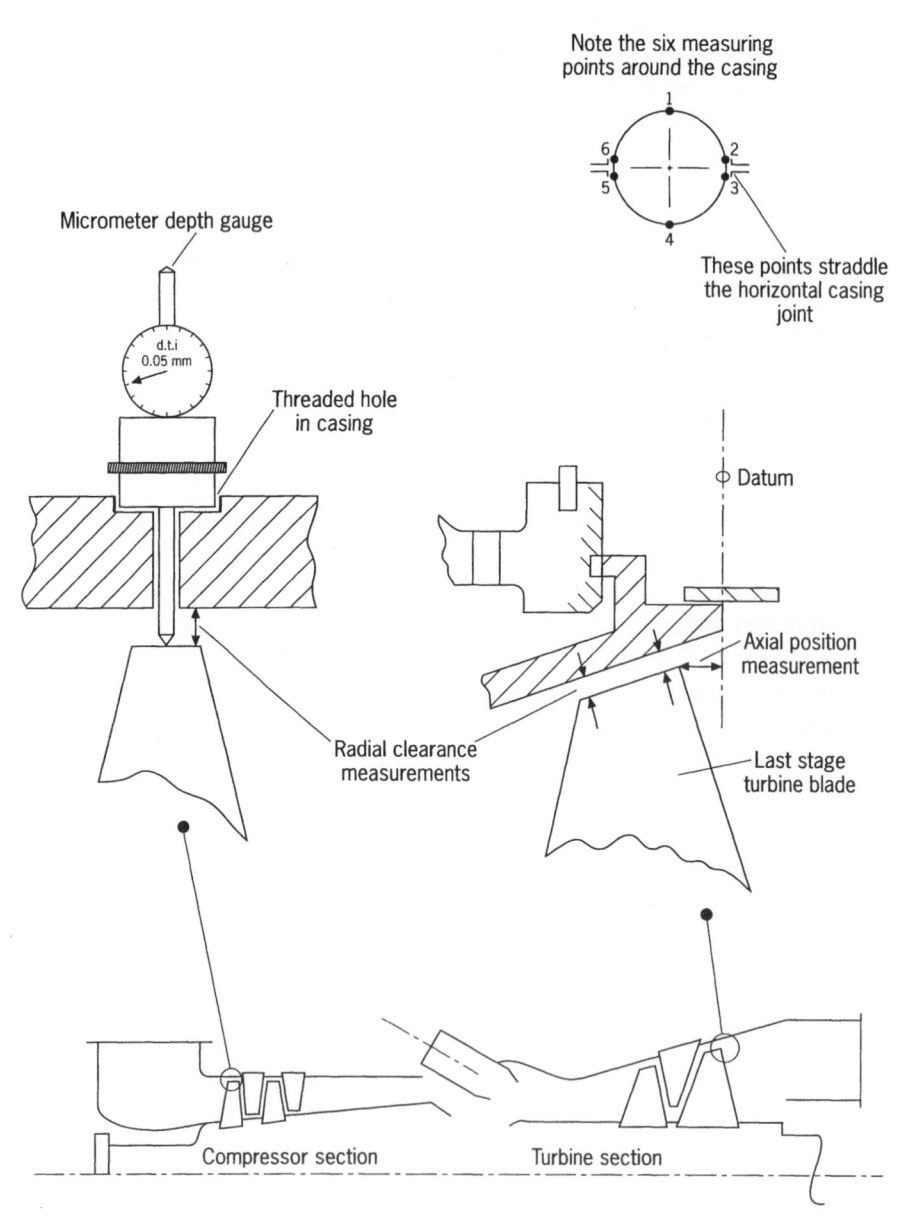

Fig 7.4 Gas turbine clearance checks

using slip gauges. They will vary between designs but indicative figures are:

– Compressor end stage 1–5	1.6 mm to 2.0 mm
– Compressor end stage 9–16	1.8 mm to 2.4 mm
– Compressor end stage 16+	2.0 mm to 2.4 mm
– Turbine end	4.0 mm to 4.5 mm
– Rotor axial position (end clearance of last blades)	7.0 mm to 8.0 mm

Accessory base test

Think in terms of *functional testing*. The accessory base test is effectively an interim performance test on the complete set of auxiliary systems that enable the gas turbine to function. It is carried out towards the end of the manufacturing procedure – at the stage where the main auxiliary components (but not the gas turbine) have been fitted to the fabricated skid. The test relies heavily on simulation. The gas turbine drive is simulated by a large electric motor. For the fluid loops, lubricating/hydraulic oil, fuel, and air, the presence of the turbine is simulated by piping loops, fitted with suitable orifices and restrictions to act as system resistance. Figure 7.5 shows the arrangement. To carry out an effective inspection during the accessory base test, you need an appreciation of the two key objectives of the test (note how these fit in with the general FFP criteria), which are:

- To make sure all the items function under (simulated) full-load conditions. This saves time during site installation.
- To detect any early failures that are going to occur in the initial stages of turbine operation. These are often due to incorrect assembly.

Preparation

Following the general principle I have suggested in other areas of this book, your first task must be to *check the circuits*. Do not start the test until you have traced the oil, fuel, and air systems – as a minimum – and have identified the system components. Once again this is not a place for a 'broad-brush approach' – you will need to review the schematics in detail to understand how the systems work. Make sure also that you identify the pipe runs on the skid – it's not wise just to look at the drawings, you are there to identify any mistakes.

The test itself

Step 1: Assembly. (refer to Fig. 7.5). The manufacturer will normally assemble the accessory base components before you arrive. Frankly, this does not matter much – the electrical wiring checks are done at this stage so you can witness them if you are present, but the rest is straightforward mechanical assembly.

Step 2: Pre-running tests. First check for static leaks, then witness the flushing and operation of the lubrication oil pumps and pipework. The other important test at this stage is to check the operation of the components of the gas fuel system.

Step 3: Cranking tests. This involves driving the starting motor, torque converter, auxiliary drive gear and coupling at cranking speed. Check again for leaks. Witness the calibration of the liquid fuel servo system and check the operation of the fuel pump.

Step 4: Rated speed and performance test. The accessory drive gearbox is driven at 100 percent rated speed by the electric drive motor. A full inventory of readings (as outlined in Fig. 7.5) is taken under these conditions. This is the time to verify that the fluid loops are simulating correctly the presence of the turbine. The manufacturer will have calibration curves for volume throughput, pressure drop, and so on – review these and compare the actual systems readings with the design values. Make sure they match. Make a thorough check that all the instrumentation is working correctly. During this test it is beneficial to check all readings manually, rather than relying on remote indications. It is vital that any uncertainties about the validity or accuracy of readings are eliminated at this stage – do not carry any doubts forward to the turbine no-load running test. You should not let the test conclude until you are satisfied. Check that all the results have been accurately recorded on the set of datasheets.

There are two basic mechanical checks to be performed during this run:

- Check vibration levels of the rotating equipment – particularly the auxiliary drive gearbox and the auxiliary air compressor. Make sure the specified type of anti-vibration mountings are being used. Check also that the drive couplings and any V-belt drives are accurately aligned.

- Check the accessibility of pipe, valves, drains, and traps. They should all be accessible by hand and not obstructed by other pipe runs or cable trays.

The GT accessory base test is used to verify systems function – this is one of the key FFP criteria.

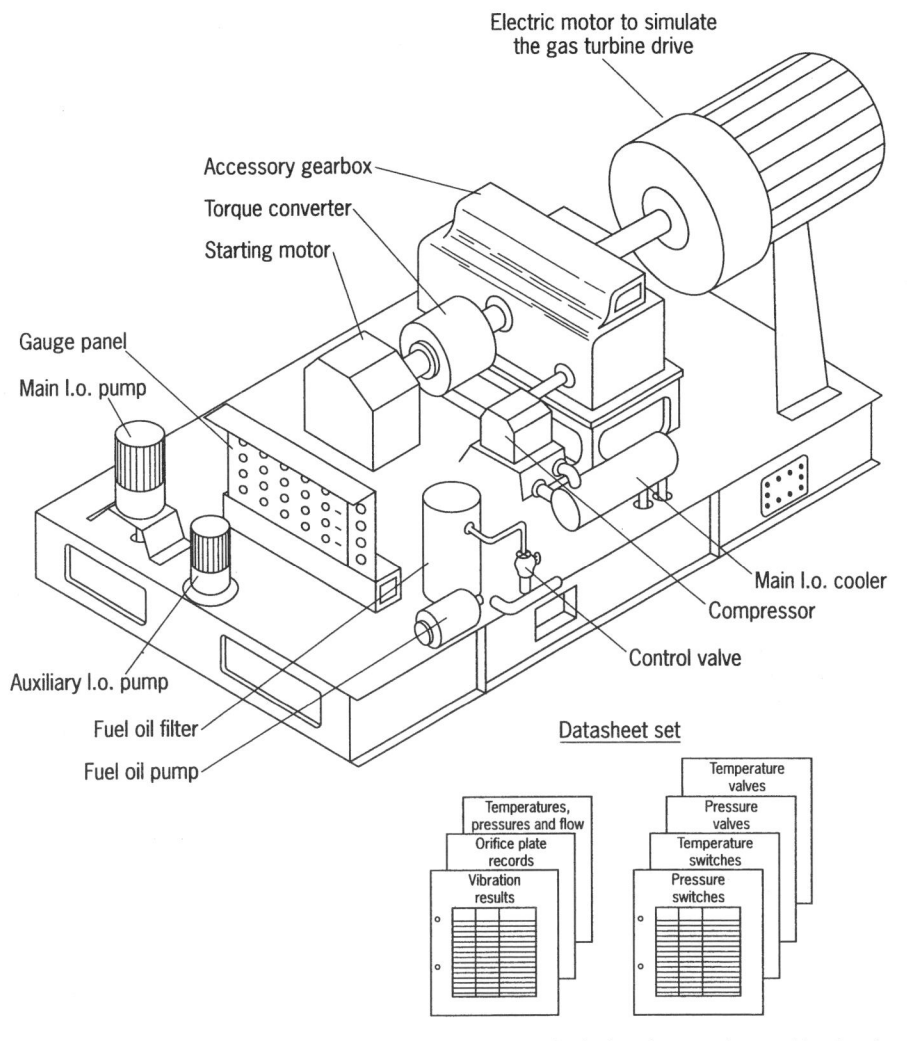

- Check that the datasheet set is completed accurately during the rated speed/no-load test.
- Verify the results against the design values.

Fig 7.5 The GT accessory base test

No-load running test

Gas turbines of standard design are not normally tested on-load in the manufacturers' works. Some manufacturers do not feel the need to do *any* running tests prior to the full performance test after commissioning on site and this attitude is becoming increasingly common as the technology develops and the number of well-proven units increases. Prudent purchasers specify that their turbines should be subject to a works running test – this is particularly the case if the contract structure is such that the roles of power plant 'packager', GT licensor, and GT manufacturer are in any way fragmented. Under these circumstances it is wise to take all possible precautions to eliminate any FFP problems *before* site installation starts. The usual mechanism for this is the no-load works running test.

The no-load running test is primarily a test for mechanical integrity of the turbine and for the cumulative function of its on-base support systems. It is not a test which will enable you to verify the acceptance guarantees described earlier in this chapter. Notwithstanding that its role is to act as a performance guideline only, it can nevertheless help greatly in verifying the other FFP aspects of the machine. To do this properly a good no-load running test needs to be a carefully planned and accurately documented series of events. To witness this test properly you must adopt an equally structured approach. The test is carried out over a period of four or five hours and involves running the turbine throughout its speed range. Allow at least one full day inspection time to witness this test and to make all the necessary checks and reviews. Don't be hurried – one half day will not be enough.

Setting-up checks

There are a number of important setting-up checks you should do before the test starts. I have itemized these below. Note that these assume that you have already witnessed, with satisfactory results, the full functional checks on the accessory base.

- Do an equipment inventory check. The purpose of this is to be completely sure which equipment items involved in the test are *contract* and which are *shop* (i.e. will not be supplied with the turbine under the purchase contract). You may find that the following are *shop* equipment (they usually are):
    - the inlet and exhaust system
    - the cooling water pumps and coolers

- the instrument air supply
- the batteries and rectifiers
- the control panels.

None of these matter very much to the validity of the test. What *is* important is that the remainder of the equipment should be the contract equipment – otherwise the test may not be meaningful in some areas. If you don't test with the contract equipment, then you are not really testing for FFP. Good GT manufacturers will have a proforma checksheet showing the equipment inventory and whether it is the contract supply. Ask for this list and check some of the serial numbers. One discrepancy may be an error – two shows a trend.

- Check the lubricating oil. Gas turbines do not like contaminated oil. Ask the manufacturer to demonstrate that the oil has been filtered to 8 μm, making sure that you are really happy with the evidence. If in doubt ask that the oil system be circulated for a further two hours with a new 8 μm filter. Check the filter element again after the two hours. It should be absolutely clean.
- The settings data record: have you seen this? It is a record of all the valve adjustments, orifice sizes, controller set-points, calibration levels, and trip settings for the gas turbine control equipment and circuits. It should be a separate document. Reviewing these data, particularly the trip points, will give you a good indication of how the system parameters behave during the test.

The test procedure

Following preliminary checks, the test procedure itself starts. This may vary slightly between manufacturers but the major steps and objectives should be common. Figure 7.6 shows a typical routine.

- *Step 1: Starting.* The unit is started up, normally on liquid fuel, and run up to cranking speed. Check for smooth starting operation and for early oil, air, or gas leakage. The rate of speed increase should be constant (this will have been pre-programmed into the GT control system software). All auxiliaries should be operating without excessive mechanical noise or vibration.
- *Step 2: Run-up.* The sequence is continued and the unit run up to 100 percent speed. This should be a constant speed increase – there is normally no need to stabilize at a lower speed unless problems are evident. Keep a close visual check on the vibration measurements

224 Handbook of Mechanical Works Inspection

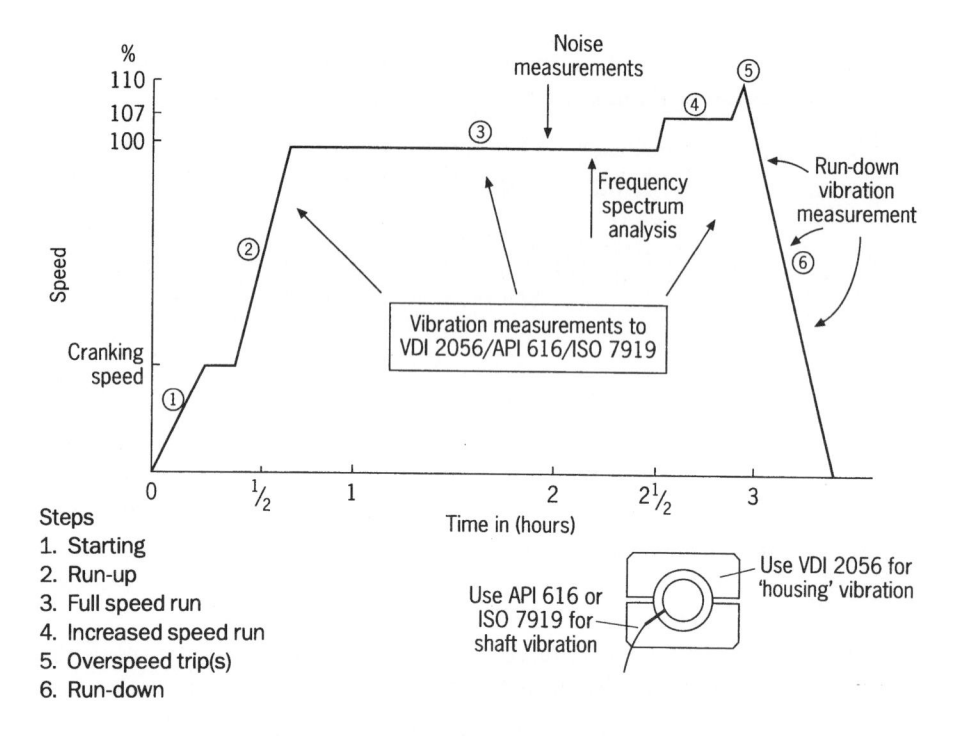

Fig 7.6 Gas turbine no-load test run

Steps
1. Starting
2. Run-up
3. Full speed run
4. Increased speed run
5. Overspeed trip(s)
6. Run-down

during this run-up, you should see some indication of the first critical speed.

- *Step 3: The two-hour full speed run.* This gives all temperatures and pressures the opportunity to stabilize. Measurements of all parameters should be recorded by the datalogging system at regular intervals – note that ANSI/ASME PTC 22 specifies a maximum recording interval of ten minutes. Vibration levels are monitored continuously during the two hours and noise measurements taken at a suitable time. Towards the end of the period, expect to see a full vibration frequency analysis carried out. It is important to witness this activity carefully.
- *Step 4: Increased speed run.* The turbines is increased to 107 percent of nominal speed and run for 15 minutes. Make sure the speed is not increased at a greater rate than 1 percent per five seconds to avoid over-stressing the rotating components. Pay particular attention to temperatures and vibration levels during this period.

- *Step 5: Overspeed trips.* The speed is slowly raised from 107 percent to the point where the first electronic overspeed trip operates. This should trip at 110 percent of nominal (rated) speed with a tolerance of $\pm$ 0.5 percent. Be precise in your assessment of this – it means $\pm$ 0.5 percent of nominal (rated) speed. Note that gas turbines have two separate overspeed trip mechanisms. The second, which may be mechanical or a separate electronic type, should be tested to 112 percent $\pm$ 0.5 percent. This is not shown in Fig. 7.6, but is normally carried out after the routine shown and should still be witnessed.
- *Step 6: Run-down.* The unit is allowed to run down naturally from its trip speed. It will take quite a few minutes to stop. Watch the speed indicator to make sure the run-down is smooth. An erratic or very short run-down is an almost certain indication of mechanical component failure (at overspeed) or shaft bearing problems. A critical part of the test is the recording of vibration data during the run-down.

Assessing vibration results

Remember that vibration levels for gas turbines and other large rotating machines can have two different bases. It is worth re-stating these:

- Bearing *housing* vibration. Housing vibration in turbines is governed by VDI 2056 group T. This specifies an rms velocity of 2.8 mm/sec as the deciding criterion and assumes a sinusoidal vibration form. It is easy to convert readings taken in other units – look now at Fig. 7.7, in which I have included an example to try and make it clear.
- *Shaft* vibration. This is a completely different concept of measurement from that of housing vibration. It is mentioned in API 616 but explained in more detail in ISO 7919/1 (BS 6749 Part 1), which both use amplitude (peak-to-peak) as the deciding criterion. A commonly used acceptance limit is a maximum of 38µm for GTs in the speed range 4000–8000 rpm and 50 µm for these operating below 4000 rpm. Note that:
    - You *cannot* make a sensible conversion between housing and shaft vibration or vice versa.
    - Fewer manufacturers measure shaft vibration than use a housing vibration measurement. This is because it is more complicated – if you look at ISO 7919/1, you will see why. It is also practically more difficult, needing specially adapted bearing housings to hold the non-contacting probes.

Before assessing the vibration test results, it is essential therefore to check which basis you are working to. Figure 7.7 shows the most common methodology used by gas turbine manufacturers but *be prepared* for different approaches. There are two further vibration aspects that should be reviewed during a witnessed test. These can also be confusing, so Fig. 7.8 shows them in simplified form.

- *Vibration spectrum analysis.* Normally these readings are taken towards the end of the two hour 100 percent speed run. Expect to see two sets of readings (horizontal and vertical) per shaft bearing. Vibration measurements are recorded across the frequency spectrum up to a multiple of six times the rotational frequency – this is termed the sixth order. The purpose is to identify any excessive vibration levels at non-synchronous frequencies, particularly at the higher end of the spectrum. As a guide, any discrete, non-synchronous vibration that is greater than about 25 percent of the acceptance limits needs to be investigated – you should issue an NCR to make this clear.
- *Run-down vibration analysis.* Most test routines on large rotating plant involves measurement of vibration information during the run-down. Again, separate readings are obtained from the horizontal and vertical sensors fitted to each bearing. Vibration velocity (or amplitude) is plotted against rotational speed as the unit slows to a stop. Expect to see, also, measurements of phase lag. Figure 7.8 shows typical plots.

Noise measurement

Most contract specifications require that the GT be subject to a noise measurement check. This is a common works inspection test. Noise measurement principles and techniques are common for many types of engineering equipment, so the following general technical explanations can be applied equally to diesel engines, gearboxes, or pumps.

The main technical standard relating to GT noise testing is ISO 10494 on measurement of airborne noise (equivalent to BS 7721) **(6)**. This is referred to by the GT procurement standard ISO 3977 and contains specific information about measuring GT noise levels. Note that there is a more general standard ISO 1996, Parts 1 to 3 – it works to the same basic technical principles as ISO 10494 but does not mention any specific machinery types (or give noise limits). You may see it used, therefore, in relation to all types of rotating equipment – it is certainly not excluded for use with GTs, but ISO 10494 is easier to work with for works

Gas turbines

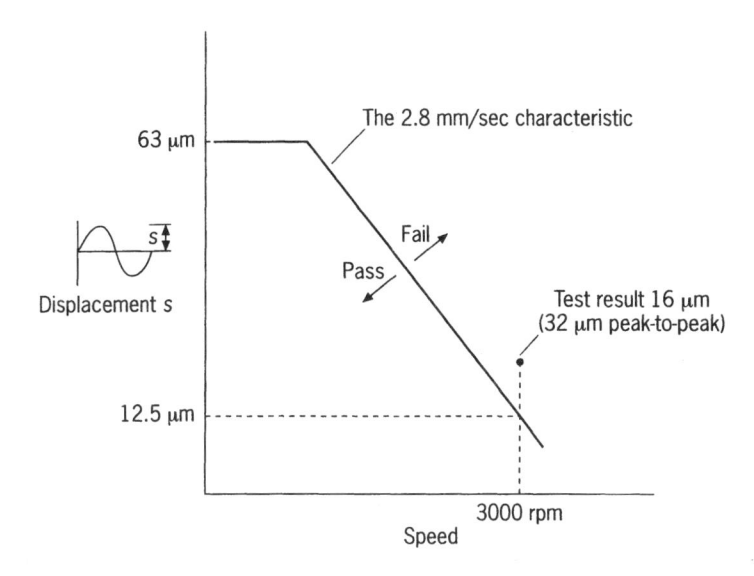

[Figure: Graph showing displacement s vs Speed, with 2.8 mm/sec characteristic line. Points marked: 63 μm, 12.5 μm, 3000 rpm, Test result 16 μm (32 μm peak-to-peak), Pass/Fail regions]

Bearing cap vibration measured as		Conversion required
A velocity (rms)	V(rms) mm/sec	Compare directly with the 2.8mm/sec Vrms acceptance level as specified by VDI 2056 groupT.
A velocity (peak)	V(peak) mm/sec	Convert to V_{rms} using $V_{rms} = V_{peak}/\sqrt{3}$ then compare as above.
A displacement (peak to peak)	s microns	First convert the 2.8mm/sec criteria to an acceptable displacement (using VDI 2056 Table 7) at the rated speed of the machine. Then compare the actual reading with this (see below).

An example:
1. A machine rated at 3000 rpm is specified to comply with VDI 2056, 2.8mm/sec criteria.
2. The test vibration reading of a bearing is 32μm peak-to-peak displacement.
3. From VDI 2056, the acceptable maximum displacement for 2.8mm/sec at 3000 rpm is 12.5μm (×2 to convert to peak to peak) = 25μm.
4. The actual reading of 32μm is greater than 25μm hence it does not comply.

Fig 7.7   How to interpret vibration readings

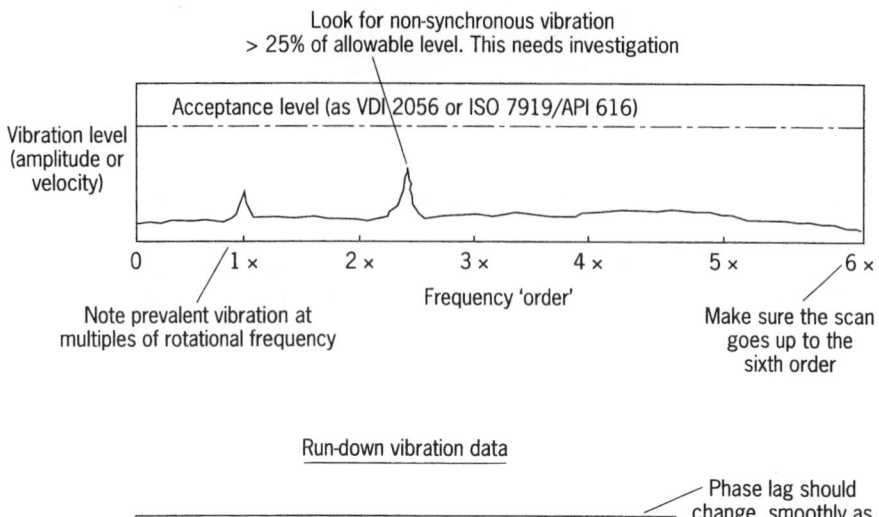

Vibration frequency spectrum analysis

Look for non-synchronous vibration > 25% of allowable level. This needs investigation

Acceptance level (as VDI 2056 or ISO 7919/API 616)

Vibration level (amplitude or velocity)

Frequency 'order': 0, 1×, 2×, 3×, 4×, 5×, 6×

Note prevalent vibration at multiples of rotational frequency

Make sure the scan goes up to the sixth order

Run-down vibration data

Phase lag should change smoothly as speed is reduced

Phase lag (degrees) — Speed from 100% → 50% → 0

Vibration level (amplitude or velocity) — Acceptance level — Speed from 100% → 50% → 0

Watch for vibration increases caused by 'compressor surging' or critical speeds

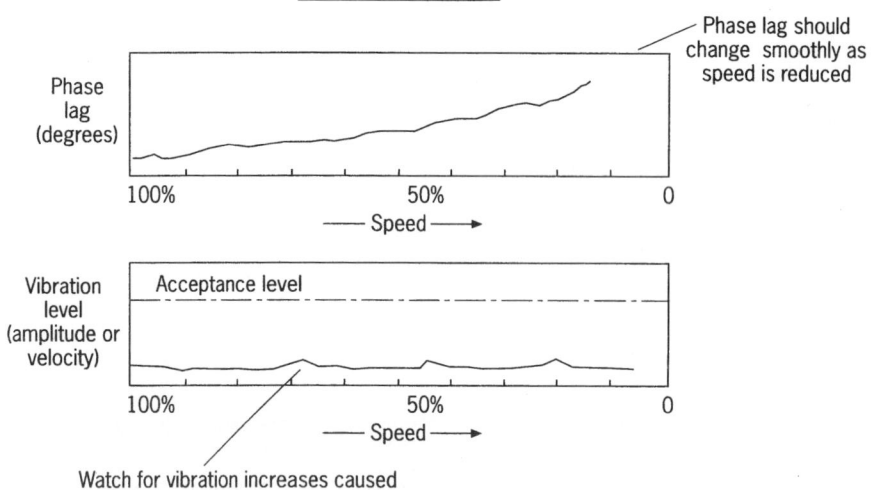

Fig 7.8 Vibration frequency and run-down analysis: gas turbines

---

inspection purposes. Expect to find allowable noise limits given in the contract specification, rather than the technical standards, which tend to concentrate more on testing methods.

Principles

Noise is most easily thought of as air-borne pressure pulses set up by a vibrating surface source. It is measured by an instrument which detects these pressure changes in the air and then relates this measured sound pressure to an accepted 'zero' level. Because a machine produces a mixture of frequencies (termed 'broad-band' noise), there is no single noise measurement that will fully describe a noise emission. In practice, you will see two common ways used – note that they are complementary. They are:

- The *'overall noise'* level. This is often used as a colloquial term for what is properly described as the 'A-weighted sound pressure level'. It incorporates multiple frequencies, and weights them according to a formula which results in the best approximation of the *loudness* of the noise. This is displayed as a single instrument reading expressed as decibels – in this case dB(A).
- *Frequency band* sound pressure level. This involves measuring the sound pressure level in a number of frequency bands. These are arranged in either octave or one-third octave bands in terms of their mid-band frequency (see Fig. 7.9). The range of frequencies of interest in measuring machinery noise is from about 30 Hz to 10 000 Hz. Note that frequency band sound pressure levels are also expressed in decibels (dB).

The decibel scale itself is a logarithmic scale, a sound pressure level in dB being defined as:

$$dB = 10 \log_{10} (p_1/p_0)^2$$

where p_1 = measured sound pressure
p_0 = a reference *zero* pressure level

GT noise characteristics

Gas turbines are inherently noisy machines, producing a wide variety of 'broad-band' noise across the frequency range. Although the noise has its origin in many parts of the machine there is inevitably some simplification required when considering a practical measurement programme in a works environment. There are three main emitters of noise: the machine's total surface, the air inlet system, and the exhaust gas outlet system. In practice, the inlet and outlet system noise is considered as 'included' in the surface originated noise. This is a simplification but one which is necessary. The machine bearings emit noise at frequencies related to their rotational speed, whilst the combustion process emits a wider, less predictable range of sound frequencies. Many industrial turbines are installed within an acoustic

230 Handbook for Mechanical Works Inspection

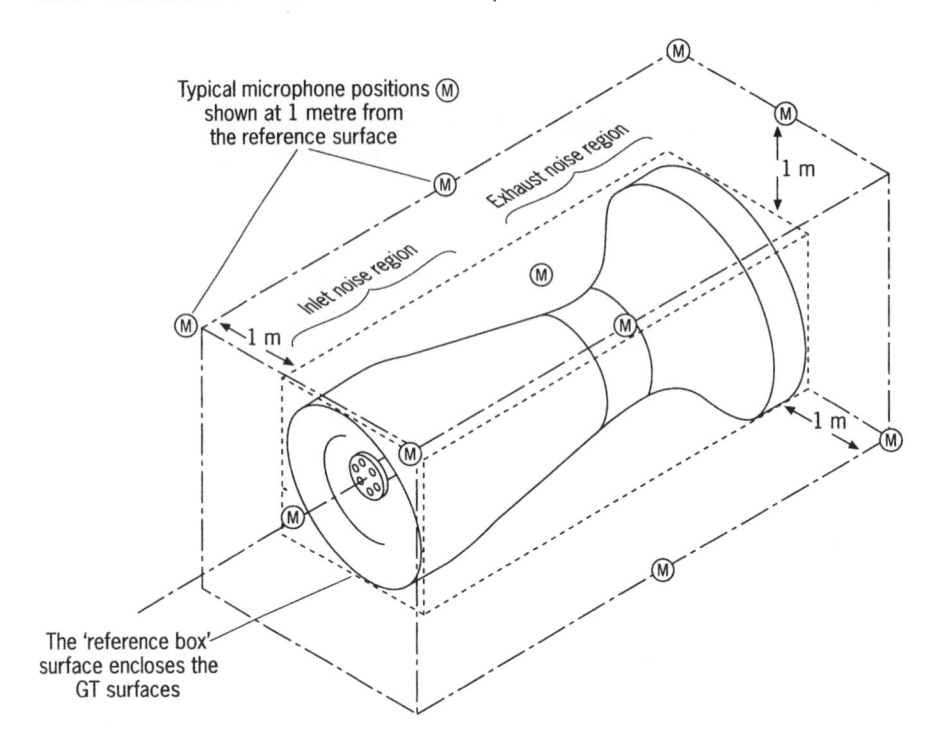

Typical microphone positions (M) shown at 1 metre from the reference surface

The 'reference box' surface encloses the GT surfaces

Commonly-used 'octave' mid-band frequencies are:

| 63 Hz | 125 Hz | 250 Hz | 500 Hz | 1000 Hz | 2000 Hz | 4000 Hz |

Background noise correction (grade 2 accuracy ISO 10494)

Difference between 'running' and 'background' noise	Subtract this correction from the 'running' noise
6dB	1dB
7dB	1dB
8dB	1dB
9dB	1dB
10dB	0
>10dB	0

A noise test comprises the following steps:

1. Define the 'reference surface'.
2. Position the microphones 1 metre from the reference surface as shown.
3. Measure the background noise (at each microphone position)
4. Measure the 'running noise' at each microphone position – take A-weighted and discrete mid-band frequency readings.
5. Calculate the background noise correction – see the above table.
6. Remember that there is an 'uncertainty' tolerance of about ±2dB in most noise tests.

Fig 7.9   Gas turbine noise tests

Gas turbines 231

enclosure to reduce the levels of 'near vicinity' and environmental (further away) noise.

The noise test

Although inherently a straightforward process, noise tests on rotating machines are sometimes marred by problems of uncertainty about the test techniques or misunderstandings about the assessment of measured results. A proper test divides neatly into the following series of steps:

- *Define a reference surface.* This is an imaginary rectangular *box* which encloses all significant surfaces of the GT unit. To all intent, this box is assumed to act as the noise-emitting surface of the GT.
- *Position the microphones.* The microphone positions are located at one metre distances from the 'reference box' surface (this one metre is a standard distance used in all noise tests). A good test should use 10–12 microphone positions (at least), located to take into account the main noise emission sources; bearings, inlet and outlet regions and the combustion chamber.
- *Measure background noise.* A separate background reading should be taken at each microphone position before the unit is started. Theoretically, the GT auxiliary base components should be running but practically this may not be possible as they are normally driven by the turbine itself. Record the background values.
- *Measure the noise with the GT running at steady-state conditions.* First measure the 'A-weighted' sound pressure level in dB(A) at each microphone – you need to allow at least 30 seconds for the reading to stabilize after switching on the microphone in each new position. Make sure it is fixed securely in a stand. Record all the results. The second step is to record the sound pressure level in the various frequency bands. There are no absolute rules as to which bands to measure. I have shown those from a 'typical' test in Fig. 7.9. Selected one-third octave band readings could be taken as well as the octave band readings across the spectrum – the main purpose of the one-third octave measurements is to look closely at frequencies which, by experience, have a tendency to exhibit higher-than-expected sound pressure levels. You have to rely a little on the GT manufacturer's experience here. Record all the results.
- *Calculate background noise corrections.* Because of the characteristic of the decibel scale this is not a straight subtraction. Figure 7.9 shows the corrections to be subtracted from the 'running' noise, depending on the recorded difference between this running noise and

the background noise measured earlier. There are limits on what is acceptable – the concept of acceptability is based on what is termed 'grade 2 accuracy' (this is an ISO 10494 parameter). To achieve this accuracy there must be a recorded difference between running and background noise of *at least* 6dB. A smaller difference means that the test results will probably be unreliable. I think it is fair for a works inspector to request that a noise test be able to achieve ISO 10494 grade 2 accuracy to be considered acceptable.

- *Assess the results against the requirements of the contract specification.* Expect different contracts to have different specified noise limits – this is quite common. Regarding tolerances, technical standards vary on the overall 'uncertainty' level they ascribe to noise measurement techniques. As a guide, under controlled test-shop conditions, with good equipment, expect a tolerance of perhaps $\pm$ 1.5 to 2dB to be applicable. If you issue an NCR for a machine that has not met its noise limits but is within this tolerance, you could find that a new test may yield a better (or worse) result. In such situations, look carefully at the frequency band readings to help you reach a conclusion – this information needs more analysis but in the right hands *can* give definitive answers.

The test report

The no-load running test must be properly documented by a complete and detailed manufacturers' test report. There is a lot of information here which will be of use during site performance testing or if any performance problems occur in the early life of the unit. You should review the test report, comparing the reported readings with the personal notes that you took during the test. It is not sufficient just to sign the test results sheet. Effective inspection is about taking care to ensure that you report on any uncertainties – and make them easy for your client to understand. As a minimum I think you should address at least the following points in your report. Make them stand out:

- Comment on the vibration measurement philosophy that has been used. Was it based on bearing housing measurements or relative shaft vibration measurements? Give a precise explanation of how errors were eliminated – you must demonstrate that you have a good working knowledge of this issue – some inspectors don't. Note that it is not acceptable to draw conclusions or inferences about vibration levels by looking back at the dynamic balance results – there are too

many complex technical factors involved. Consider balance results and vibration results as two discrete issues – each of which should be reported separately and accurately.
- Make clear statements about the observed spread of GT operating temperatures. Mention journal and thrust bearing temperatures, wheel-space temperatures, and exhaust gas temperatures. If you have witnessed bearing metal temperatures above 70°C, wheel-space temperatures above 300°C, or exhaust temperatures above 250°C during steady state no-load operation, then you should provide an explanation of how you satisfied yourself that these temperatures were acceptable.
- Give your considered opinion on the no-load performance data. These comprise mainly air/gas flow, pressure ratio, flow and fuel consumption. It takes a lot of experience to be able to project these no-load readings to the designed site conditions, even using the manufacturers' correction curves, so do not try to predict site performance. Stick to accurate reporting within the terms of reference of the no-load test.
- Because of the limitations of those international standards relating to gas turbines, a purchaser is often unsure as to whether turbine manufacturers are constrained by these standards, or whether they have free rein to set their own acceptance standards for operating temperatures, vibration levels, etc. A good inspection report will help explain the situation to the purchaser. Provide simple explanations as to whether the machine complies with the requirements of API 616, VDI 2056, and ISO 7919/1. If the machine was not specified to these standards, but it is better, or worse, then say so.

Most gas turbine witnessed test programmes end on completion of the no-load running test. Unlike a diesel engine, there is not usually a stripdown examination, unless it is specifically requested by the contract. Manufacturers often do their own post-test borescope inspection of the combustion spaces, but there is normally only a stripdown if any problems have been evident during the running test.

Common non-conformances and corrective actions: gas turbines

Non-conformance	Corrective action
During no-load test	
Leaking air or gas from flange faces.	Don't be too hasty in issuing an NCR for this – the leakage often seals as the casing reaches steady-state operating temperature.
High journal or thrust bearing metal temperatures.	You must first eliminate simple faults so: – verify the lubrication oil flow rate. – check the measuring instrument calibration. – re-check the bearing clearance records. If there is no obvious reason for the high measurement, the *only* solution is a stripdown to inspect the bearing.
Excessive mechanical noise.	The test must stop immediately. Do *not* allow a re-start until the manufacturer has made a clear statement that the problem has been identified and rectified. It is unlikely to be solved without a stripdown.
Excessive vibration of auxiliary equipment mounted on base.	Check the accessory base test results. This is nearly always a monitoring problem.
Excessive vibration of turbine bearing housings or shaft.	The action is: you must describe correctly the exact nature of the excessive vibration – this is critical to assist in defining the problem. Categorize your observations as follows: • Is the excessive vibration constant across the speed range or does it vary? • Do a scan to find the most prevalent frequency. Record this as a multiple of rotational frequency. Make a specific note if it is prevalent at second order (double) frequency. • Make reference to the lowest natural bending frequency of the shaft. • Check the vibration wave form – is it regular or fluctuating? These will help with categorizing of the problem. Your corrective action (CA) report must describe the necessary actions you have agreed with the manufacturer.
A cautionary note Vibration of gas turbines is a specialized field. Use your experience to describe accurately the problem, but don't try to solve it. Accept the superior experience of the GT licensor or manufacturer, then make sure that the manufacturer *does* address the problem.	

KEY POINT SUMMARY: GAS TURBINES

Fitness for purpose

1. The two explicit FFP criteria which you can verify during a works inspection are *running integrity* and *systems function*. It is not possible to verify guarantees such as thermal performance or efficiency.
2. Other 'inferred' FFP criteria which you can check are:
   - governor characteristics
   - overspeed settings
   - vibration, critical speed, and noise levels.

Standards

3. GT licensors'/manufacturers' own standards are much more comprehensive than published technical standards. Treat GT standards such as API 616, ISO 3977, and ASME PTC 22 as a framework only, rather than as a set of prescriptive requirements.

ITP steps

4. The most important ITP steps are the pre-assembly rotor tests (runout, balancing, and overspeed), blade clearance checks, the accessory base functional test, and the no-load running test of the completed machine.
5. GT rotors are dynamically balanced to the requirements of ISO 1940 or API 616 using two correction planes.
6. The accessory base functional test relies heavily on simulation to verify the performance of the oil, air, and fuel *loops*.
7. A no-load running test is used to check general operation, noise, and vibration levels and the GT's safety mechanisms. You need to adopt a structured approach to witnessing this test. It is not sufficient just to sign the manufacturers' test sheets – your report should provide *good engineering observations* about:
   - vibration measurement techniques (and the results)
   - operating parameters (temperatures and pressures)
   - whether the machine exhibits any *non-compliance* with published technical standards
8. *Be careful*. Use your experience to describe accurately any problems that you identify but do not try to solve them. GTs are complex items – accept the superior experience of the manufacturer.

References

1. BS 3863: 1992 (identical to ISO 3977 : 1991). *Guide for gas turbine procurement.*
2. BS 3135: 1989 (identical to ISO 2314). *Specification for gas turbine acceptance test.*
3. ANSI/ASME Performance Test Code 22: 1985. *Gas turbine power plants.* The American Society of Mechanical Engineers.
4. API 616 *Gas turbines for refinery service*, 1989 American Petroleum Institute.
5. ISO 1940/1: 1986 (identical to BS 6861 Part 1: 1987). *Balance quality requirements of rigid rotors: Part 1, method for determination of permissible residual unbalance.*
6. ISO 10494: 1993 (identical to BS 7721: 1994). *Gas turbines and gas turbine sets. Measurement of emitted airborne noise-engineering/survey method.*
7. VDI 2056: 1984. *Criteria for assessing mechanical vibration of machines.* Verein Deutscher Ingenieure.
8. ISO 7919/1: 1986 (identical to BS 6749: Part 1: 1986). *Measurement and evaluation of vibration on rotating shafts – Guide to general principles.*
9. VDI 2059 Part 4: 1981. *Shaft vibrations of gas turbosets – measurement and evaluation.* Verein Deutscher Ingenieure.

Chapter 8

Steam turbines

Steam turbines are genuine heavy engineering – so they bring with them the range of technical problems related to large components and heavy material sections. Whilst they are undeniably mature technology, there is an increasing widening between the activities of design and manufacture. The majority of turbines of 20 MW and above are now manufactured by companies with vastly different cultures and practices to those that designed them. This brings technical risk. Technical risk causes problems, so do not expect to find that the inspection of steam turbines is trouble-free.

Fitness-for-purpose criteria

From an inspector's viewpoint, FFP is a more elusive concept for steam turbines than for many other types of equipment. Part of this stems from the fact that although the machine is a complex prime mover, it can only be tested properly for function when it is connected up to its steam system. It is necessary to accept the corresponding restriction on what can be achieved by test activities at the source of manufacture. The same problem exists with acceptance guarantees – power output and efficiency are not normally tested until site installation. To deal with this you need to take a slightly different stance on FFP. The criterion of 'running integrity' that I put forward in the parts of this book discussing gas turbines, diesels and gearboxes is still relevant, but for steam turbines you need to look *back* one step, return to the manufacturing process to make sure that FFP has been built-in to the machine. This is one way to add value to your FFP verification. I think it is the best way.

This means that you need to know more about *how to manufacture* a steam turbine than you would about, for instance, a diesel engine.

Basic technical information

It is not possible in this book to provide a detailed explanation of the techniques of steam turbine design and manufacture. There are many excellent references that you can consult. From an inspection viewpoint it is not necessary to have a detailed understanding – we must be selective in choosing the knowledge that we need to acquire. You should find the following three manufacturing 'approaches'; of some value – treat these as an essential simplification of the wide background of steam turbine experience. Their only purpose is to be useful – try to build them into your thinking.

Look at the materials

Steam turbines can be classified broadly into those designed for saturated steam and those designed for superheated steam. The only *real* difference from a works inspection viewpoint concerns the materials of construction that are used for these two categories – in particular that for the pressurized casing components. These are commonly ferritic steel castings, designed so as to have specific tensile properties at elevated temperatures. Most are weldable, so that defects can be excavated and repaired and so that the turbine casing can be attached permanently to its valve chests. A typical cast steel used is DIN 17245 grade GS – 17 Cr MoV 511 (you will also find this listed under the German 'Werkstoff' classification No. 17706). It has a low carbon content (0.15 to 0.2 percent) and obtains its properties predominantly from its manganese, Mn (0.5 to 0.8 percent), chromium, Cr (1.2 to 1.5 percent), and molybdenum, Mo (0.9 to 1.1 percent) alloying elements. UTS is in the range 590–780 N/mm^2, with 0.2 percent proof stress being used instead of yield (R_e) to define strength at operating temperatures up to about 550 °C.

As for other types of castings, steam turbine components are normally classified into a 'quality grade' (also referred to by some standards as a 'severity level'). This concept was introduced in Chapters 4 and 5. The particular standard in use will depend upon the individual contract requirements – the DIN 17245 alloy mentioned would typically reference severity level T2 of DIN 1690 **(1)**, whilst a steam turbine specified to API 611 **(2)**, would reference ASME/ASTM – recognized

Steam turbines 239

standards. Defect levels in steam turbine casings are often a closely – guarded aspect of a manufacturer's practice and are not always as transparent and well defined as the technical standards may infer. Most of the technical standards allow for the provision of a manufacturer/purchaser agreement to define acceptable defect levels. Note that it is not unusual for two separate levels to be specified, tight acceptance criteria for critical areas such as weld-end and flange–face regions and a more relaxed set of criteria for the rest of the casing.

Because of the importance of material integrity in ensuring FFP, material specification, test results, and the documentation aspects of material traceability and certification are key areas for the works inspector. It is worth concentrating on those areas that are common sources of problems with steam turbine castings – you can soon learn to recognize them. The main ones are:

- *Variation in chemical analyses.* Expect to find variations in chemical analyses of large castings. They can be caused by the way in which the analysis is taken (i.e. whether from a 'cast-on' test piece, or from the body of the casting itself).
- *Variation in mechanical properties.* These can vary both with the position (orientation) of the test piece, and with the thickness of cast material from which they were taken. The final heat treatment of the component has a significant effect on properties. Most steam turbine castings leave the foundry in a 'quenched and tempered' condition, but there *can* be variations. Check the material standard carefully – it should link the required mechanical properties to a particular heat treatment and 'finishing' condition.

Consider the size of the machine

Steam turbine units can be very large: there are machines installed of 1500 MW capacity with LP (Low Pressure) rotor diameters of several metres. Strictly, the size of a turbine (and the complex design considerations that arise from it) do not affect directly the fundamental content of works inspection activity. What does happen, though, is that increased size results in a increased potential for a range of practical problems that *do* have an effect on the way that you should approach some of the major witness points during manufacture.

Expect very large machines to have:

- High superheat temperatures and pressures – with the corresponding high specification material choices.

- Thicker material 'ruling' sections in the casing parts. This attracts a number of particular material defects more likely to occur in thick, cast sections.
- Longer unsupported rotor lengths. This gives a greater tendency for bending and subsequent vibration, particularly on single-shaft machines.
- Larger diameters, particularly of the LP rotors, in which most of the stress on a blade is caused by centrifugal force rather than steam load. Higher stresses mean a greater sensitivity to defect size, requiring more searching NDT techniques on the rotating components.

All of these points means that greater inspection vigilance is required for a large steam turbine than for a small one.

Try to think like a manufacturer

An effective inspector can think like a manufacturer. If you accept the FFP issues suggested earlier, it follows that you need to obtain, at least, a better than average understanding of the techniques used in turbine manufacture. If you can understand the methods, then it is a short step to understanding the problems that are likely to occur. I have heard it said that the opposite is probably also true. There are three quite different sets of parts of a steam turbine – note how the methods of their manufacture differ.

The stator parts

These are heavy castings, normally produced by a sub-contracted foundry. Foundry practice is not easy to understand, as much is based on cumulative experience over many years. Some of the techniques have a strong empirical basis (which means it is difficult to analyse exactly what is happening), and some are governed by trial and error. Perhaps because of this, heat treatment and material analysis assume great importance. Machining operations for stator parts are just the opposite, they are straightforward, consisting mainly of milling. Dimensional accuracy has relatively relaxed tolerances.

The rotor parts

Rotor parts are made from forgings in which greater emphasis is placed on *directional* material properties. Design stresses are high and machining accuracy is to very tight tolerances. Balancing is a key step (fortunately this has a rigorous theoretical origin). Rotor manufacture is nearly always closely controlled by the manufacturer holding the main licence for manufacture and assembly of the turbine – and so carries less technical risk.

The fitted parts

These are the inventory of components which may be bought-out from specialist suppliers rather than made by the steam turbine manufacturer. They include labyrinth glands, bearings, valve internals, and (surprisingly perhaps) some of the blades. The level of precision is high, but the turbine manufacturer relies entirely on the supplier for the fitness for purpose of these components.

Do you see how these three manufacturing 'categories' differ?

The purpose of this approach is to show you how, when considering the FFP of steam turbine parts, you should first stand back and decide which of these three categories you are dealing with. Remember that the level of technical documentation available to you will vary – some parts will have been subjected to much better manufacturing control than others – and for some parts (I am thinking here of the sub-contract items) you will be much more likely to find breaks in the chain of traceability. If you think in these terms it will give you a framework on which to hang experience. It should also help you to keep the *correct focus* on FFP.

Guarantees

It is not easy to show you a 'typical' set of steam turbine acceptance guarantees. You will need to accept that a less direct form of guarantees (typically the information and parameters included in the technical schedules of the purchase specification) will form the basis of your FFP assessment. You will also have to rely on the various technical standards to provide clear guidance on some technical parameters that may not be stated explicitly in the contract acceptance guarantees.

Specifications and standards

I have made the statement in the earlier chapter on gas turbines that I believe gas turbine standards are less well developed than manufacturers' practices. This is *not* the case for steam turbines – perhaps because steam turbine technology is more mature and does not change as quickly.

International standards that you are likely to meet fall rather neatly into three categories.

The generalized technology standards

These provide a broad coverage of design, manufacture and testing. By now you should be starting to recognize this scope as being the domain of the API standards – the two relevant ones are API 611 *General purpose steam turbines for refinery services* (2), and API 612 *Special purpose steam turbines for refinery services* (3). As with other major API standards, their application is much wider than the titles suggest – they provide a good overall view if you have the time to work methodically through the explanations. ASME/ANSI PTC (Performance Test Codes) No. 6 (4) is a related document group which complements the API standards. It is not in very common use for large power generation turbines.

Performance test standards

These cover *only* the performance testing of turbines under steaming conditions. They are used for performance verification after commissioning on site but not for works testing. On the very rare occasion that a steam test is performed in the manufacturing works these are the standards to use. They are BS 5968 (5) (similar to IEC 46-2) and BS 752 (6) (similar to IEC 46-1) *Test code for acceptance tests*. It is most unlikely you will need to use them.

Procurement standards

This is the most useful category for works inspection purposes. The predominant document on steam turbine procurement is BS EN 60045-1 (7) – identical to IEC 45-1. It has been recently updated and encompasses many of the modern practices governing the way steam turbines are specified and purchased. It is worth mentioning some of the more important aspects that it covers:

- It provides clear guidance on governor characteristics and overspeed levels.
- Vibration is addressed in the two ways that I have introduced as a general approach in this book, i.e. bearing housing vibration using VDI 2056/ISO 2372 (using mm/sec as the guiding parameter) and shaft vibration using ISO 7919 and the concept of relative displacement measurement.

Steam turbines 243

- Definitive requirements are stated for hydrostatic tests on the pressurized components of the turbine.

Make sure you have access to this standard – you should note down some of its key points in your inspection notebook. I have already mentioned (in other sections of the book) the all-important balancing and vibration standards that are used. There is one final standard which will prove useful if radiography is used for the volumetric NDT of the turbine casings. This is BS 2737 *Terminology of Internal Defects in Castings* **(8)**. The identification and classification of defects is an important early step in determining the FFP of the cast turbine components.

Inspection and test plans

ITPs for steam turbines follow a well developed format. The underlying philosophy is one of very tight control of materials, both for the large individual cast components and for components such as blades, which are produced by batch manufacturing methods. Material test and analysis standards generally follow those described in Chapter 4. Figure 8.1 shows the common extent of the various material tests for the main turbine components. Note the predominance of NDT activities to ensure material integrity of both rotating and static components.

Also included in the ITP will be post-assembly checks, which are mainly concerned with clearances and alignment. Try to witness some of these tests (perhaps more than you would on a gas turbine) – most are straightforward measuring techniques using feeler gauges. Expect to work to the manufacturer's drawing tolerances.

The main post-assembly checks you will see on the ITP are:

- blade clearances (axial and radial)
- gland seal clearances (axial and radial)
- nozzle casing and balance piston clearances (if fitted)
- position check of the rotor in its bearing pedestals
- journal bearing radial clearances
- clearances of overspeed, lubricating oil (l.o.) pump and turning gear mechanisms
- bearing pedestal/casing alignments
- stop valve and control valve stroke and clearance measurements
- prestretch of bolts for the blade carriers, nozzle casings and outer casings.

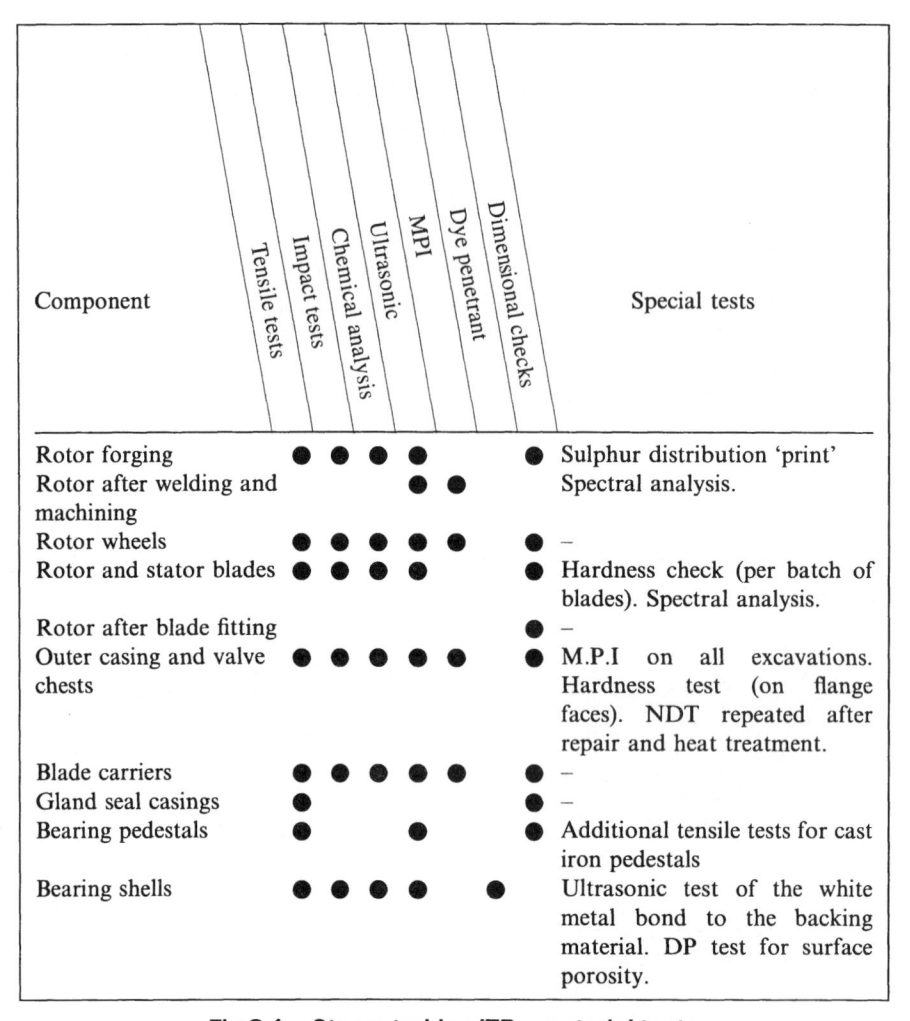

Component	Tensile tests	Impact tests	Chemical analysis	Ultrasonic	MPI	Dye penetrant	Dimensional checks	Special tests
Rotor forging	●	●	●	●			●	Sulphur distribution 'print'
Rotor after welding and machining				●	●			Spectral analysis.
Rotor wheels	●	●	●	●	●		●	–
Rotor and stator blades	●	●	●	●			●	Hardness check (per batch of blades). Spectral analysis.
Rotor after blade fitting							●	–
Outer casing and valve chests	●	●	●	●	●		●	M.P.I on all excavations. Hardness test (on flange faces). NDT repeated after repair and heat treatment.
Blade carriers				●	●	●	●	–
Gland seal casings				●			●	–
Bearing pedestals				●		●	●	Additional tensile tests for cast iron pedestals
Bearing shells				●	●	●	●	Ultrasonic test of the white metal bond to the backing material. DP test for surface porosity.

Fig 8.1 Steam turbine ITP – material tests

Test procedures and techniques

A good witnessed inspection programme for a steam turbine will involve at least three separate visits, spaced several weeks apart. These coincide neatly with the three main categories of test that you should witness.

Aim for continuity on these visits – it is preferable to witness all three stages, this way your assessment of documentation will become cumulative and you will obtain a better overall view of the fitness for purpose of the machine.

Casing tests

The collective view of steam turbine standards is that all components which experience a positive gauge pressure should be subject to comprehensive metallurgical and non-destructive testing followed by a hydrostatic test. This approach fits in well with the realities of manufacture of cast components, in which defects are often unavoidable. A key aspect of foundry practice is the identification and repair of these defects. As a works inspector, one of your most important tasks in ensuring FFP is the monitoring and verification of this set of activities. First, note the chronological series of events for a steam turbine casing.

- After solidification, the casing is removed from the mould, cooled, and the risers and 'moulding parts' rough-cut off. It is then fettled to a semi-finished condition.
- A visual surface examination is performed, supplemented by preliminary volumetric NDT (radiograph and ultrasonics are both commonly used). At this stage the casing will have defects – there will almost certainly be several major ones, possibly even 'holes' extending through the full casing wall thickness in some places. The important point is that these defects are identified and recorded using a *defect map*.
- The defects are 'ground-out' (excavated) back to sound parent metal ready for repair. At this point, a *repair procedure* is established – this is analagous to the WPS/PQR regime described in Chapter 5 and it is normal practice for the PQR element to be subject to third party approval in the same way. Figure 8.2 shows a typical defect map prepared for a hp (high pressure) turbine casing. Note the outline repair procedure shown, designed to utilize a GTAW root and SAW 'fill' technique with specified pre-heat and pwht (post weld heat treatment) requirements.
- Following repair and final heat treatment, the casing is subject to final NDT. This is a comprehensive surface and volumetric examination of all areas, not just those that have been repaired. A typical ultrasonic examination standard would be ASTM A609 **(9)** –

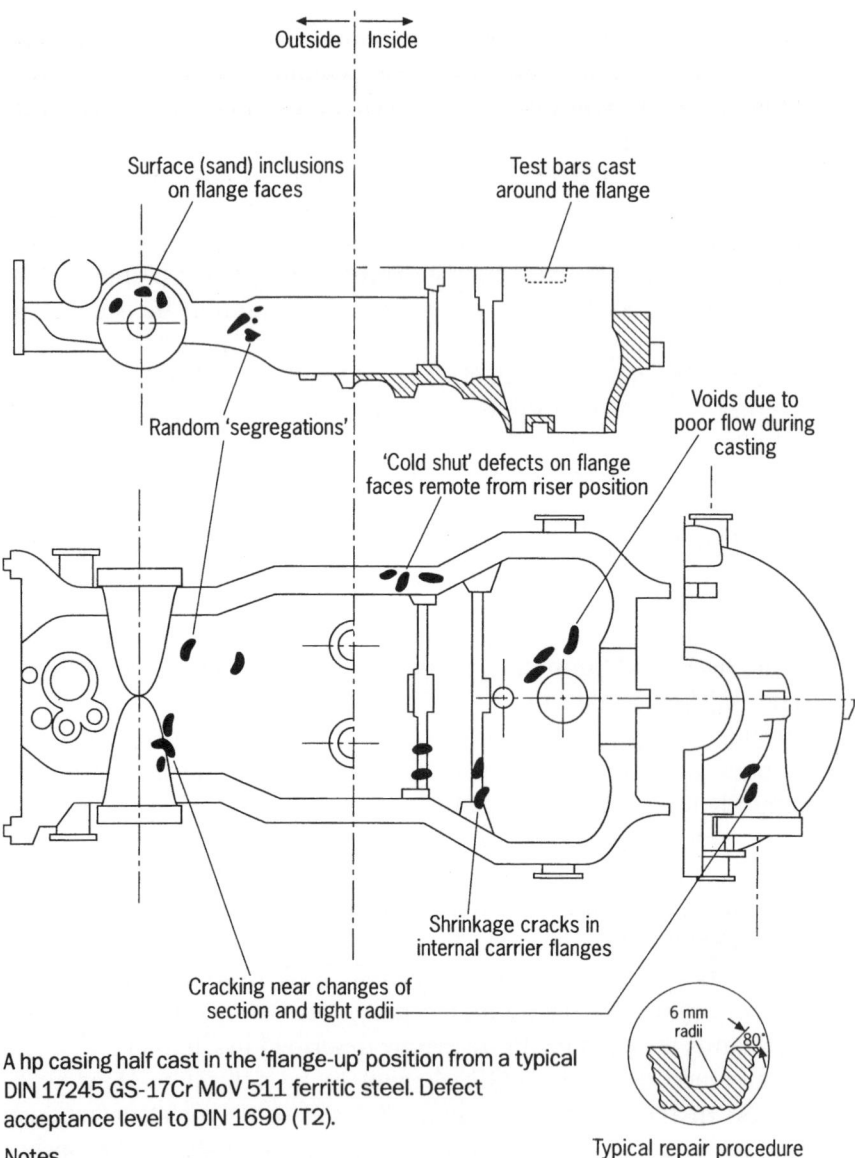

Labels on figure:
- Outside | Inside
- Surface (sand) inclusions on flange faces
- Test bars cast around the flange
- Random 'segregations'
- Voids due to poor flow during casting
- 'Cold shut' defects on flange faces remote from riser position
- Shrinkage cracks in internal carrier flanges
- Cracking near changes of section and tight radii
- 6 mm radii, 80°
- Typical repair procedure

A hp casing half cast in the 'flange-up' position from a typical DIN 17245 GS-17Cr MoV 511 ferritic steel. Defect acceptance level to DIN 1690 (T2).

Notes
1. Record all defect sizes (length × width × depth) in mm.
2. Defects frequently extend for the full wall thickness.
3. Excavate all defects and use an approved repair welding procedure.
- Preheat temp. 300°C
- Interpass temp. 400°C
- Full stress relieve after welding repair
- GTAW root, SAW fill

Fig 8.2 Common turbine casing casting defects

note that the ASME VIII Div I 'referenced' standards are commonly used for this application. The corresponding surface examination would be to MSS SP 55 **(10)** (visual) and ASTM E709/E125 (for MPI).
- Radiographic testing: although some steam turbine manufacturers use only ultrasonic testing of casings, radiography is still the more frequently used for most turbine sizes, except the largest designs. Typically this will be based on ASTM E94 and ASME Section VIII Div I UW–51 and UW–52.

Treat the radiographic test as the most important – this is an area where you need to have a clear idea about what to look for. It needs some preparation. Figure 8.3 show the outline of events that I recommend. Note the additional qualifying information that will be useful to you. Although the radiographic technique used is important, in order to obtain good results, from an inspection viewpoint the more relevant

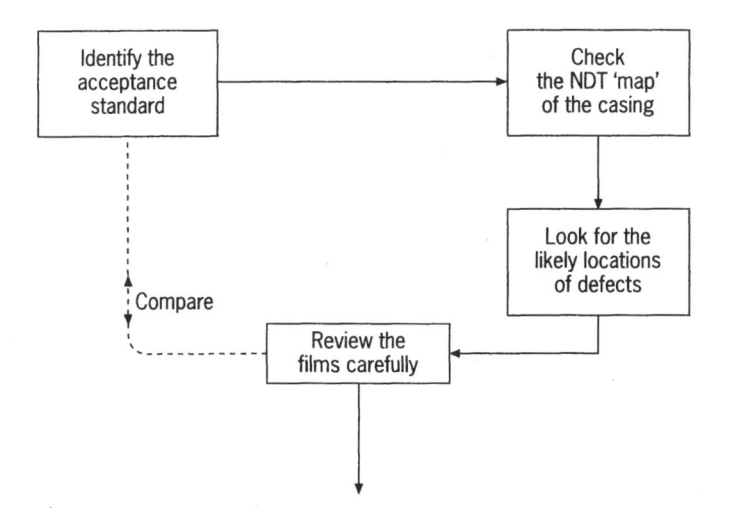

The 5 categories of defects are:
- Voids
- Cracks
- Cold shuts
- Segregations
- Inclusions

Use this 'qualifying' information to help you:
1. ASTM E280 and E446 'reference radiographs'.
2. BS 2737 'Terminology of defects in castings'.
3. MSS SP-55 'Visual standards for steel castings'.

Fig 8.3 Assessing turbine casing radiographs

standards are those which define acceptable defect levels. Manufacturers' comments that castings should comply with 'acceptable practice' or 'commercial quality' are frankly just not the basis of a good product. The best way of ensuring FFP is to make sure that recognized defect acceptance standards *are* incorporated as part of the contract requirements between manufacturer and purchaser. If you find that they are not, you will have to use these standards as background 'qualifying' information only, to complement your engineering judgement.

The hydrostatic test

The predominant design criterion for turbine casings is the ability to resist hoop stress at the maximum operating temperature. For practical reasons, a hydrostatic test is carried out at ambient temperature – this is only one of a list of real practical constraints which limit what a hydrostatic test can tell you. Watch out for the most common misconception, which is:

- 'The size of defects in a casting do not matter too much, if a casing can withstand hydrostatic test it must be acceptable'.

  This is simply *untrue* – a hydrostatic test imposes principal stresses under non-operating conditions whereas casings actually fail by fatigue, creep, or crack propagation mechanisms – or a combination of all three.

Treat the hydrostatic test as an important and valid part of your assessment (refer to Fig. 8.4) but be aware that it does not, in itself, confer fitness for purpose.

Check these important points when you do witness the test:

- The test pressure is normally 150 percent of the maximum allowable pressure the casing will experience in service. Note that it is necessary to apply a multiplying factor to compensate for the difference in tensile strength of the steel between ambient and operating temperature. Practically, the manufacturer may refer to the ASME code (Section VIII Division 1 for casing or B31.1 for piping) to determine material stresses and the corresponding test pressure. You can ask to see these calculations. Be wary of testing at an artificially low pressure.
- Some types of casing (typically those that have been designed to very 'tight' stress criteria) are tested by sub-dividing the casing with steel diaphragms held in place by jacks – this enables the various regions

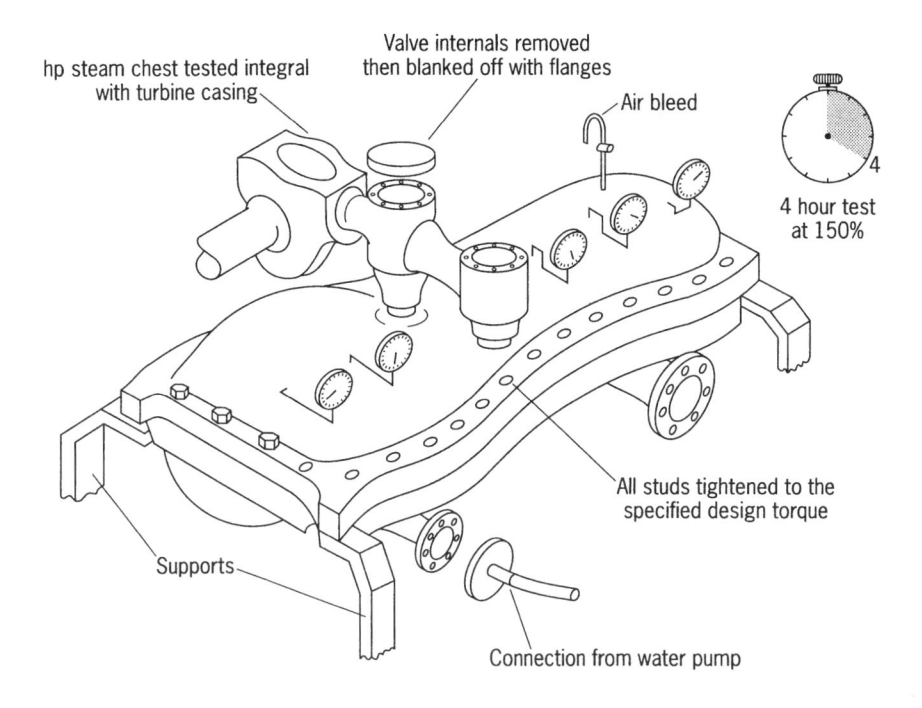

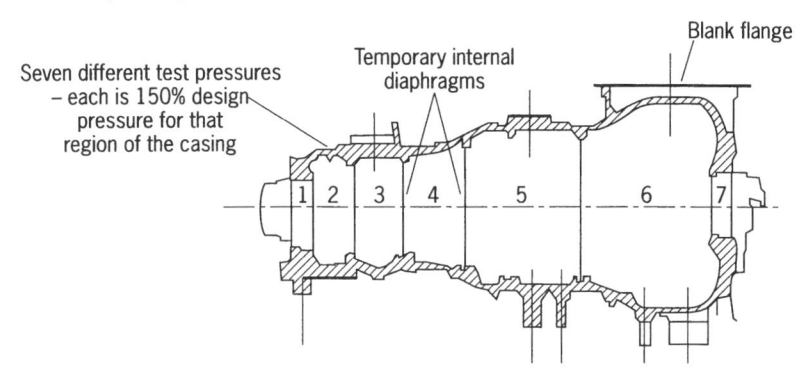

Steam turbines 249

Fig 8.4 Hydrostatic tests of a HP/IP turbine casing

(Upper diagram labels:)
- hp steam chest tested integral with turbine casing
- Valve internals removed then blanked off with flanges
- Air bleed
- 4 hour test at 150%
- All studs tightened to the specified design torque
- Supports
- Connection from water pump

(Lower diagram labels:)
- This shows an advanced combined hp/ip turbine casing tested at multiple test pressures
- Seven different test pressures – each is 150% design pressure for that region of the casing
- Temporary internal diaphragms
- Blank flange

of the casing to be tested at individual pressures which are more representative of the pressure gradient the casing experiences in use. Figure 8.4 shows such an arrangement.

- Pressure should be maintained for an *absolute minimum* of 30 minutes. Several hours is better. Keep a close check on the pressure gauges (there must be two) to identify any pressure drops – do not isolate the gauges using their shut-off cocks. The most accurate way to test is to use gauges where the test pressure is at 70–75 percent of the full-scale deflection.
- It is acceptable to use a sealant compound on the casing flange faces but be aware that this can mask flange distortion or a poor milled finish on the mating surfaces. If you see sealant, look carefully at the flange faces after disassembly.
- Casual observation of the pressurized casings is not good enough. You need to dry the exterior surfaces completely with compressed air and then examine all areas under a strong light. Pay particular attention to areas near changes of section and tight radii – try and anticipate where shrinkage defects and tearing would occur during cooling. It is also worthwhile to check the foundry repair records to see where major excavations and repairs were located.
- It is good practice to examine the inside of the turbine casing for visible defects after the hydrostatic test is completed – it normally takes about three hours for the casing to be drained and disassembled. You should include the results of this examination in your inspection report. This will form a useful future reference if defects are subsequently discovered on the casing internal surfaces during the operating life of the turbine. Make sure you also address the following points:
    - *Flange faces*. The manufacturer should re-check the *flatness* of the flange-faces (using marking blue) after the hydrostatic test to make sure no distortion has occurred. Try to witness this check, particularly if it is a long casing such as a combined hp/ip (high pressure/intermediate pressure) design. Pay particular attention to the inside edges, which is where distortion often shows itself first. Any 'lack of flatness' means that the faces must be skim-milled.
    - *Bolt holes*. Visually check around all the flange bolt-holes for cracks.
    - *Internal radii*. Check that small radii inside the casing have been well dressed and blended to minimize stress concentrations.
    - *General surface finish*. This assessment can be a little subjective – look for a good 'as cast' finish on the inside of the casing without signficant surface indentations. Use the visual inspection standard

MSS SP 55 as a broad guide. At this stage of manufacture all visible surface 'defects' should have been removed. Use a good light to check the surface carefully.
- Report your findings carefully and accurately using the guidelines given in Chapter 15.

Rotor tests

Steam turbine rotors are subject to dynamic balancing, overspeed and vibration tests and these are common works inspection witness points. The techniques used are similar to those for gas turbine and gearbox rotors – you can refer to Chapters 7 and 10 for the basic explanations. Acceptance limits can vary – this is another of those areas where manufacturers sometimes set their own limits. In general they should be based on the commonly used standards.

Dynamic balancing

This is carried out after the blades have been assembled, normally at low speed (400–500 rpm). Smaller hp and ip rotors will have two correction planes for adjustment weights whilst large lp rotors have three.
    API 611/612 specify a maximum residual unbalance (U, measured in g mm) per plane of :

$$U = \frac{6350\ W}{N}$$

where W = journal load in kg
N = maximum continuous speed in rpm

ISO 1940 specifies its balance quality grade G2.5 for steam turbine rotors. A similar approach is adopted by VDI 2060.

Vibration

API 611/612 specify vibration as an amplitude. The maximum peak-to-peak amplitude A (microns) is given by: A (μm) = 25.4 $\sqrt{(12\ 000/N)}$ with an absolute limit of 50 μm. BS EN 60045-1 adopts the same approach as other European gas turbine standards. Bearing housing vibration follows ISO 2372 (similar to VDI 2056) using a velocity V (rms) criterion of 2.8 mm/second. Shaft vibration is defined in relation to ISO 7919-1 (I have explained in Chapter 7 how this is a more complex approach).

Overspeed

The overspeed test may be carried out in a vacuum chamber to minimize problems due to windage.

API 611/612 infers that a steam turbine rotor should be overspeed tested at 110 percent of rated speed. BS EN 60045-1 places a maximum limit of 120 percent of rated speed for the overspeed test. In practice, this is more usually 110 percent. A close visual examination is required after the overspeed test to check for any breakage or yielding. Pay particular attention to the blade shroud wires and tip restraints – they are normally the first areas to break.

Assembly tests

During final assembly of the turbine, the manufacturer will perform a detailed series of clearance and alignment checks. The dimensions recorded at this stage will be used as an important reference point when the condition of the turbine is assessed at various inspections during its working life. They are the main indicator of wear and alignment problems. Unlike gas turbines, most steam turbine clearances are measured before fitting of the outer turbine casing top half. This is an effective witness point – try to find the time to witness the blade and gland clearances. This will add a little to your design knowledge – you can compare the clearances at each stage and make useful notes for comparison between manufacturers' designs. Figure 8.5 shows the locations at which the main clearances are taken and gives indicative values for a double-casing type hp turbine. Note the following points when taking the readings and recording the results:

- *Gland clearances.* Radial and axial clearances are normally *larger* at the low pressure (condenser) end. The readings should be confirmed at four diametral positions as shown in Fig. 8.5.
- *Nozzle casing and balance piston seals.* Expect the axial clearances to be *approximately* three times the radial clearances.
- *Blade clearances.* Use long (300–400 mm) feeler gauges to take clearance measurements at the less accessible radial locations. Note how the radial and axial clearances (and the allowable tolerances) *increase* towards the low pressure end. Radial clearances for the rotating blades tend to be broadly similar to those for the stationary blades, however lower temperature turbines in which the fixed blades are carried in cast steel diaphragms may have smaller clearances for the labyrinth seal between the diaphragm and the rotor (this is due to

Steam turbines 253

Clearances shown are indicative for a double-casing type hp turbine – exact dimensions will depend on the turbine design.

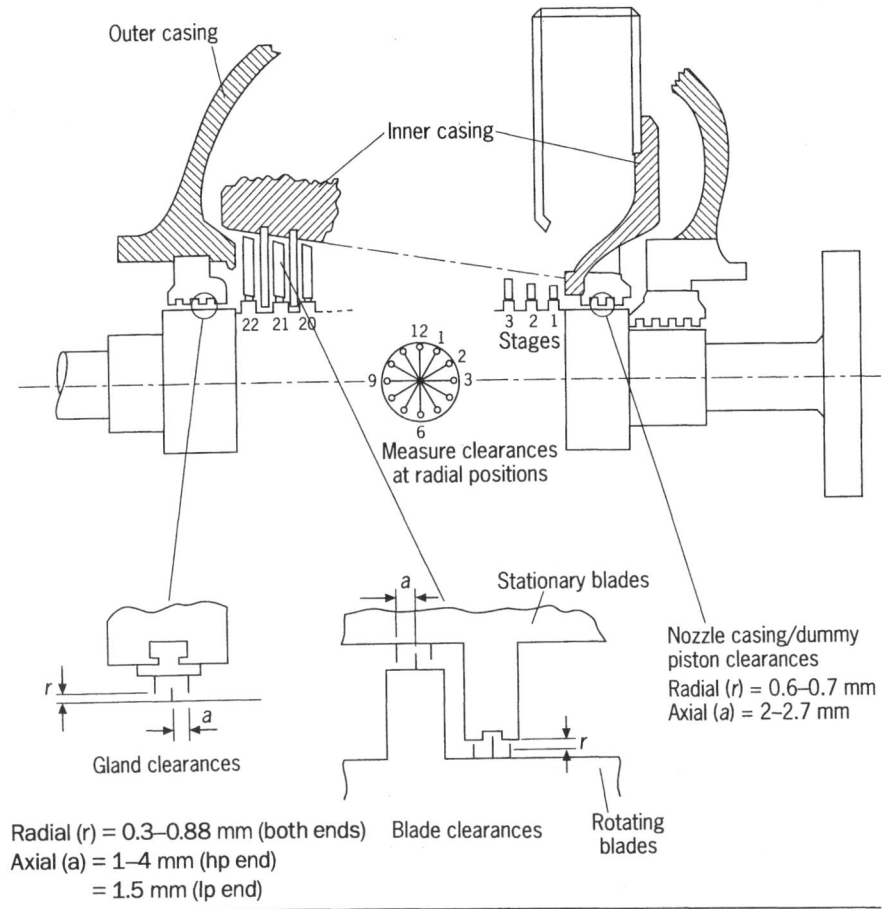

Nozzle casing/dummy piston clearances
Radial (r) = 0.6–0.7 mm
Axial (a) = 2–2.7 mm

Gland clearances
Radial (r) = 0.3–0.88 mm (both ends)
Axial (a) = 1–4 mm (hp end)
     = 1.5 mm (lp end)

| | Rotating/stationary blades | |
Stage	Radial (r)	Axial (a)
1–4	0.9 ± 0.3 mm	4 ± 1.5 mm
5–9	0.9 ± 0.3 mm	5 ± 1.6 mm
10–14	1.0 ± 0.35 mm	5 ± 1.6 mm
15–18	1.2 ± 0.35 mm	6 ± 1.6 mm
19–22	1.4 ± 0.40 mm	6 ± 1.6 mm

Fig 8.5 Typical HP turbine clearances

the high pressure drop across the impulse stages). Always check the manufacturers' data-sheets to see what the clearances should be.
- Take all blade clearances at a minimum of four positions (normally 3, 6, 9 and 12 o'clock as shown in Fig. 8.5. It is good practice to take additional readings at interim radial positions, particularly if there is any suggestion of shaft misalignment. Record all the results – it is easier to use the manufacturers pro-forma for this.

Common non-conformances and corrective actions: steam turbines

Non-conformance	Corrective action
Lack of material traceability of the rotor or stator.	This is a serious problem. The correct course of action is: • Make a full document review – go back to the foundry or forge using the heat-number as a basis for investigation. • Carry out a 'metalscope' analysis of the component. It is best to agree beforehand what analysis will be acceptable, then do the test. This is only a partial solution – if mechanical test pieces are not available for full analysis you may be obliged to reject the component. • Make sure you agree a written course of action with the manufacturer.
Failed mechanical tests on turbine casing(s)	Do not jump to conclusions. First categorize fully the failure. Do this by addressing the following points in your report: • How many of the specimens (normally 8–10 test bars per casing) failed the test? • What was the orientation (direction) of the failed specimen? Is it in a 'ruling section'? Obtain the turbine designer's comments on this report, then • *Repeat the tests* (witnessed) using left-over specimens from the original test bars. Check the laboratory records to ensure the correct specimens are used. If the problems are reproduced it is advisable not to accept the casing. Your NCR should recommend replacement. Be wary of allowing concessions. • In marginal cases it may be possible to increase the hydrostatic test pressure – but do remember the limitations of this test.
Casting defects observed by radiography of the casing. (See Fig. 8.2 for some common defects)	You must categorize the defects. Refer to BS 2737 and describe them fully. Loose description will be meaningless. Compare with the recognized defect acceptance levels. If they are outside accepted levels you should not accept the casing. Ask the foundry to propose a written excavation and repair procedure. Repeat all NDT and testing after the repair.
Excessive rotor unbalance	Do not accept *any* concessions on balance limits. Double-check the balance procedure for compliance with API 611/612 or ISO 1940 – check the machine accuracy as API describes.
Leaking flange faces during casing hydrostatic test	• Check bolt torques. • Do not allow any further jointing material or sealant to be applied *until* you have visually checked the flange faces for defects. Use an accurate straightedge to detect warping, then • Repeat the test.

KEY POINT SUMMARY: STEAM TURBINES

FFP criteria

1. FFP is a difficult concept to verify for steam turbines because they are not normally subject to a running test in the manufacturers' works.
2. Works inspectors need to know a little more about *how to manufacture* a steam turbine than they do for other types of equipment.

Your technical understanding

3. Try to understand the effect of the following factors on your activities when you are inspecting a steam turbine:

   - the materials of construction (particularly the cast parts)
   - the size of the machine – because larger machines have thicker material sections
   - the different manufacturing methodologies used for stator parts, rotor parts and bought-out fitted parts, which carry different levels of technical risk.

Test procedures

4. The identification and classification of casting defects is vitally important. You should witness and verify these activities.
5. Most large turbine castings *will* contain defects. Only properly approved and documented repair procedures are acceptable.
6. The casing hydrostatic test is important but it is not a replacement for the correct NDT activities.
7. *Don't accept* poor defect acceptance criteria based on vague 'commercial quality' arguments. Agree proper criteria that are based on proven technical standards. If in doubt, start with DIN 1690, ASTM E186 or BS 4080 and BS 6208.

References

1 DIN 1690: 1985. *Technical delivery conditions for castings made from metallic materials: steel castings, classification into severity levels on the basis of non-destructive testing.*
2 API 611: 1989. *General purpose steam tubines for refinery services.* American Petroleum Institute.
3 API 612: 1987. *Special purpose steam tubines for refinery services.* American Petroleum Institute.
4 ANSI/ASME. Performance Test Code No. 6: 1982. American Society of Mechanical Engineers.
5 BS 5968: 1980. *Methods of acceptance testing of industrial type steam turbines.*
6 BS 752: 1974. *Test code for acceptance of steam turbines.*
7 BS EN 60045-1: 1993. *Guide to steam turbine procurement.*
8 BS 2737: 1995. *Terminology of internal defects in castings as revealed by radiography.*
9 ASTM A609/A609M: 1991. *Practice for castings, low carbon, low alloy and martensitic stainless steel – ultrasonic examination.*
10 MSS SP 55: 1986: *Quality standards for steel castings – visual method.* The Manufacturers Standardisation Society of valves and fittings industry.

Chapter 9

Diesel engines

It is by no means easy to inspect a diesel engine competently. There are many different types which, particularly in the large sizes, can display a bewildering array of complex design features. Inspectors cannot easily check the design of diesel engines but even as a non-specialist it is worth adopting a professional approach to their inspection.

Fitness-for-purpose criteria

We are fortunate that FFP criteria for diesel engines can be clearly defined. There are two key criteria:

Integrity

As with the other prime-movers, the core notion of integrity incorporates the mechanical integrity of the mechanical components and the effective operation of the engine's safety and protection features. Because of the design features unique to reciprocating machinery, we must also recognize the importance of correct *running clearances* in ensuring correct operation of the engine. The brake test and stripdown examination details given later will look at this in more detail.

Performance

Engine performance is the main issue that drives the ongoing design and development of diesel engines. If you look at competing engine manufacturers' literature you will see evidence of the continual striving for increased power output and fuel efficiency. The prime performance criterion is *specific* fuel consumption (sfc). Expressed in grams (of fuel) per kW (brake power) hour, this is a direct way of assessing and comparing engines – a good FFP criterion. It is important not to underestimate the role of assessing performance criteria closely during the witnessed load (brake) test of a diesel engine – it is all too easy to be diverted to other matters (turn back to Chapter 2 and look again at Fig. 2.3 if you need reminding of how this can happen).

Basic technical information

Speed classification

The three main diesel engine types are classified broadly by speed. Slow speed diesels, the largest two or four stroke designs, operate below 300 rpm with piston speeds less than 9 m/s and are used for power generation and ship propulsion. Medium speed engines range from 300–800 rpm. Speeds above 800 rpm are broadly designated 'high speed' – these engines are almost exclusively of 'Vee' configuration and are widely used for driving locomotives, fast marine craft, and smaller auxiliary generators.

These three types of engine differ in size and power output and have followed different development routes. It is also true, however, that modern developments have been such as to blur the strict distinctions between the types – features such as heavy fuel capability, formerly found only in slow speed engines, are now commonplace in the medium speed ranges. Conversely, lighter construction methods originally used for medium speed designs are influencing the design of larger slow speed engines. From the inspection viewpoint, this convergence of technology is an advantage as it results in an increasing level of commonality of inspection activities.

Some technical considerations

Listed below are some of the technical differences between the three types of diesel engine:

- Piston speeds and the methods of piston lubrication differ: this means that you have to look for potentially different wear regimes on the piston/liner surfaces.
- Crankshaft construction: large slow speed engines tend to have 'built-up' (rather than one-piece forged) crankshafts. The key inspection activity of measuring crankshaft deflections, however, is essentially the same.

- Turbocharging and scavenging arrangements are different. This is purely a design consideration and does not affect the method or techniques used during works inspection.
- Larger 'trunk' engines have crossheads and sometimes more complicated torsional vibration damping arrangements. These are additional items to be assessed during the stripdown inspection.

Notwithstanding design features, you can see that from the inspector's viewpoint there are no great differences in the task of verifying fitness for purpose of the various types of diesel engine. The engineering techniques used are really quite similar. Experience shows that there is also a lot of commonality of the types of faults and non-conformances that regularly occur. It is reasonable to expect, therefore, that a non-specialist inspector can become equally competent in witnessing tests on all types of diesel engine.

Acceptance guarantees

Engine guarantee schedules follow the well defined pattern shown in Fig. 9.1. I have shown indicative parameters for an advanced medium speed engine – to give you an idea of what to expect. Note that *not all* of the parameters may be shown in this schedule – some may be cross referenced to a 'technical particulars' schedule included elsewhere in a contract specification. Note these points about this guarantee schedule:

- A number of key design requirements, such as critical speed range, governor characteristics, acceptable bearing temperatures and peak cylinder pressures are not stated *explicitly*. That does not mean that they are not a legitimate part of the requirements of the specification.
- The guarantee schedule cannot therefore stand alone as a complete statement of FFP requirements, it is necessary to use a technical reference standard to provide 'qualifying information'.

Vibration

The vibration characteristics of a diesel engine are very different from those associated with turbomachinery and similar rotating plant. The existence of large reciprocating masses means that some of the mass

GUARANTEES

- Nominal site power output shall be 12MW (note the general statement here, to be qualified by reference to ISO 3046/2).
- Engine to be capable of 10% greater output for a period of one hour.
- Maximum continuous speed 500 rpm.
- Specified fuel: e.g. BS 2869 Class 'F'.
- Governing class: Type A1, single speed, accuracy class A1.
- The test standard shall be ISO 3046.
- **Specific fuel consumption** (units:grammes per (brake) kilowatt hour; g/(b)kWhr)

MCR (%)	Guarantee (typical)	Guarantee +2.5%
50	220g/(b)kWhr	225g/(b)kWhr
60	216g/(b)kWhr	221g/(b)kWhr
80	214g/(b)kWhr	219g/(b)kWhr
90	212g/(b)kWhr	217g/(b)kWhr
100*	210g/(b)kWhr	205g/(b)kWhr

  *Note that the 100% MCR specific fuel consumption guarantee is often qualified by a maximum allowable exhaust back-pressure.

- **Emission levels**

  Maximum emission levels with the specified fuel grade at stated load (90%) shall not exceed:

NO_x	:	1400 mg/m^3
CO	:	450 mg/m^3
Particulates	:	100 mg/m^3
Non-methane hydrocarbons (NMH)	:	100 mg/m^3

- **Engine adjustment factors:**

  It is common to assess the engine design by evaluating closely the effect of retarding the fuel injection timing. You may see it expressed like this:

  Effect of 2 degrees injection retard
  : Maximum +3% increase in sfc.
  : Minimum decrease of -12% in NO_x
  : Maximum increase of +3% in CO
  : No measurable effect on Particulates or NMH.

  Remember: These are good indicative figures but it depends on the particular engine design.

- **Lubricating oil consumption**

  Guarantee l.o. consumption shall be less than 1.5 g/(b)kWhr ($\pm 5\%$) after site running-in for a period of 1000 running hours.

- **Site rating**

  At site conditions of : p = 1 bar (sea level), T = 300K
  relative humidity = 40 to 80%
  The rated power shall be 12000 kW.
  Site load condition will be 90% MCR.
  No percentage de-rating of the engine for site conditions.

- **Noise levels**

  At floor level and 1m distance from the engine the maximum allowable decibel level shall be a maximum of 115dB.

Fig 9.1 A typical diesel engine guarantee schedule

effects simply cannot be balanced. There are two different types of vibration to be considered. Radial and axial 'housing' vibration (in the x, y, z planes) is of the same type as that discussed in the chapters on steam and gas turbines – it can be measured during engine testing using relatively simple techniques. Diesels also exhibit *torsional* vibration which is a different specialized subject. Works inspection activities are normally limited to the measurement and assessment of radial and axial vibration – you are unlikely to become involved much with torsional vibrations when witnessing engine manufacture and brake testing in the works.

The balancing of diesel engines is also a difficult issue. The rotating element is the crankshaft and flywheel assembly, together with the rotating parts of the connecting rod bottom end bearings. This assembly has complex geometry and there is often uncertainty as to whether it should be classed as a 'rigid rotor', to allow an analysis under ISO 1940/1, or whether it is considered a 'flexible rotor', in which case the specific standards ISO 5343 and 5406 could apply. It is also not possible to link balance grades with vibration performance due to the effect of resonant frequencies. In practice, the balance grade of engine crankshafts is a design issue rather than a works inspection one. You shouldn't need to get too involved.

The levels of 'housing' vibration that you will see in diesel engines will be several orders of magnitude *higher* than in turbomachinery. Typical vibration velocity (rms) levels can be up to 50 mm/s during normal running with possibly up to 500 mm/s at critical (resonant) speeds. Large engines will pass through one, possibly two resonant speeds between starting and reaching full speed. It is not easy to find definitive guidance or acceptance levels in technical standards – the best one to use is probably VDI 2056 (the same one used for turbomachinery) which specifies reciprocating engines in its group D and S machines. It does not, however, provide any graphs or data that you can use to calculate acceptable vibration limits for these groups. Note that the concept of relative shaft vibration (ISO 7919/1) has no relevance to reciprocating engines.

Specifications and standards

Technical standards for diesel engines divide into two well-defined categories. The first, manufacturing standards, tends to concentrate on the major forged components of the engine – the crankshaft, connecting rods, camshaft and major gears, and the major castings such as cylinder heads, liners, pistons, and entablature/crankcase assembly. In the second category, there may also be general fabrication standards which are relevant to the bedplate and crankcase, depending on the design of the engine. These standards are mentioned in Chapters 4 and 5 – use them to monitor construction of the relevant engine components, as you would for other fabricated equipment items covered by similar standards. The main body of technical standards that you will encounter are those relating to *testing* of the engine. The most widely accepted standard is ISO 3046: Parts 1 to 7, *Reciprocating internal combustion engines: performance* (identical in all respects to BS 5514 Parts 1 to 7) **(1)**. This is a wide-ranging and comprehensive standard – it is worth looking at the various parts (they are separate documents) and the technical areas that they cover.

- ISO 3046/1: *Standard reference conditions*. This defines the standard ISO reference conditions which qualify engine performance. It is also useful in explaining the way in which various power measurements such as rated power, service power, etc. are defined. As inspectors, it is essential that we understand these if we want to determine properly whether an engine has met its performance guarantees.
- ISO 3046/2 *Test methods*. This part is definitive in describing the principles of acceptance guarantees during the load (brake) test. It explains how to relate engine power at actual test conditions to ISO conditions and site conditions (both of which normally feature in the engine's guarantee specification). A second useful, but more complex, topic covered is that of power *adjustment* and power *correction*. You may need this in some engine tests. I strongly recommend that you review the 'List A' schedule which is included in this part of the standard. It is a comprehensive list of measurements that should be taken during a brake test. By using this, you can confirm whether an engine manufacturer's measurement schedule is complete.
- ISO 3046/3 *Test measurements*. Use this part if you have any doubts whether a test procedure is being carried out at the correct temperatures, pressures and rotational speed. It gives permissible deviations.
- ISO 3046/4 *Speed governing*. The most frequently used content is that defining the five possible classes of governing accuracy. An engine specification and acceptance guarantee will quote one of these classes – so it is best to know what to look for.

- ISO 3046/5 *Torsional vibrations*. This is a complex part of the standard and not particularly useful in a works inspection situation.
- ISO 3046/6 *Specification of overspeed protection*. This is a very short document but it usefully defines the overspeed levels. It does not add a lot of information to the requirements of a good engine specification, so is perhaps not essential reading.
- ISO 3046/7 *Codes for engine power*. Manufacturers frequently classify the power output of their range of engines using a string of code numbers. This standard describes how they might do it. You might find the document useful – I have never needed to use it.

One final important standard is BS 2869 **(2)**. This covers classes of fuel. The main parameters are chemical composition and (lower) calorific value (lcv) – it is worth checking the test fuel carefully against this standard. It will *not* always comply.

Inspection and test plans

ITPs for diesel engines can be slightly different from those for some other equipment items. One major difference is the possible influence of a classification society. Both slow and medium speed engine designs, because of their potential marine applications, tend to be built to 'class', i.e. to the survey requirements of one or more of the major classification societies. For this reason, material specifications, traceability, and testing requirements are well controlled and specified – this is normally reflected in the content of the ITP. Because a diesel engine is an independent prime mover (unlike a steam turbine which needs a boiler) compliance with acceptance guarantee requirements can be proven in the manufacturer's works – you can expect therefore that performance test standards will be well specified in the ITP. A typical ITP for medium speed engine assembly is shown in Fig. 9.2.

Test procedures and techniques

During the manufacturing process of a diesel engine there are several key stages which you may find defined as witness points. I have included an outline explanation of the important stages, with a more detailed description of the brake test.

Activity	Checks
• Crankshaft documentation:	Classification Society approval of materials and manufacture.
• Piston Rings:	Metallurgical checks.
• Crankcase:	Crack detection to BS 6072 (MPI) and metallurgical test on test bars. Alignment of bearing bores checked by wire or laser measurement. Main bearing bore sizes.
• Cylinder heads:	Metallurgical tests.
• Cylinder liners:	Dimensional checks. Surface finish check to BS 1134 or DIN ISO 1302. Hydraulic test.
• Connecting rods:	MPI crack detection to BS 6072 before machining. Dimensional checks.
• Large end blocks:	MPI crack detection on forgings before machining. Dimensional checks.
• Cylinder heads:	Metallurgical checks on castings before machining. Hydraulic test of cooling water passages. Hydraulic test of indicator bores.
• Assembly:	Record crankshaft thrust bearing clearances and web deflections.
• Flushing check:	Cleanliness check and record.
• Running-in check:	Manufacturer's procedure.
• Load test:	Test to ISO 3046 (Parts 1 to 7).
• Post-test crankshaft alignment:	
• Stripdown examination:	
• Reassemble and protect from corrosion:	
• Paint engine.	
• Document review and certification.	

Fig 9.2 Diesel engine manufacture – typical ITP content

Alignments

Alignments are particularly important for large slow-speed engines but are also performed for the smaller types. There are three key alignment parameters to be measured:

Bedplate alignment

This is the first alignment operation to be carried out. Its purpose is to achieve absolute concentricity of the main bearings about the crankshaft centreline after assembly of the bedplate sections – larger engines often have the fabricated bedplate manufactured in two or three sections for ease of fabrication and transport. For the purposes of this check, the bedplate sections are subject to a trial alignment in the works, often before final machining of the important datum faces. This is a key witness point – accurate alignment at this stage will greatly reduce the risk of later problems during the engine assembly operations.

The most common method used is a combination of water-level and taut-wire measuring techniques. Figure 9.3 shows the principles for a two-section bedplate of a large slow speed engine. Note that there are *two sets of measurements* to be taken; those from wire A will determine the relative concentricity of each main bearing about a datum centreline, whilst those from wires B and C will determine the accuracy of the bedplate in the vertical plane. The test steps are as follows:

- The bedplate sections are assembled on chocks and their relative position adjusted until they are level. This is checked by filling the channels machined longitudinally along the upper faces with water. Final adjustment is then made to achieve an equal depth of water over the length of the channels.
- The 'centreline' wire A is positioned along the line of the crankshaft axis, passing between the saddles that will hold the main bearing keeps. Measurements are taken from the wire to the milled abutment faces on each side of the saddle (these are the faces that locate the main bearing keeps accurately in position relative to the bedplate). These measurements therefore indicate the accuracy of location of the bearing centreline in relation to wire A (see Fig. 9.3). Final skimming of the abutment faces can then be carried out to bring all the bearing centres in line. Note that alignment is required in both the x and y planes to give the correct 'truth' of the main bearing axis.
- Wires B and C are positioned longitudinally above the bedplate flange and across the diagonals as shown. Accurate *vertical*

measurements are taken from the wire to the machined horizontal datum faces of the bedplate. Variations in these dimensions indicate whether further skim-milling of these faces is required in order to achieve an accurate flat surface. If these surfaces are not flat, they will pull the cylinder axes, and hence the reciprocating running gear parts, out of alignment once the crankcase and entablature are assembled. This will result in severe wear and probably failure.

Any taut-wire measuring technique suffers from sag of the wire between its suspension points. This is a well-understood phenomenon and is documented in catenary tables which list the corrections that must be added to the vertical dimensions (depending on the overall length of the wire and the position along its length from which a vertical dimension is taken). Dimensions taken during the measurement checks will therefore need to be *corrected* before they will give a true representation of the alignment. This is a quick procedure, good manufacturers will have the catenary corrections already printed on their bedplate alignment report sheet. The acceptable tolerances for misalignment depend on the physical size and stiffness of the bedplate and the type of engine. All diesel engine licensors and manufacturers include a bedplate alignment drawing as part of their design package – compare the test results carefully with the stated tolerances on this drawing. Report accurately what you find – don't just make general statements that 'the bedplate alignment was witnessed'.

Crankshaft alignment (deflections)

The purpose of measuring crankshaft deflections is to check the 'as installed' alignment of the main journal bearings. The check is first performed towards the later stages of engine assembly, after installation of the crankshaft. Expect to see it repeated several times, including after the brake test procedure (and again after site installation). Figure 9.4 shows the technique. A dial test indicator (dti) is placed horizontally between the crank webs directly opposite the crank pin. The crankshaft is slowly rotated through one complete revolution and gauge readings taken at the five points shown. This is repeated for each crank web pair (i.e. one set of readings per cylinder) and the results tabulated. Figure 9.4 also shows the sign conventions that are normally used when referring to deflection readings. Note particularly the $+/-$ convention relating to the direction of distortion of the crankshaft webs and the way

Diesel engines 269

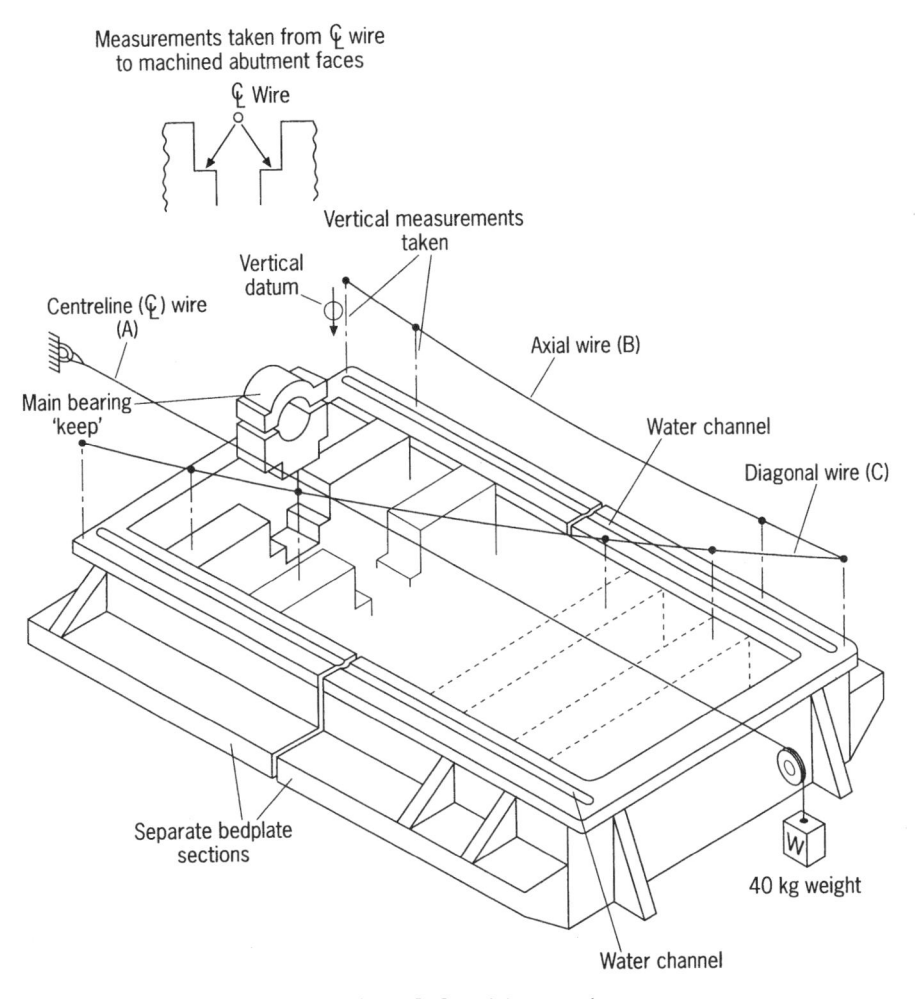

- Large slow speed engines may have 2–3 bedplate sections
- The objective is absolute concentricity of the bearings shells around the crankshaft centreline
- Measurement accuracy is ($\pm$) 0.01mm
- Vertical measurements need to be corrected to compensate for the wire catenary. Tables are available

Fig 9.3   Diesel engine bedplate alignment check

in which the sides of the diesel engine are usually referred to: i.e. exhaust (E) side and camshaft (C) side.

The measured values that we are interested in are the vertical misalignment (shown as T–B) and horizontal misalignment (shown as C–E). There will be clear acceptance limits for deflection readings set by the engine licensor – they may be shown in the form of a graph or a table. The figure shows a typical set of deflection results – note how the values are interpreted to diagnose which bearings are misaligned. Vertical misalignment can be corrected (to a limited extent) by using shims to adjust the position of the bearing keeps.

There are a few key points to bear in mind when witnessing deflection checks:

- Before taking deflections, check the main bearing clearances using feeler gauges. The bottom clearances directly underneath the journals should be zero. If they are not, then the bearing train is *definitely* out of alignment.
- Because of the connecting rod, it is not possible to take a deflection reading when the crank is at bottom dead centre. The accepted way of getting a reading here is to take readings as near as possible either side (positions x and y) and then take an average.
- The engine staybolts should be tightened to their design torque, slack staybolts will cause errors in the deflection readings.
- Make sure that before each reading is taken, the turning gear is reversed and backed off slightly to unload the flywheel gear teeth – this helps to avoid inaccuracies.

Reciprocating parts alignment

For large engines it is necessary to check various dimensions that indicate the alignment of the pistons and other reciprocating parts. These dimensions include: piston inclination (axial and transverse), crosshead clearances, and crosshead pin clearances, all taken in several piston positions to indicate the truth of the engine assembly. They are important dimensions which must be carefully taken in defined positions. Ask to review the engine manufacturer's alignment procedure document – this is the best (and only) document to use for guidance.

Diesel engines 271

Position the gauge as shown:

Crankpin
Web
dti

Convention
⊕ ‾_/‾
⊖ _/‾_

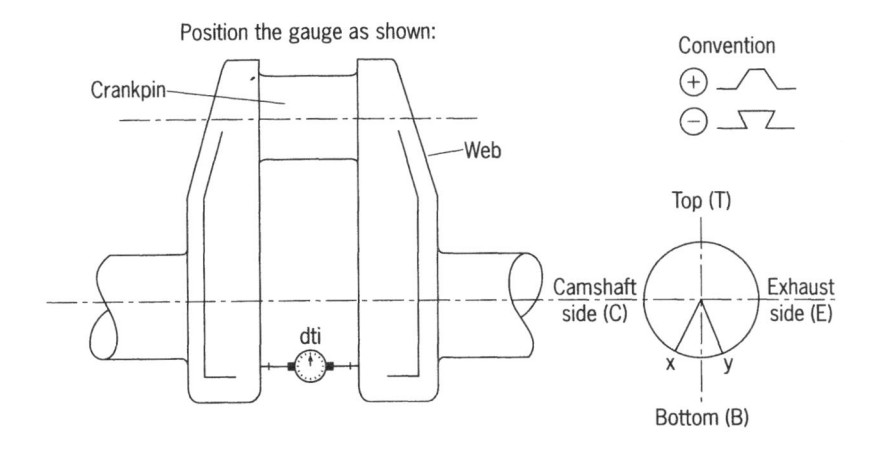

Top (T)

Camshaft side (C) | Exhaust side (E)

x   y

Bottom (B)

1. It is convention to set the gauge using x = 0 as a datum.
2. Don't confuse the algebra: e.g. (– 2) – (– 4) = + 2
3. Diagnosis. In this example the main bearing between cylinders 3 and 4 is <u>high</u>.

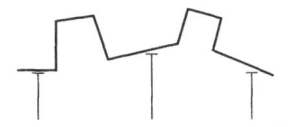

Typical readings (in 0.01mm) for a 6-cylinder engine

Crank position	Cylinder number					
	1	2	3	4	5	6
x	0	0	0	0	0	0
C	+4	+1	+3	-6	-2	+1
T	+8	+3	+10	-12	-6	+3
E	+4	+2	+5	-6	-4	+2
y	-2	+2	-2	0	0	-2
B=(x+y)/2	-1	+1	-1	0	0	-1
VM=T-B	+9	+2	+11	-12	-6	+4
HM=C-E	0	-1	-2	0	+2	-1

VM - Vertical misalignment
HM - Horizontal misalignment

Fig 9.4 Crankshaft web defelections

The brake test

Most diesel engines larger than 1–2 MW are factory-tested under load. This is called the brake test and is a common inspection witness point. The brake test is the key activity to prove fitness for purpose. Once again, it is best to think of the test procedure in terms of discrete steps and try to witness them all – this is not a good area for a half-hearted or 'cost-saving approach'. There are four steps:

Step 1: Flushing

Before a new engine is started for the first time it is *absolutely essential* that the lubricating oil piping system and sump is properly flushed out. The objective is to get rid of all scale, dirt, and weld spatter. There are, unfortunately, numerous examples of engines that have failed in early service because of inadequate flushing.

Check the following points:

- Before filling with the flushing oil charge, the crankcase and sump should be thoroughly cleaned by hand. Pay particular attention to the inside of welded pipework and ensure that all weld scale and spatter is properly loosened by hammering and then scraped off – otherwise I can guarantee to you that it will come off later, probably when the engine is running on site. Be wary of welded steel pipework that has not been acid pickled to remove slag and scale – insist on a visual inspection of the pipe internals if you are in any doubt.
- You must make sure that debris from the crankcase and sump is filtered out before it has a chance to reach the important parts of the engine. This is done by isolating the engine lubricating oil feed galleries (using blank flanges) whilst circulating the flushing oil through the supply system until the filters have removed the debris. It is not sufficient to do this quickly – a long circulating period without evidence of filter contamination is necessary to prove that the oil is clean. Try to *witness* the flushing process and verify the flushing log sheets.

Step 2: Running-in

It is normal for diesel engines to undergo a running-in period before the witnessed brake test. As well as the traditional objective of bedding-in the bearing and wear surfaces (the relieving of high spots) this also serves as a period for carrying out various adjustments necessary for efficient operation of the engine. Check that the running-in period lasts 18–24 hours and that the following activities form part of the procedure:

Diesel engines 273

- A short period of running below 20 percent load to check starting, control and emergency systems.
- Running for 2–4 hours at 30–50 percent load: checking of indicator diagram cams by taking sample indicator cards (or electronic recording, if fitted).
- Measurement of crankshaft deflections after 8–10 hours total running-in. Ask to see the results. Compare them with the pre-test deflection readings.
- Running for 4–6 hours at 50–70 percent load: testing of emergency stop and minimum rpm.
- At least 3 hours at 80–90 percent load. It is normal to make a detailed check of the injection timing under these conditions, particularly if the engine has a variable injection timing (VIT) facility.
- The final running-in step is typically a 3–4 hour period, increasing steadily from approximately 90 to 100 percent load, normally defined as the maximum continuous rating (MCR). A full test-log should be taken during this period. Look for evidence of a preliminary governor test at this stage.

It is not uncommon for engine manufacturers to perform several running in-tests, particularly on large engines which can need more adjustments. It is important to look at the results of these tests – you can learn a lot about the engine by really investigating what happened during repeated pre-brake-test running. This is where potential future problems will show themselves.

> – *A review:* a good engine manufacturer may understandably be reluctant to discuss in detail the reasons for repeated running-in tests. I introduced a basic tactical approach to you in Chapter 2 of this book. I would now like to suggest that you turn back to Chapter 2 and re-read the short section entitled 'Asking and Listening'. This may help you.

Step 3: The brake test

An engine brake test should follow a well-defined programme. If a written procedure is not available it can be difficult to follow what is happening. The exact format of the test depends on the requirements of the purchase contract, but you will find that generally it follows a quite well-defined set of rules. The engine will be run for the longest test period at a load corresponding to the planned site load, which will be

274 Handbook of Mechanical Works Inspection

stated in the specification or acceptance guarantee schedule. Shorter runs will be made at other loads and there will always be a period at 100 percent rated load. Figure 9.5 shows a typical load/time characteristic – in this case it refers to an engine designed for a specified 90 percent site load. You should always check the validity of a manufacturer's proposed running test programme to make sure it relates properly to the specification and guarantee requirements.

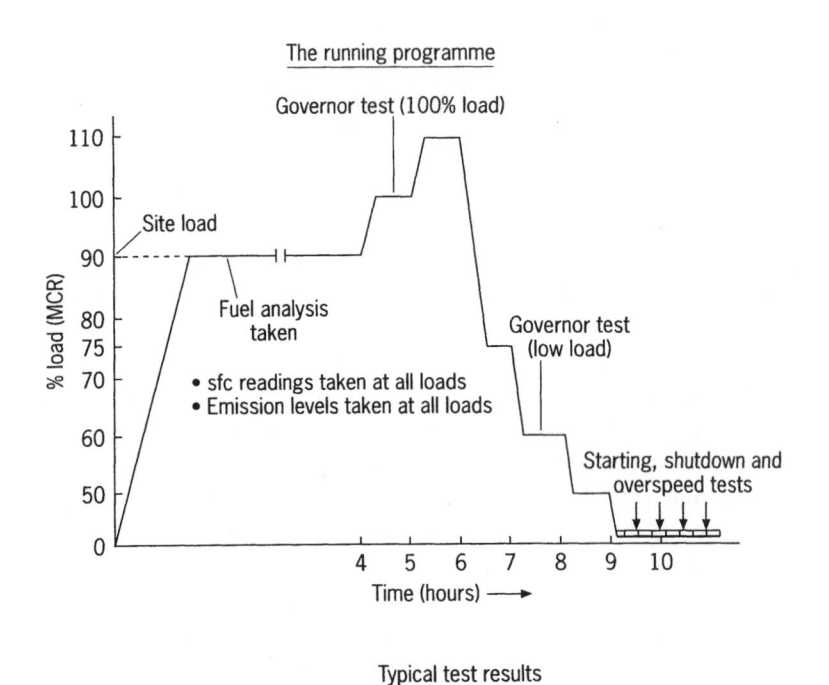

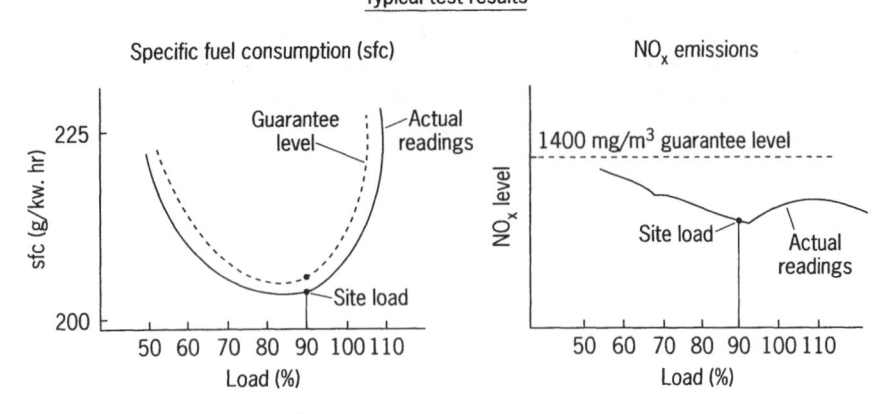

Fig 9.5 Diesel engine brake test

Here are some further useful checks to make before the test commences:

- *Flushing.* You must, repeat *must*, make sure this has been done correctly, as I have described previously. It is not wise to accept manufacturers' broad assurances – ask to see the flushing log. If in any doubt remove a crankcase door, take a lubricating oil sample from the crankcase (with the main l.o. pump on) and check the oil for metallic and oxide debris.
- *The data logger.* Most engine manufacturers will use a data logger to collect and process the multiple temperature and pressure measurements during the test. Check the 'zero-readings' of these measurements before the engine is started. The data logger should then be set to provide a print-out of the parameters at least every 15 minutes.
- Pay particular attention to the method of measuring brake power. This will almost certainly be derived from a torque measurement taken from the brake shaft – you have to rely therefore on using the correct 'brake constant' in order to make an accurate power calculation. Make sure the brake calibration certificate is current, and that it verifies the correct brake constant to be used.
- Check the set-point for engine inlet air temperature (this is controlled via the charge air cooler circuit). There should be a very closely controlled air temperature range – essential as a baseline for assessing the exhaust emission readings.
- *Engine adjustments.* You have to watch out for this one. Before the test starts, a manufacturer should be able to *demonstrate* to you the state of the engine adjustments – this is essential if the test performance and emission readings are to have real meaning. Remember that adjustments are commonly made during the pre-brake-test running-in periods – so you can ask to see these results.

Adjustments which should be recorded for each cylinder are:

- exhaust cam lead (degrees)
- fuel cam lift (mm)
- VIT settings (if fitted).

Adjustment characteristics will frequently be included in the guarantee schedule so they are an important part of the verification process. An engine which passes an acceptance test on its limit of adjustment will probably *not* meet emission limits after a few thousand hours of running. Check this area carefully – remember that ISO 3046 provides further detail if you need it, but a good engine manufacturer will be able to explain it to you. An engine manufacturer should also not make unplanned and unrecorded engine adjustments *during* the brake test.

When the brake test commences, parameters are normally monitored from a remote measuring station or control room. Assuming that the data logger is doing its job properly, there is no real need for you to spend too much time watching it print. Once you have established that each load setting is correct, the best place to be is near the engine where you can monitor the local instrumentation and develop a feel for noise and vibration levels.

Pay particular attention to these points:

- Double-check the *vibration levels* at the site load condition (90 percent MCR in our example) and at 100 percent MCR. Feel round the engine, as well as looking at the instruments.
- Listen for *turbocharger surging*. This is very audible and will also show itself as significant surges in turbocharger speed (exceeding the allowed deviation of ± 2 percent rpm), particularly at the 110 percent MCR condition. For a new engine, turbocharger surging is an indication either of incorrect adjustment or of more fundamental design 'matching' problems.
- *Bearing temperatures* are a key FFP point. They will not all be identical (a spread of 10–15°C is acceptable) but they should fall into a pattern which remains fairly constant throughout the brake test. Be wary of any white metal bearing operating at *above* 85°C oil drain return temperature. The white metal surface temperature will be higher and may be approaching the safe limit.
- Exhaust *gas temperatures* give the best indication of what is happening in the engine cylinders. Expect a 25°C or so spread of temperatures but be vigilant for excessively high temperatures – they are a sign of fuel injection or exhaust valve problems.
- *Governor operation* is normally tested at two loads (MCR and 40–50 percent MCR 'low load'). At each load, a test is made with the two governor settings at zero droop (this condition is known as isochronous) and at a higher droop level. You are looking for two things:

Diesel engines 277

- The time it takes the speed to settle down after a load change (expect 8–15 seconds).
- The change in speed (known as droop) between zero load and 100 per cent MCR.

If in doubt check with ISO 3046 part 4.
- Trips: at the end of the brake test the various trips and safety features will be tested. Make sure that you witness the test on these which are:
  - overspeed trip, using the control air circuit to close the fuel racks
  - emergency (mechanical) overspeed trip
  - low lubricating oil (l.o.) pressure trip, which should operate at approximately 25 percent below normal l.o. supply pressure.
  - low turbocharger l.o. pressure (for slow speed engines)
  - oil mist level high (on some engines)
  - emergency stop lever
  - turning gear interlock (on *all* engines).

Step 4: Stripdown inspection

The stripdown inspection is an integral part of the brake test, although ISO 3046 is weak in this area – there is only a one-line reference to post-brake-test inspection in Part 2 of the standard. A comprehensive stripdown examination, however, is an excellent way to prove running integrity, which is a key FFP criterion. It is also the best way to evaluate the particular lubrication and future wear regime of an individual engine.

Stripdown inspections help to build up your inspection experience. You will find that evaluating and diagnosing the condition of components is not a simple pass or fail exercise – it is more about interpretation and experience. As I have suggested in other chapters of this book, it is important to use a checklist. You must know what you are looking for – if you do not you may find yourself feeling able only to nod in approval at the decisions of the engine manufacturer. Diesel engines are also a good area in which to practise concise technical reporting – in particular the technique of *describing* what you find (take a quick look forward to Chapter 15, which discusses this).

The examination involves dismantling one complete cylinder liner/piston assembly (sometimes termed an 'upper line strip') and the corresponding large-end bearings and main crankshaft journal bearings, (the 'bottom-end strip'). Together, you may hear these referred to as a 'complete one-line strip'. Some manufacturers perform a borescope inspection on all cylinders before deciding which cylinder to strip down, whilst others seem to choose one almost at random. It is always worth having an input into this decision; weight your decisions based on cylinder pressures and exhaust temperatures recorded in the brake test results, making sure that you also take into account the results of the pre-test crankshaft deflection readings. Cylinders with high exhaust temperatures, and in which corresponding crank webs have 'on the limit' deflection, are good candidates for stripdown.

Figures 9.6 and 9.7 show which areas to examine and what to look for. Note the various measurement and record sheets that should be filled in – you can structure your checklist around these. I have already mentioned that the interpretation of stripdown observations is heavily reliant upon experience. There are, however, several observations which under all circumstances give cause for serious concern. Watch for these and have your non-conformance report ready. They are:

- Any *active* wear mechanism such as scoring or scuffing on the sliding surfaces of the piston, piston rings and liner. This signifies that the lubrication regime is inadequate. You must be careful not to confuse this with mild 'bedding-in wear' which is normal and actually desirable in developing a good lubrication regime.
- Exhaust valve burning or erosion of the valve seat or stem.
- Poor 'bedding' of any of the hydrodynamic rotating bearings to the extent that the white metal backing layer is exposed. This is indicative of serious misalignment.
- Signs of deep scoring or gouging of *any* of the bearing surfaces. This is a sure sign of lubricating oil contamination of some sort. It is a sorry fact that numerous diesel engines have failed on site because of this.
- Look particularly at the gudgeon pin and its bush. Gudgeon pins are a very highly loaded component with relatively tight clearances and only intermittent hydrodynamic lubrication. A lot of potential problems with oil quality or distribution will first show themselves here. Treat gudgeon pin and bush wear seriously, with less tolerance for marginal cases than you would the bottom-end or main journal bearings.

In common with other rotating equipment, diesel engine failures in service are relatively common. If this happens, the pre-commissioning works inspection reports become an important baseline to work from when trying to diagnose the reason for the failure. Accurate and

Diesel engines 279

Check valves
- stem clearance
- sealing face condition

Check the valve rotators

TDC

Valve 'cage'

Check gudgeon pin, bush clearances, and condition

Top ring position
Second ring position

Liner

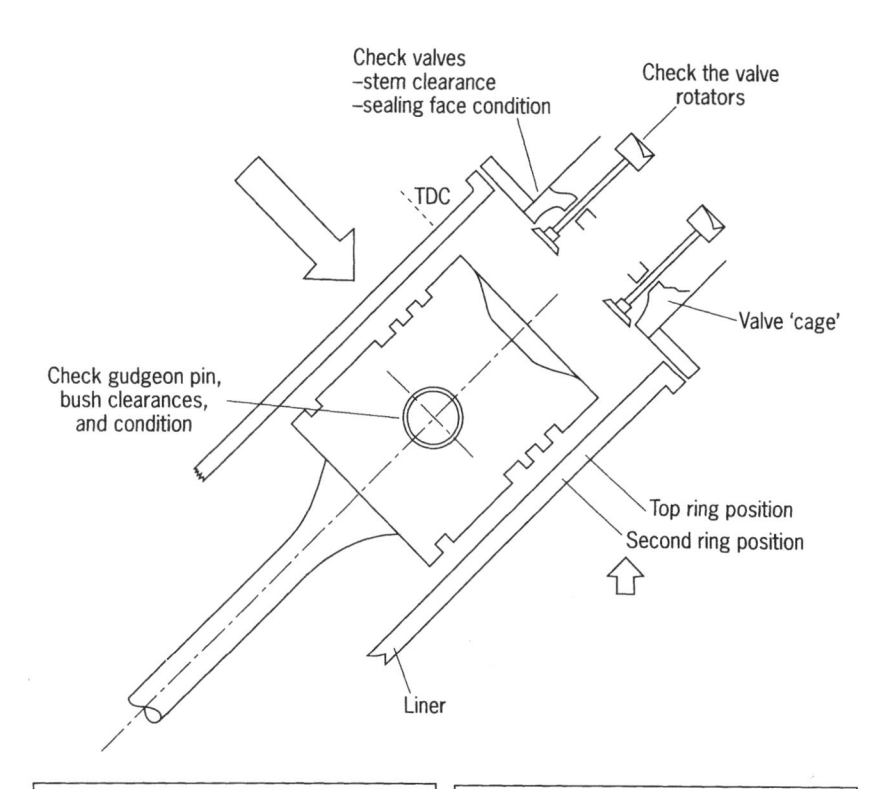

Piston, ring and liner observations

Record any:

- scratches
- microseizure
- active microseizure
- wear ridges
- scuffing
- 'cloverleaf' wear

ALSO

- check the ring gaps
- measure ring clearances
- check the ring action – watch for sticking

Measure liner wear (to 0.01mm) at

A - top ring (axial)
B - top ring (transverse)
C - 2nd ring (axial)
D - 2nd ring (transverse)
Mean dia. = (A+B+C+D)/4
Record the mean wear

During the stripdown:
- take photographs of the running surfaces
- describe accurately what you see

Fig 9.6   Diesel engine – upper line strip

280 Handbook of Mechanical Works Inspection

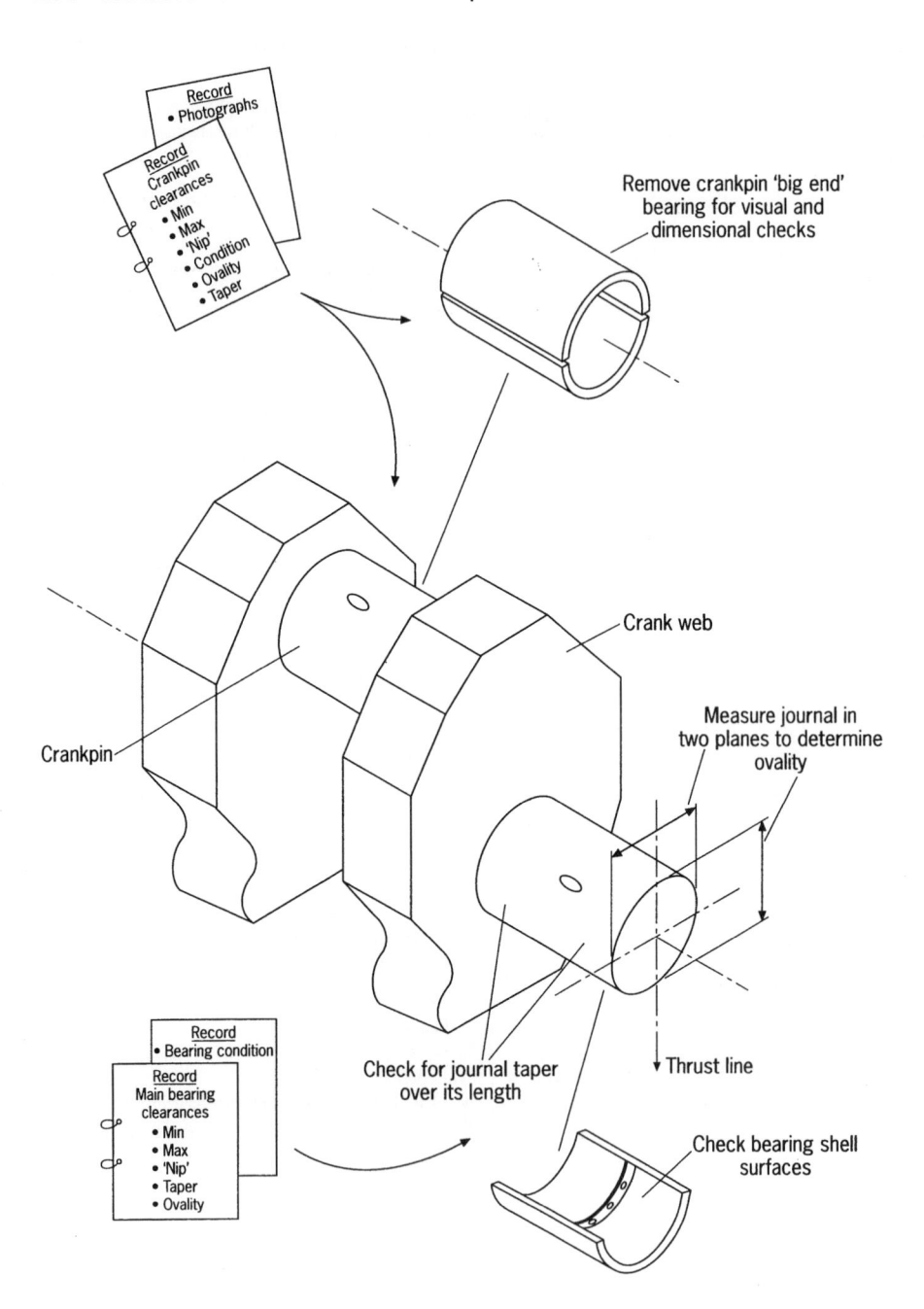

Fig 9.7 Diesel engine – bottom end strip

Common non-conformances and corrective actions : diesel engines

Non-conformance	Corrective action
Peak cylinder pressures too high, or low, or varied.	This is not unusual. The key point is to check whether these correspond with unacceptable emission analyses – this may indicate fuel rack settings are poorly adjusted. Also, feel for leaking cylinder head joints. A re-test is required.
Excessive crankshaft web deflections.	First, double-check the web deflections whilst making sure that the crankshaft journals are sitting correctly in their bearings. Then check that all main bearing shells are correctly fitted and tightened. If this is not the cause then there is a major problem, Do not run the engine until the cause has been fully investigated and rectified.
Valve seat burning or erosion.	This is most likely a fuel problem. Re-check the fuel analysis and the injection timing – but the damaged valve(s) must be replaced before a re-test.
Fretting or metal 'pick-up' (or a polished appearance) on the backs of bearings shells.	Check the serial numbers of the bearing keeps and shells to ensure they are standard size. Check the tightness of bearing studs. Do not allow random machining of the bearing keep surfaces.
Excessive vibration.	First make sure the nature of the vibration is identified – a frequency analysis will be required to help identify the source. A check of all bolt torques inside the crankcase is advisable – and then a careful re-test. If vibration continues, a more detailed investigation is required – there is unlikely to be a quick solution.
Bearing surface wear or scoring.	Don't try to solve this by replacing the worn bearing and repeating the test. The correct action is to re-measure all crankshaft web deflections and reciprocating parts alignment/clearance measurements. Repeat *all* of them, not just the offending item. Make sure that you understand the results before agreeing to a re-test. For scoring, flush out the l.o. system, check all the bearings and start again.
The engine will not reach its power rating.	Check you are not confusing test, ISO and site power ratings. Check with ISO 3046 and review the allowable measurement deviations. Note: It is, frankly, unusual for diesel engines not to reach their power rating. Make sure you really understand the calculations before issuing an NCR in this area.
Engine exceeds emission limits.	Almost always caused by poor fuel or valve timing adjustment. The best action is to request a complete adjustment log and review this carefully before the re-test. Don't forget to take all measurements again during the re-test. Check that the sfc (specific fuel consumption) does not suffer from attempts to minimise the emission levels. Retarding injection timing will increase sfc.
Any maloperation of trips or safety features.	*Do not* disconnect or 'gag' any trips to enable the brake test to continue. Stop the test immediately. If it is an excessive overspeed problem, record the terminal speed accurately to enable a designer to evaluate potential damage due to over-stressing.

descriptive reporting is therefore essential if your works inspection activity is to realize its full and proper potential. Remember the basic rule, which I have mentioned in several sections of this book – *say* what you did but *describe* what you found.

KEY POINT SUMMARY: DIESEL ENGINES

Fitness for purpose

1. The two main FFP criteria for a diesel engine are mechanical *integrity* and (brake) *performance*. Both can be verified by a programme of works inspections.
2. Engines can be classified into slow, medium or high speed types. Works inspection activities are much the same for all types.
3. The format of performance guarantees is well defined but the guarantee requirements do not form a complete statement of the fitness for purpose of an engine.

Standards

4. The predominant standard you will meet is ISO 3046. It deals with engine performance testing and consists of seven separate parts. The manufacture of a diesel engine follows general engineering standards. A classification society may be involved if the engine is built 'to class'.

Test procedures

5. During manufacture there are three important alignment checks you should witness. These are:
   - bedplate alignment – this is performed using a water level and taut wire technique
   - crankshaft (deflections) – the purpose is to ensure the main bearings are in alignment
   - reciprocating parts alignment.
6. A *brake test* is used to demonstrate an engine's performance. Oil flushing and running-in are important preliminary steps. Make sure you understand the test procedure before you witness a brake test.
7. Vibration levels are much higher then for turbomachinery. Levels of up to 50 mm/s (rms) are acceptable.
8. A *stripdown* examination is performed after the brake test. Treat the stripdown as:
   - the most important part of the test procedure
   - a diagnostic exercise (look for lubrication and running clearance problems)
   - an opportunity to practise *clear descriptive reporting*.

References

1 BS 5514 Parts 1 to 7: *Reciprocating internal combustion engines: performance.*
2 BS 2869 : Part 2: 1988. *Specification for fuel oil for agricultural and industrial engines and burners (classes A2, C1, C2, D, E, F, G and H).*

Chapter 10

Power transmission

GEARBOXES

Gearboxes range in application from small pump drives to the large and heavy-duty sets used for marine propulsion or generator drives. Gearing is a detailed technical subject involving complex design criteria and advanced manufacturing techniques. Fortunately, from an inspection viewpoint, it is not necessary to have a detailed understanding of these aspects – it is sufficient to have an overall appreciation of the technology, and to know how to apply the common test techniques.

Gearboxes do fail in use – not infrequently. Experience shows that perhaps 70 percent of mechanical failures are due to operational factors (such as external shock loading or problems with the lubricating oil supply) whilst the remaining 30 percent can be linked to 'manufacturing activities'. Because gearboxes are precision items, manufacturing and testing techniques are well developed. As with the other machinery types, we must first look at the requirements of fitness for purpose.

Fitness-for-purpose criteria

The purpose of any gearbox is to transmit drive efficiently and there are three clearly defined FFP criteria that we need to consider. They are:

- A correctly machined and aligned gear train. This will minimize the possibility of excessive wear and subsequent failure.
- Correctly balanced rotating parts, to keep vibration to acceptable levels.
- Mechanical integrity of the components, particularly of the highly stressed rotating parts and their gear teeth.

Basic technical information

Gearbox and gear train designs are carefully matched by the designer to their intended application. For general power and process plant use, the category that we are concerned with is that of high speed units which are in continuous use. These are used for turbomachinery (mainly gas turbines), diesel engines, pumps, compressors, centrifuges, and similar auxiliary equipment drives. They may act as speed reducers or increasers depending on their application. Within this broad category there are three basic types.

Spur gears

Spur gears are the simplest type of parallel-axis gears and are still used in basic non-reversing high speed gearboxes. Works testing is particularly concerned with evaluating the high noise levels that are often a feature of this design. For high speed and high reduction ratio applications, spur gears are often arranged in an epicyclic format – this results in a very compact, low weight gearbox.

Single helical gears

These gears have a single set of helical teeth. They have lower noise and vibration levels than spur gears as they are designed to have several sets of teeth in mesh at any moment in time. A key inspection issue for single helical gearboxes is successful performance of the bearings, in particular the thrust bearing necessary to absorb the residual end thrust from the gear-teeth helix.

Double helical gears

These have double helices on the main wheels in the gear train, the helix directions opposing each other to eliminate end-thrust. Alignment can be a problem – double helical gears require extreme precision in their manufacture. Works inspection is therefore oriented towards the verification of alignment.

The principles of works testing are basically similar for these three gearbox types, although there may be slight changes in emphasis of the elements of FFP. In most contracts the requirements will be clearly expressed in the content of the guarantee schedule. I have shown below a typical schedule and acceptance values that you are likely to see – in this case for a gas turbine generator drive.

The design standard	e.g. API 613
Rated input/output speeds	5200/3000 rpm
Overspeed capability	110 percent (3300 rpm)
No-load power losses	Maximum 510 kW (this is sometimes expressed as a percentage value of the input power)
Oil flow	750 litres/minute (with a tolerance of ±5 percent)
Casing vibration	VDI 2056 group T: 2.8 mm/second rms (measured as a velocity)
Shaft vibration	Input pinion 39μm Output shaft 50μm peak-to-peak (both measured as an amplitude)
Noise level	ISO 3746:97 dB(A) at 1 metre distance

In addition to these explicit criteria there are several inferred requirements which must be borne in mind. These are expressed clearly in the applicable design standard.

Specifications and standards

There are standards from the API/AGMA, BS, VDI/DIN and ISO ranges which address high-speed gearboxes. From the viewpoint of determining fitness for purpose in the manufacturers' works, we can make the situation clearer by seeing these as being in two categories: gear design standards and inspection/testing standards.

Gear design standards

Gear design standards are defined at the specification stage and relate to the *application* of the gearbox. The main items they address are design (strength) factors, tooth forms, materials, and manufacturing accuracy. As a works inspector you do not have to have a detailed understanding of all of these standards – but you do need to understand their scope, and the way in which they specify other complementary inspection and testing standards. I have listed them below:

API 613 **(1)**. This is a broad and comprehensive standard which covers design criteria and testing aspects. It has direct relevance to works inspection and you will find it a useful and practical standard. It is used in many industries (not just refinery service as the title suggests). For further technical details, API 613 cross-references the American Gear Manufacturers Association (AGMA) range of standards.

ISO 1328 **(2)** is a detailed standard on manufacturing tolerances, using ISO Q-grade classes of accuracy. There are a further ten or so ISO standards which impact on gearing design but you are unlikely to meet them.

BS 1807 **(3)**: this covers gears for marine propulsion drives. It is concerned specifically with tooth design factors and shock loading considerations.

Gear inspection/testing standards

API 613 is a good broad guide for works inspectors but you will also find the ISO, VDI and BS ranges very useful in providing detailed information for use during the works inspection. The main contents of these 'supplementary' standards are methods of determining and measuring acceptable levels of noise and vibration.

VDI 2056 **(4)** covers criteria for assessing mechanical vibration of machines. Be careful to note that it is really only applicable to the vibration of gearbox bearing housings and casings – not the shafts. This is an easy-to-follow standard – machinery is divided into six application 'groups' with gearboxes clearly defined as included in group T. Vibration *velocity* (rms) is the measured parameter. Acceptance levels are clearly identified.

ISO 2372 (equivalent to BS 4675) **(5)** covers a similar scope to VDI 2056 but takes a different technical approach.

ISO 8579 (equivalent to BS 7676) **(6)** is in two parts, covering noise and vibration levels. It provides good coverage but is not in such common use as other ISO and VDI standards.

ISO 7919/1 (equivalent to BS 6749 Part 1) **(7)**. This relates specifically to the technique of measuring shaft vibration and is quite complex. It is perhaps best to view this standard as a complement to others but you may find it being used alone for some large gear units. Expect to see it specified also for large turbomachinery prime movers where the test procedure needs to identify several modes of vibration within the service speed range. I have provided further details in Chapters 7 and 8 covering gas turbines and steam turbines – the principles are the same. API 613 provides a simpler method of evaluating gearbox shaft vibration using amplitude as the measured parameter.

ISO 3746 **(8)** and AP1 615 **(9)** are relevant noise standards. Other parts of test procedures which are addressed by standards are dynamic balancing and tooth contact tests. You will find that API 613 explains these in a straightforward way.

Inspection and test plans (ITPs)

An ITP for a gearbox follows a well defined format, the activities included being closely linked to the verification of the three main FFP criteria. A typical example is shown in Fig. 10.1.

Activity	Standard/Requirement
• Technical datasheets	API 613 requirements
• Material test certificates	
– all rotating components	EN 10 204 (3.1B)
– bearings	EN 10 204 (3.1B)
– gearcase	EN 10 204 (2.2)
• NDT of rotating parts	100% US test (ASTM A388)
	100% MPI test (ASTM A275)
• NDT of bearings	US test on white metal adhesion
• NDT of gearcase	MPI on welds and critical areas
• Runout measurements	Maximum runout 8μm.
	After installation combined mechanical/electrical 'slow roll' runout $\leq$ 25% of maximum vibration amplitude.
• Dynamic balancing	API 613 (Section 2.6.2) or ISO 1940 G6.3
• Tooth contact checks	> 85% contact along pitch line
	> 75% contact on pitch height
• Running test and overspeed	120% overspeed test
• Vibration test	VDI 2056 or API 613
• Noise test	ISO 3746 : 97 dB(A) at 1 metre
• Certificate of compliance	–

Fig 10.1 Gearbox ITP content

Test procedures and techniques

Most gearbox works inspections involve witnessing a no-load running test followed by a stripdown examination. It is also common for earlier dynamic balancing and sometimes tooth contact checks to form part of the witnessing schedule. Taken in order of manufacture, the first of these is the dynamic balancing test.

Dynamic balancing test

In most gearbox designs, dynamic balancing is carried out *after assembly* of the gear wheels and pinions to their respective shafts. The assembly is then commonly referred to as the gear 'rotor'. Dynamic balancing is carried out with the rotor mounted by its journals in a specialized machine. The rotor is spun at up to its rated speed and multiphase sensors mounted in the bearing housings sense the unbalanced forces, relaying the values to a suitable instrument display. The purpose of dynamic balancing is to reduce the residual unbalance to a level that will ensure the vibration characteristics of the assembled gearbox are acceptable. In practice, gearbox manufacturers will attempt to reduce this unbalance to zero, or as low a level as practically possible. What can be achieved depends mainly on the size of the gear rotor and the skill of the manufacturer.

You can find a useful first approximation for maximum permissible unbalance in API 613 (see Fig. 10.2). It is important not to confuse the units of unbalance. The correct compound unit is g mm (gram millimetres), i.e. an unbalanced mass operating at an effective radius from the rotational axis. Nearly all modern balancing machines will provide the unbalance reading on a digital readout, so no manual calculations are required. Make sure you check the calibration certificate. Any residual unbalance is corrected (after stopping the rotor) by adding weights into threaded holes. The test is then repeated to check the results.

Some useful points to remember when you are witnessing balancing tests are:

- Check the test speed. It is normally the rated speed of the rotor but may be less for high speed units, owing to the limitations of the balancing machine. Don't test it at above the rated speed.
- The manufacturer should be attempting to obtain zero unbalance. A rotor which just achieves the 'acceptable' levels can likely be improved. If not, it is wise to pay particular attention to vibration measurements during the subsequent gearbox running test. Unbalanced rotors can cause a forcing frequency which may reduce one of the critical speeds, bringing it near the rated speed. Be wary.
- When recording the results of a dynamic balance test you must be accurate in the way that you record the results. Figure 10.2 shows the API 613 method. Check particularly that the correction planes (i.e. where any balance weights are added) are actually those shown on the test result sheets.

Power transmission 291

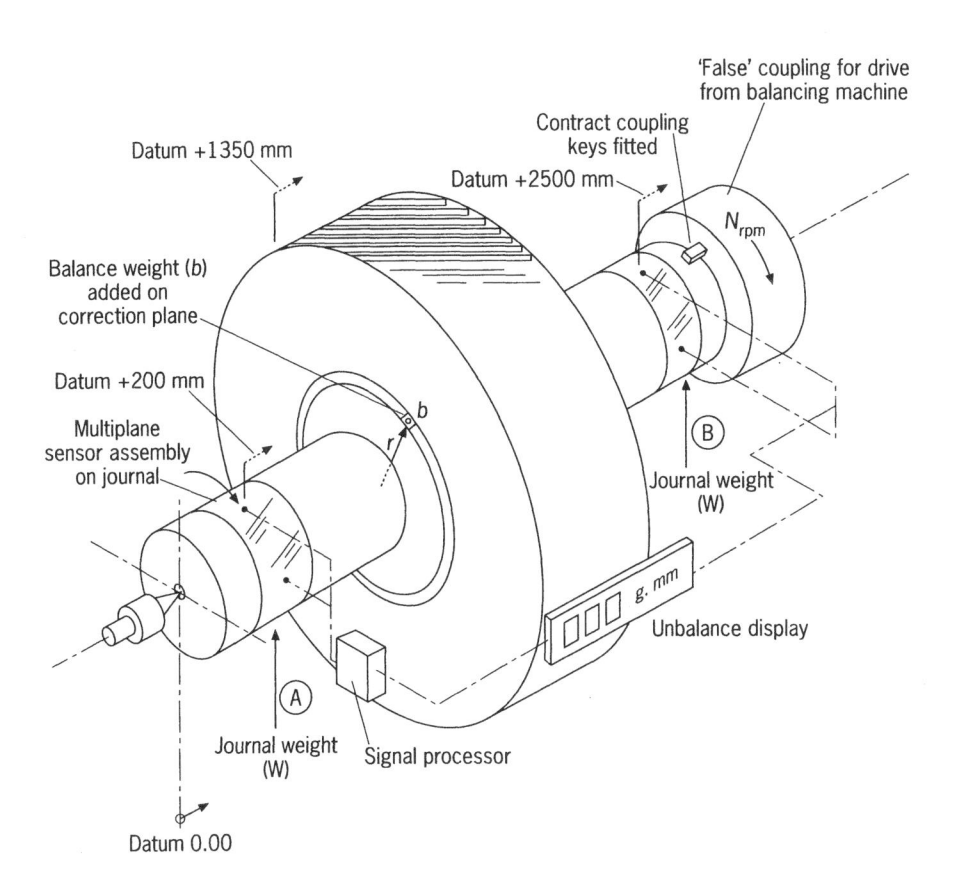

Record the results like this:
High-speed shaft. Rated (N) = rpm. Rotation anticlockwise (ACW) seen from coupling.

	Journal A	Correction plane	Journal B	
Location from datum 0.00	200mm	1350mm	2500mm	
Static loading (kg)	W	–	W	Specimen calculation (API 613)
Balance wt added	–	b	–	$\hat{U} = 6350\ W/N$
Residual unbalance (U) g.mm	U_A	–	U_B	

Fig 10.2 Dynamic balance of a gear rotor

- A practical point: check if there are any shaft keys or shaft collars which will be fitted to the rotor but were not there during the balance test. Extra items such as these should be either fitted, or allowed for in the unbalance calculation.
- You will see that some testing standards require a sensitivity check on the balancing machine – by adding known weights to the rotor and checking that the results tally. Frankly, a modern and properly calibrated balancing machine shouldn't need this, but follow the standard, if it is specified.

Contact checks

Contact checks are a simple but effective method of checking the meshing of a gear train. The results can provide a lot of information about the machined accuracy of the gear teeth, and are also a measure of the relative alignment of the shafts. The test consists of applying a layer of 'engineers blue' colour transfer compound to the teeth of one gear of each meshing set and then rotating the gears in mesh. The colour transfer then shows the pattern of contact across each gear tooth.

These are frequently two stages of contact checks. Contact stand checks are carried out during the early assembly stage of the gearbox (see Fig. 10.3). The stand should be accurately constructed with gear centrelines accurately positioned to within 25μm. It is not very common for these tests to be witnessed by clients' inspectors, unless the gears are of a 'one-off' or particularly specialized design. The results should be recorded, though, and available for review. The best way to record the contact pattern is to lift off the blue compound using cellulose tape, transferring the print to a sheet of white paper. You can also take photographs. 'In-situ' contact checks are carried out *after* the gear trains are mounted in the gearcase. The technique is the same.

In order to obtain meaningful results when doing a tooth contact check it is necessary that some load be placed on the teeth – this helps to ensure that clearances and alignments adopt their 'under load' condition. This is particularly important with double helical gear trains where manufacturing and alignment accuracy is critical. The most common method is to apply a drag torque to the output shaft using a band-brake. There is an extreme version of this test where the full load torque is applied, but you will normally only see this on large heavy-duty gear sets. For gear units in which the shafts are mounted on rolling element bearings it is necessary to remove any 'end float' before

Power transmission 293

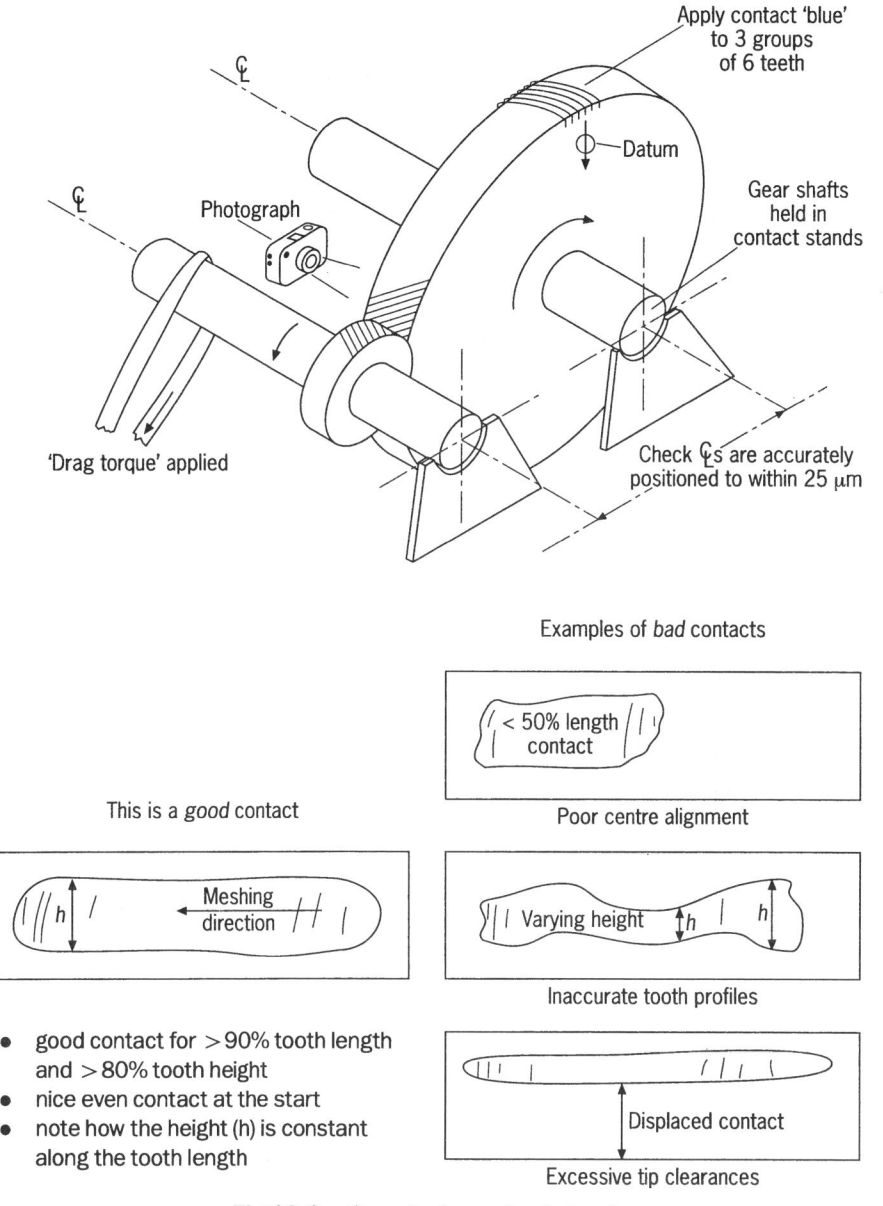

- Apply contact 'blue' to 3 groups of 6 teeth
- Datum
- Gear shafts held in contact stands
- Photograph
- 'Drag torque' applied
- Check ℄s are accurately positioned to within 25 μm

This is a *good* contact

- good contact for > 90% tooth length and > 80% tooth height
- nice even contact at the start
- note how the height (h) is constant along the tooth length

Examples of *bad* contacts

< 50% length contact

Poor centre alignment

Meshing direction

Varying height

Inaccurate tooth profiles

Displaced contact

Excessive tip clearances

Fig 10.3   Gear train contact checks

performing contact checks. Some useful points to check when witnessing contact checks (see also Fig. 10.3) are:

- Make sure the gears are dry and degreased before applying the blue compound. You can't carry out proper contact checks with oil in the gearcase.
- Do not apply the blue compound randomly. Three groups (at 120 degrees spacing) of six teeth on the largest gear of the meshing train is sufficient. The blue should be rubbed on (giving a thin 20μm film) rather than painted. During the test, the blue is transferred from the wheel to the pinion, and then back onto the wheel. Contact markings are then taken from those teeth *between* the original blued teeth.
- When recording the results, refer to the position of the teeth in question. Some gear sets have a so-called 'hunting tooth combination', where teeth change their meshing pattern with each revolution. Gear wheels often have a zero datum marking and the teeth are numbered clockwise from this.
- Pay particular attention to double helical gears – and make sure contact tests are carried out on both sets of helices.
- Most gearboxes are non-reversible, but if you encounter one that is reversible, it is necessary to carry out contact checks on both sides of the tooth flanks.
- Remember the need for a drag torque in order to obtain meaningful results.

Running tests

The mechanical running test is the key proving step for the gearbox and is the most commonly specified witness point. Although some manufacturers have the facility to perform the running test at full or part load, this is becoming less common. Most purchasers rely on a no-load running test. You may also see this referred to as a 'proof' test (although, strictly, it isn't). The key objectives of the mechanical running test are to check that the oil flows, vibration, and noise levels produced by the gearbox are within the acceptance levels. Remember that these levels are 'guarantee' items, so as an inspector you need a very clear view of these requirements and the ways in which they will be verified during the test. First, review the test procedure – it normally contains the following steps (refer to Fig. 10.4).

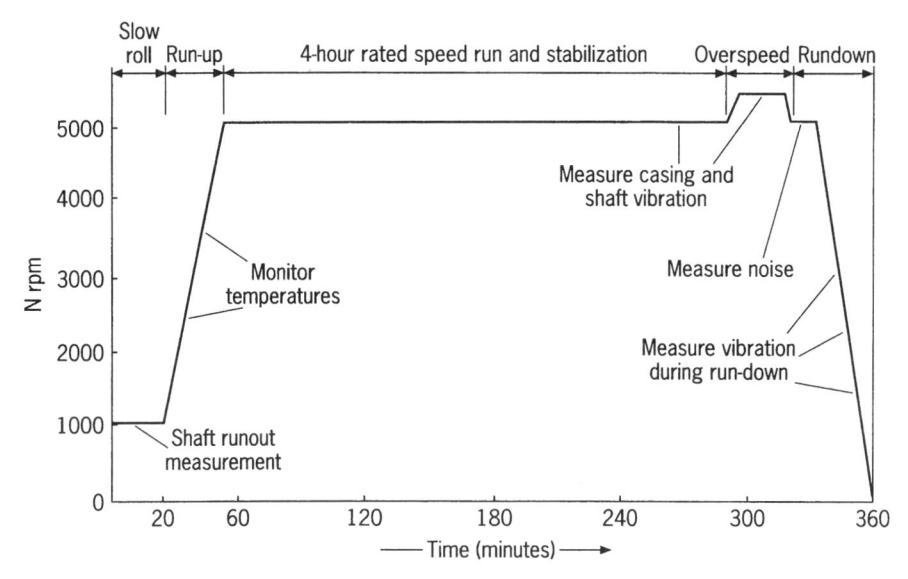

Fig 10.4 Gearbox no-load running test

Step 1: Slow speed run

Run the gear unit at slow speed; shaft runout is measured at this stage. This gives a value for slow-roll mechanical and electrical runout to be used when interpreting shaft vibration measurements. Check the lubricating oil (l.o.) supply. Watch for oil leaks.

Step 2: Run-up

Gradually increase to 100 percent rated speed, taking l.o. pressure and temperature readings at least every ten minutes.

Step 3: Rated speed run

Maintain at 100 percent rated speed for four hours. Continue to record l.o. pressures and temperatures at regular intervals. Pay careful attention to the l.o. discharge (or thermocouple) temperatures of the journal and thrust bearings – this will help indicate the thermal balance of the gearbox and show which bearings are the most highly loaded. Combined journal/thrust bearings should be monitored very closely, particularly if they are fed by a single l.o. distribution pipe. The bearings and their operating temperatures must be correctly identified. If it is not clear (or won't be if someone reviews your report in two years' time), give each bearing a number and record it on the general arrangement drawing.

The maximum *surface* operating temperature of a proprietary white metal gearbox bearing is around 125 °C. You need to allow at least 15 °C for measurement inaccuracies and margin, so a monitored bearing oil *drain* temperature of 110 °C is a cause for concern. Watch for any bearing temperatures that vary for no apparent reason. As a general rule, be very vigilant of bearing temperatures during a no-load running test. The situation will not improve with load – it will *always* get worse. Address any problems now.

Vibration should be measured once the l.o. temperatures and pressures have stabilized. The general principles follow those for other rotating equipment – however, there are some specific practices and techniques relating to gearboxes which are important. Looking back to earlier in this chapter, we can see that different standards are used for vibration assessment of the gearcase and the shafts.

For the gearcase, VDI 2056 uses the parameter of 'vibration intensity' – defined as a velocity (rms) value in mm/second. This definition is convenient for inspection purposes because it approximates vibration at multiple frequencies down to a single V_{rms} reading. This makes checking against the acceptance criteria relatively straightforward as long as the correct measuring instrument is used. Check that you are using a velocity-sensing mechanical/electrical transducer with a square-law characteristic. This will give a direct rms readout, independent of frequency. If the instrument performs mechanical displacement measurement only, then you need to make multiple measurements across the frequency range – so you will need a frequency analysis as well. Gearboxes fall under Machinery Group T of VDI 2056, so for a typical gas turbine/generator gearbox (50 MW load capability), look for around 2.8 mm/second as an acceptable maximum vibration level (check Fig. 7 of this standard if you are in any doubt).

Shaft vibration is not covered by VDI 2056. API 613 or ISO 7919/BS 6749 use maximum peak-to-peak displacement (amplitude, *s*) as the measured parameter. Vibrations are sensed across the frequency range (typically 0–1000 Hz). The calculation is very simple – API 613 uses this formula:

Maximum peak-to-peak displacement $(s) = 25.4 \sqrt{(12\,000/N)}$ μm
where N is measured in rpm.

This is subject to a maximum of 50 μm, which corresponds to that for a 3000 rpm synchronous speed drive. Before comparing readings with the acceptable limit it is necessary to subtract the slow roll combined mechanical and electrical runout, to get a true reading.

The location of the vibration sensors on the gearbox is important. On the gearcase, sensors should be mounted at those points at which vibration energy is transmitted to the mountings. This is normally taken as at the mounting feet or flanges. For bearing vibration, non-contacting shaft vibration sensors are usually located on the horizontal or vertical

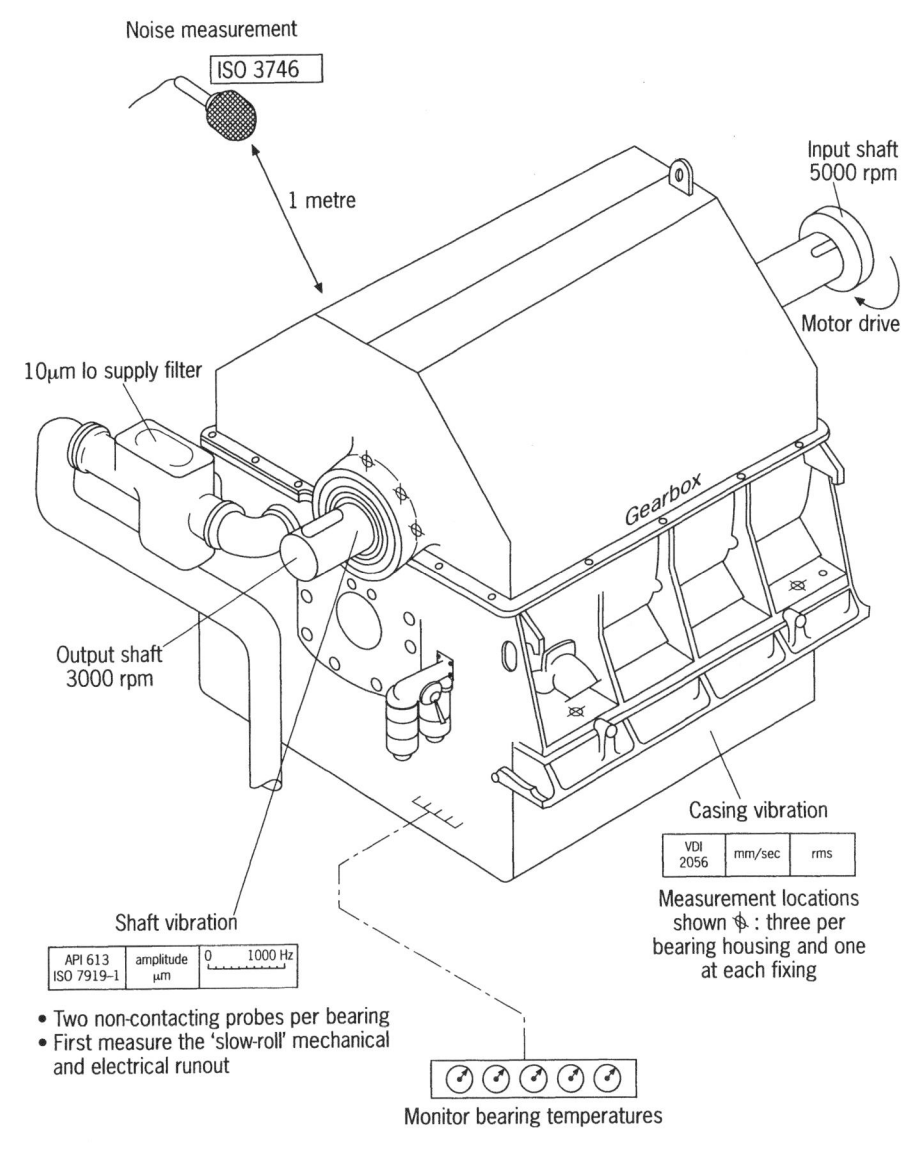

Fig 10.5 Gearbox test: monitoring

axes in the bearing housings, as near the journal location as is possible. Figure 10.5 shows sensor location for a single helical, single reduction gearbox.

Step 4: Overspeed

Most gearboxes are subject to an overspeed test. This is to verify the strength and integrity of the rotating parts, and to make sure that there is not a critical speed region too near the continuous rated speed. A typical overspeed level is 120 percent rated speed. The speed should be increased gradually from rated speed over a minimum of 15 minutes, with regular temperature and pressure monitoring. Maximum 120 percent speed is then maintained for a continuous period of three minutes. Vibration *can* be measured at the overspeed condition, (although this is not always specified) but expect the readings to be up to 40–50 percent higher than at rated speed. Any greater is cause for concern. Shaft vibration is also recorded during the rundown, towards the completion of the running test.

Step 5: Noise

Speed is reduced gradually back to rated speed to enable noise tests to be carried out. Noise level is a guarantee parameter – often measured in accordance with the requirements of ISO 3746 or API 615. Most guarantees will simply specify an overall value in decibels dB(A) – look for a level of 97 dB(A) at 1 m distance for an average set (refer to Fig. 10.5). Remember that the dB scale is not linear though, it is logarithmic. Occasionally, guarantees may require levels at specific frequencies. Here, you have to scan the frequency range in octave bands (approximately 31 Hz to 16 000 Hz) and compare the results with those maximum values specified. Frequency distributions vary significantly for different types of gears and hence are often closely specified for very specialized applications.

Step 6: Stripdown

The final step is the stripdown. Some manufacturers do not do this unless it is specifically requested (API 613 requires it). You should take care to make a thorough stripdown examination. For gearboxes the emphasis is on visual observation of the *condition* of the rotating components, rather than looking at design features inside the gearcase.

A good checklist is a necessity if you want to carry out the inspection properly. It is easy to have a cursory look inside the gearbox and proclaim that 'everything seems all right'. Be forewarned that many of the common manufacturing defects and problems that can later result in failure of the gearbox will *not* be identified by a cursory inspection. Make a thorough and structured examination. Use your engineering knowledge – be diagnostic. Here is the basis of a stripdown check list:

- Firstly, keep your fingers *out* when the gears are moving.
- Check the integrity of every tooth. During the first complete revolution, inspect the 'top corners' of the teeth – looking for chips or small visible cracks at the tooth tips.
- Rotate the gears a second full revolution, this time checking only the root radii at the bottom of the teeth. Look carefully for ridges due to pitch errors (particularly on double helical designs); any problems that do exist may not necessarily be visible all around the gear. Take your time.
- *Tooth contact*. It is possible to draw useful conclusions on tooth contact by observing the visual signs. On a well-cut set of nitrided gears, the 'as-machined' surface finish of the teeth will normally be 0.6 to 0.8 μR_a, reducing to better than 0.4 μR_a after a period of running – a comparator gauge is always a useful tool to carry. During a five or six hour running test, provided the unit has not been repeatedly tested beforehand, you should find that the contact surface will dull slightly. Under a good light, this will give a good visible indication of the tooth contact pattern. It is essential to identify any 'scuffing' of the teeth – this can indicate a poor lubrication regime. Scuffing is characterized by the existence of dark bands on the tooth surface. They are often wavy, with a more discontinuous appearance than that of a normal contact surface. Look particularly for scuffing where sliding velocities are the highest. If you do find scuffing, look at the affected areas around the gear to see whether there has been any more extreme oil-film breakdown causing microseizures. Microseizure is characterized by a very dull, fibrous-looking surface – it may be visibly speckled and feel rough to the touch. This is normally much more than 'running-in' wear and is often accompanied by other lubrication problems, usually in the bearings.
- *Examination of bearings*. Bearings are the most susceptible components of a gear unit and warrant a detailed examination during the stripdown inspection. Once the gearcase top half is lifted it is a relatively straightforward task to remove the journal bearing top shells and then to 'turn-out' the lower shells using a special tool. Thrust bearing pads may be more difficult to access as they are often

enclosed in a ring or cage but it is still a straightforward maintenance task. There are four main problems that you should look for. Between them, they are responsible for just about all bearing failures and are often a precursor to further consequential damage to the gear teeth.

- *Scoring of the surfaces.* This can occur on journals or thrust bearings. Look for it first on the white metal surfaces – it can be detected visually and by rubbing a fingernail along the surface at 90 degrees to the direction of rotation. Scoring is caused by lubricating oil contamination and in most cases the contaminant is welding slag or debris that has not been properly flushed out of the gearcase. Check the oil filter element – it should have a maximum mesh size of 10 μm but don't be misled into thinking a clean filter means that there is definitely no contamination in the bearings. I have seen this disproved many times. If the white metal is scored, the shells or pads can be replaced. You should never accept a scored journal shaft or thrust collar – despite all the arguments it *will not* get better in service.
- *Pinholing of the white metal.* You may find small pinholes in the surface of the white metal, more often on thrust bearing pads than journal shells. This can range from a few isolated pores to a 'honeycomb' effect in extreme cases. If it occurs in isolation i.e. without other problems such as overheating or discolouration, it is almost certainly caused by a manufacturing fault in the bearing material. Repeat the running test with new shells.
- *Uneven wear.* This is easiest to see in thrust bearing pads. Check the pad thickness with a micrometer and compare with the drawing dimensions. Check that the pad tilting movement is not constrained, and that the correct pads are fitted. The origin of this type of problem normally lies *only* with the thrust bearing assembly and is not indicative of other problems elsewhere in the gearbox.
- *Overheating and discolouration.* Assuming that the journal and thrust bearings are correctly sized for their design loadings, overheating is indicative of the wrong lubrication oil film thickness. On journal bearings this is affected by diametral clearances (check the drawing again – and the shimming arrangements) whilst thrust bearing film thickness is a function of sliding speed and tilt angle. Make sure you also check the lubricating oil viscosity – the oil should have the specified 'ISO 3448 Vg' rating.

Common non-conformances and corrective actions : gearboxes

Non-conformance	Corrective action
High vibration *velocities* at casing or bearing locations	First strip down and check for : - broken gear teeth - loose mountings - cracks in the gearcase (these reduce stiffness). If no obvious causes are found, disassemble the unit and re-check the dynamic balance of rotating parts, and at the same time review critical speed calculation. You can ask to see the designer's no-load undamped response analysis. Remember: do not accept excessive vibration under a concession. Vibration will only get worse.
Irregular tooth contact pattern (after running test)	If there is clear irregularity or discontinuity in the pattern after 4–5 hours operation then there is a serious problem. Gears (strictly) are not designed to need a 'running-in' period. There is not much that can usually be done in-situ. The correct solution is to strip the unit, measure the tooth profiles and check the centreline alignments.
Discovery of broken or cracked teeth	You must remove the broken gears as a pair and carry out full surface NDT to quantify the damage. A contact stand test will help diagnosis. On new gear units, broken teeth should not be repaired in-situ. Be wary of proposals to repair broken teeth. Replacement is best. Do not just replace the gear pair without looking for other causes of the failure (alignment, etc). Then retest.
Persistently high (>110 °C) bearing temperatures	Check the gearbox loadings and thermal balance. It is possible to increase l.o. flow to hot bearings by changing the distribution pipes or orifice sizes – but be careful of the effect on the other bearings. Re-test, paying particular attention to the new temperature distribution.
High shaft vibration *displacement* readings (from non-contacting shaft probe)	First, check the 'slow roll' runout again using the same probe. It must be less than 25 percent of maximum allowed peak-to-peak displacement or 6 μm (whichever is the lower). Check how the transducers are mounted – the best way is in tapped holes in the bearing housing, if not, they will be mounted on brackets. Try the test again, this time using a separate instrument to see if the *brackets* are vibrating and giving a false reading. Check the surface of the shaft under the sensor. If the surface finish is worse than 0.4 μm R_a, it may be causing measurement errors. Finally, check there is not vibration coming from nearby machines. *Then repeat the test.* Still not acceptable? Do a stripdown.

COUPLINGS

The rotating shaft and coupling which transmits the drive to the gearbox are key components of the drive train. The inspection and testing of these items are important, although occasionally neglected, activities in ensuring the future trouble-free performance of the gearbox.

Fitness-for-purpose criteria

There are two fundamental FFP criteria for drive shafts and couplings.

- The mechanical integrity of the items must be equal to that of the gearbox rotating components.
- The design and installation must be such that they do not exert any harmful forces on the gearbox due to parallel offset, angular misalignment or axial displacement. This is particularly relevant for high-speed couplings (above 5000 rpm) – these must have a truly neutral effect on gearbox tooth loadings over the full operating speed and temperature range.

The simplest shafts or couplings are of the rigid type; a solid shaft connected to the driver and gearbox by bolted flanges. The types that we are more interested in are the flexible diaphragm type or limited end-float types of coupling. These are in common use for gas turbine and high speed diesel gearbox drives. They are flexible in the axial and radial (lateral) directions and need careful design matching and installation.

Acceptance guarantees for high speed flexible couplings should take the following form:

The design standard	e.g. API 671
Rated torque	Stated in Nm (Newton metres)
Overspeed capability	Typically 110 percent (may be higher for turbine drivers)
Axial dimensions*	Unloaded length between flange faces, static, at ambient temperature
Balancing grade	To API 671
Axial stiffness*	Typically 2 to 4 kN/m over the operating speed and temperature range
Lateral characteristic*	The ability to tolerate a specified amount of lateral misalignment per metre length

You may find that some contract specifications do not contain all of these guarantee parameters. Commonly, the three criteria shown (*) may not appear in contractors' or users' specifications – you should find however that they *are* included in the prime mover or gearbox manufacturer's specification to the coupling manufacturer. Be aware that they exist and be prepared to review them during the works inspection.

Specifications and standards

The main coupling standard you will meet is API 671 *Special purpose couplings for refinery service* **(10)**. This is orientated towards high speed gas turbine drives and, like API 613, has many applications in power plant, marine, and general industrial use. It provides good general coverage of design, manufacture, and testing. You should find the balancing information included in API 671 particularly useful – it is clearly explained and easy to understand. There are also some good engineering details on tolerances, fits, and mounting arrangements. Use appendix A of the standard as a general checklist of technical data related to coupling design.

AGMA 510 covers bores and keyways in couplings. It is predominantly a design standard but does contain some useful qualifying information (a concept I introduced in Chapter 2) which you may find useful.

The most frequently used balancing methodology and limits are those included in API 671. You may also see ISO 5406 *Mechanical balancing of flexible rotors* (equivalent to BS 5265 Part 2) specified. Note that ISO 1940/1 (BS 6861 Part 1), which relates to rigid rotors, is not normally used.

Inspection and test plans for shafts and couplings follow broadly the content used for gearboxes introduced earlier in this chapter. Remember that the material traceability and NDT requirements are equally important.

Test procedures and techniques

Ideally, all gearboxes would undergo their works running test when fitted with their contract shaft or coupling. This is termed a 'string test' and it is the best way to assess the overall vibration characteristics of the assembly as it will be installed on site. In practice this does not happen very often and whereas there are some cases where a coupling 'idling adapter' is fitted to allow both to be tested together you are more likely to be called to witness a separate test of the shaft or coupling. There are two discrete aspects to this:

Design check

Compared to most equipment that you may be called to inspect, shafts and couplings are relatively simple components. They incorporate a number of important, but definable, principles of material choice, statics, and dynamics. If you want to be an *effective inspector* you will benefit from adding to your design knowledge.

Here is the basic design check (refer to Fig. 10.6).

- *Check the unloaded length*: The coupling manufacturer will have *either* an installation drawing, showing the available space between the prime mover output flange and the gearbox input flange *or* a clearly specified 'cold' (ambient temperature) unloaded coupling length. Check the drawing or specification tolerance for this dimension. Then check the actual dimension of the coupling. It will normally be slightly shorter than the available 'between flange' length, to allow for cold pre-stretch to be incorporated. Be careful to:

  - Use a large vernier calliper – a ruler is not accurate enough.
  - Measure the length at three locations around the flange to check if the flanges are parallel.
  - Mount the coupling properly on wooden supports or vee-blocks when taking the measurements, to minimize any errors due to distortion on long (>1m in length) couplings.

- *Estimate the thermal expansion.* Using the specified operating temperature (T) of the coupling, you can calculate the likely thermal expansion using $\partial l = l \propto (T - T_{ambient})$ where $\propto$ is the coefficient of thermal expansion.

- *Estimate the axial tension in the coupling.* Do this when it is stretched to the cold pre-stretch condition. The manufacturer should have a load-v-extension design characteristic to enable you to do this. You cannot calculate this accurately yourself unless you are very familiar with plate-spring characteristics (and even then it is difficult). Make sure this pre-stretch tension is within the capability of the gearbox and prime mover thrust bearings. Look for a factor of safety of at least three at this stage, because dynamic effects will reduce it once the coupling starts to rotate (refer to Fig. 10.6 again).

- Find out the axial natural frequency (ANF) of the coupling. The

Power transmission 305

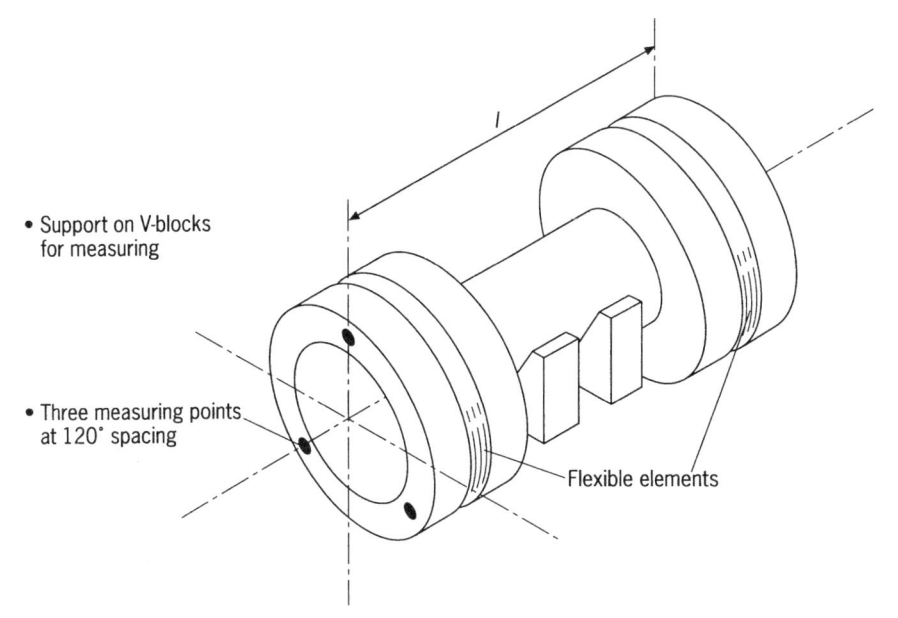

- Support on V-blocks for measuring
- Three measuring points at 120° spacing

Flexible elements

Follow these steps:
1. Check the unloaded length (at three positions)
2. Calculate the thermal expansion: $\delta l = l \propto \delta t$
3. Estimate the 'cold' tension on the gearbox using the coupling characteristics

Check these loads against the gearbox thrust bearing capability

Dynamic
Static

Extension (mm)
δl

Note the 3 'states' of the coupling:

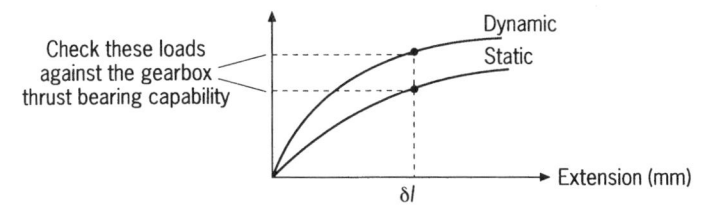

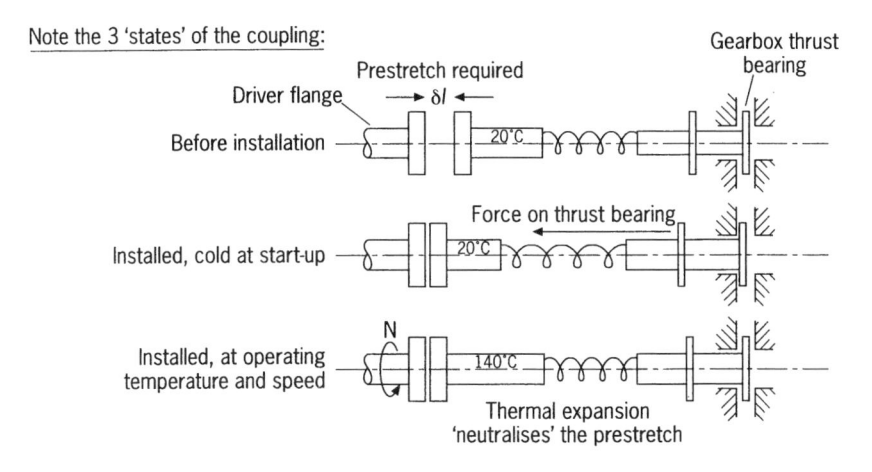

Driver flange — Prestretch required δl — Gearbox thrust bearing

Before installation — 20°C

Installed, cold at start-up — 20°C — Force on thrust bearing

Installed, at operating temperature and speed — 140°C — Thermal expansion 'neutralises' the prestretch

Fig 10.6 High speed couplings – a simple design check

manufacturer should have data showing how this varies with axial displacement. Check that the ANF is greater than 210 percent full speed – this gives a sufficient margin to avoid reasonance in use.

Record these basic design check criteria in your inspection report. Say what you did then *describe* what you found. This will help your client.

Balance

Couplings have specific runout measurement and balancing requirements. These are precise technical operations which must be performed in the correct order.

- *Runout measurement.* It is not sufficient simply to measure total indicated runout (TIR) The correct method is first to measure ovality. The calculated distance between the centre of rotation and the geometric centre of the coupling must be within 1μm per 25 mm diameter. This is a very tight tolerance. Do not accept balancing test results until you have first seen this runout properly verified.
- *Balancing technique.* Strictly, API 671 requires that all the components of a coupling should be balanced separately before assembly. Follow these steps to ensure compliance:

  –Balance each individual component to a maximum unbalance of

  $$U \text{ (g mm)} = \frac{6350 \ W\text{(kg)}}{N\text{(rpm)}}$$

  where W = weight of component. Remember that for a high speed coupling (> 5000 rpm) you should do a 'double-check' using trial balance weights (see API 671 appendix F)

  – Assemble the coupling (with match-marks) then check for an assembled maximum unbalance of

  $$U \text{ (g mm)} = \frac{63 \ 500 \ W\text{(kg)}}{N\text{(rpm)}}$$

  Note that this is ten times the allowable unbalance of the individual components.

  – If the residual unbalance is higher than this, the coupling must be disassembled and the individual components rebalanced. It is not acceptable to 'trim balance' the assembled coupling.

Common non-conformances and corrective action : couplings

Non conformance	Corrective action
Coupling is too long	This can often be rectified by machining the end flange, after disassembling the coupling. Don't machine off more than 1–2 mm thickness (or 10 percent of flange thickness), or the flange will be weakened. The reassembled coup-ling must be balanced again. Make sure the 'as-built' drawing is amended.
Coupling is too short	Reject it. It needs to be re-manufactured. Couplings which are too short cause excessive axial tension when tightened to pre-stretch length. The manufacturer may argue that the driver or gearbox can be repositioned to accept an undersized coupling. Don't accept this unless it is confirmed by the contractor for the drive 'string' – there may be other aspects to consider.
Excessive out-of-balance forces	The coupling must be rebalanced to well within the specification limits. It is not wise to issue a concession. An out-of-balance coupling or shaft can cause a gearbox failure.

KEY POINT SUMMARY: POWER TRANSMISSION

Gearboxes

1. Although gearing is a complex subject, inspectors only need an *appreciation* of design.
2. Make sure you *do* understand the tests for mechanical integrity, vibration and noise.
3. API 613 is a good 'broad' gearing standard. It references the AGMA range of detailed gear design standards.
4. There is a correct way to perform tooth contact checks.
5. There are five very well defined steps to the no-load running test.
6. Make a structured examination during the stripdown test. Having a procedure to work to helps you record observations which are useful.
7. Vibration nearly always gets *worse* in use, not better.

Shafts and couplings

8. An inspector can do a basic design check. This is *effective works inspection*.
9. Don't hesitate to reject couplings which are too short or out of balance.

References

1. API 613: 1988. *Special purpose gear units for refinery services.* American Petroleum Institute.
2. ISO 1328: 1975. *Parallel involute gears – ISO system of accuracy.* This is a related standard to BS 436 Parts 1 to 3, *Spur and helical gears.*
3. BS 1807: 1988. *Specification for marine main propulsion gears and similar drives: metric module.*
4. VDI 2056 *Criteria for assessing mechanical vibration of machines.* Verin Deutscher Ingenieure.
5. BS 4675 Part 1: 1986. *Basis for specifying evaluation standards for rotating machines with operating speeds from 10 to 200 revolutions per second.*
6. BS 7676 Part 1: 1993. *Determination of airborne sound power levels emitted by gear units.*
   BS 7676 Part 2: 1993. *Determination of mechanical vibrations of gear units during acceptance testing.*
7. BS 6749 Part 1: *Measurement and evaluation of vibration on rotating shafts – Guide to general principles.*
8. ISO 3746: equivalent to BS 4196 Part 6: 1986. *Survey method for determination of sound power levels of noise sources.*
9. API 615: 1987. *Sound control of mechanical equipment for refinery services.* American Petroleum Institute.
10. API 671: 1990. *Special purpose couplings for refinery service.*

Chapter 11

Fluid systems

CENTRIFUGAL PUMPS

Experience shows that pumps fail more frequently, and generally cause more problems, than other components in a fluid circuit. Pump tests and inspections are therefore an important part of a good inspection strategy. There are several hundred identifiable types of pump design tailored for varying volume throughputs and delivery heads, and including many specialized designs for specific fluid applications. The most common type, accounting for perhaps 80 percent of fluid transfer applications, is the broad centrifugal pump category. We will concentrate on this type.

Fitness-for-purpose criteria

The fitness for purpose of a pump is predominantly to do with its ability to move quantities of fluid. There are many pump performance parameters, some of which are complex and may be presented in a non-dimensional format. For works inspection purposes, however, you only need to consider those which normally form part of the pump 'acceptance guarantees'.

Volume flowrate (q)

Flowrate is the first parameter specified by the process designer, who bases the pump requirement on the flowrate that the process needs in order to function. This 'rated' flowrate is normally expressed in volume terms and it is represented by the symbol q, with units of metres^3/second.

... d (H)

Once the rated flowrate has been determined, the designer then specifies a total head (H) required at this flowrate. This is expressed in metres and represents the usable mechanical work transmitted to the fluid by the pump. Together q and H define the *duty point*, the core FFP criterion.

Net positive suction head(NPSH)

NPSH is slightly more difficult to understand. Essentially, it is a measure of the pump's ability to avoid cavitation in its inlet (suction) region. This is done by maintaining a pressure excess above the relevant vapour pressure in this inlet region. This pressure excess keeps the pressure above that at which cavitation will occur. Acceptance guarantees normally specify a maximum NPSH required. The unit is metres.

Other FFP criteria

- Pump efficiency (η percent): the efficiency with which the pump transfers mechanical work to the fluid.
- Power (P) in watts, consumed by the pump.
- Noise and vibration characteristics.

It is normal practice for the above FFP criteria to be expressed in the form of acceptance guarantees for the pump. The objective of the performance testing programme is to demonstrate compliance with these guarantees.

Basic technical information

A large number of pump designs fall under the general categorization of centrifugal pumps. These include radial, mixed-flow, and axial pumps and they can be tested using similar methods. In most plant specifications, you can expect to see a similar format of acceptance guarantees for the various pump designs within this wide category. The first step is to understand the set of curves that are used to describe pump performance. These are commonly known as 'characteristics', or simply 'curves'. Figure 11.1 shows a typical set.

The q/H curve

For most centrifugal pump designs the q/H characteristic looks like that shown in Fig. 11.1. The test is carried out at a nominally constant speed

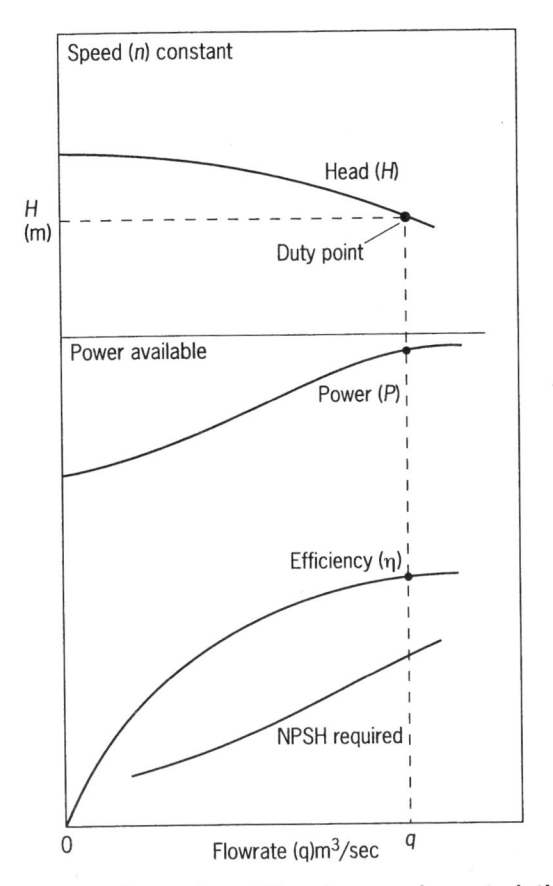

Fig 11.1 A set of centrifugal pump characteristics

and the head (*H*) decreases as flowrate (*q*) increases, giving a negative slope to the curve. Note how the required *duty point* is represented and how the required pump power and efficiency change as flowrate varies.

The NPSH (required) curve

One reason why NPSH can be confusing is because it needs two different sets of axes to describe it fully. The lower curve in Fig. 11.1 shows how the NPSH required to maintain full head performance rises with increasing flowrate, but note that this curve is not obtained directly from the q/H test – it is made up of three or four points, each point being obtained from a separate NPSH test at a different constant q (see Fig. 11.4). This is normally carried out after the q/H test. In the NPSH test you will be looking for the pump to maintain full head performance at an NPSH equal to or less than a maximum guarantee value.

Typical acceptance guarantee schedule

Pump acceptance guarantees are expressed in quite precise terms – if you look at a good specification you will see something like this (I have shown indicative values for a large circulating water pump to give you an idea of magnitude):

Rated speed (n)	740 rpm
Rated flowrate (q)	0.9 m^3/second ⎫ together, these define
Rated total head (H)	60 m ⎭ the *duty point*
Rated efficiency	80 percent at duty point
Absorbed power	660 kW at duty point
NPSH	Maximum 6 m at impeller eye for 3 percent total head drop.
Vibration	Vibration measured at the pump bearing shall not exceed 2.8 mm/sec rms at the duty point
Noise	Maximum allowable level of 90 dB(A) at duty point (at agreed measuring locations)

Now the specification states:

- Tolerances should be ± 1.5 percent on head (H) and ± 2 percent on flow (q) (these are typical, but can be higher or lower, depending on what the designer wants) *but* $+ 0$ on NPSH.
- The acceptance test standard, e.g. ISO 3555, is important – it tells you a lot about test conditions and which measurement tolerances to take into account when you interpret the curves.

Later we will see how to check whether the pump has complied with the requirements.

Specification and standards

We are fortunate in that pump performance testing is well covered by a tried and tested set of standards which relate specifically to the radial, mixed and axial flow category. These standards relate only to the pump itself. Pumps are only rarely subject to performance tests in the process system for which they are intended, normally they are tested in a specific performance test rig. The main standards are listed below.

ISO 2548 (identical to BS 5316 Part 1) **(1)** is for Class C levels of accuracy. This is the least accurate class and has the largest allowable measurement tolerances which are applied when drawing the test curves, and hence the largest acceptance tolerances on q and H.

ISO 3555 (identical to BS 5316 Part 2) **(2)** is for Class B levels of accuracy, with tighter test tolerances than for Class C.

ISO 5198 (identical to BS 5316 Part 3) **(3)** is for Class A (or precision) levels of accuracy. This is the most stringent test with the tightest tolerances.

DIN 1944 'Acceptance tests for centrifugal pumps' **(4)**. This is structured similarly to BS 5316 and has three accuracy classes, in this case denoted Class I, II or III.

API 610 'Centrifugal pumps for general refinery service' **(5)**. This is a more general design-based standard.

ISO 1940/1 (identical to BS 6861 Part 1) **(6)** is comonly used to define dynamic balance levels for pump impellers.

VDI 2056 **(7)** is commonly used to define bearing housing or pump casing vibration. A more complex method, measuring shaft vibration, is covered by ISO 7919-1 (similar to BS 6749 Part 1).

DIN 1952 and VDI 2040 are currently withdrawn standards but are still in common use to specify methods of flowrate (q) measurement.

In some inspection situations the test standard will not be quoted. This shouldn't happen. If it does I normally apply the following simple guidelines:

- If there is no guidance to the contrary use ISO 3555.
- If high *test* accuracies are necessary, i.e. if the pump guarantees are very closely specified on a reasonably 'flat' q/H curve, use DIN 1944 (Class I).
- If the pump is NPSH-critical or the NPSH available from the fluid system is in any way uncertain (check with the system designer), or if it is an experimental pump, then it is perhaps best to use ISO 5198. Its definition of NPSH testing is quite comprehensive and it specifies clear measurement tolerances of ± 3 percent NPSH or ± 0.15 metres.

There are other standards that you may meet, although frankly they are rarely used during a works inspection.

Inspection and test plans (ITPs)

Check the ITP for the pump manufacturing sequence. It should address as a minimum the following items:

Pump casing

- Material test certificates (to EN 10 204 type 2.2).
- Material identification records.
- NDT results.
- Record of casting defects, MPI and repairs.
- Hydrostatic test (normally at a maximum of 2 x working pressure).

Pump shaft and impeller

- Material test certificates (to EN 10 204 type 3.1B).
- Heat treatment verification.
- NDT tests as specified.
- Dynamic balance certificate (a common level is ISO 1940 grade G6.3).

Assembled pump

- Completed technical data sheet.
- Guarantee acceptance test results and report.
- Painting records and report.
- Pre-shipping documentation review.

Some manufacturers will exceed these minimum requirements, others will not.

Test procedures and techniques

The pump acceptance test is carried out in a purpose-built test circuit in the manufacturer's works. In practice the layout of the circuit may be difficult to see as some of it is often underneath the test bay floor plates. Luckily most test circuits follow a similar pattern. Figure 11.2 shows what to look for. Note that there are effectively two different parts, the basic circuit for the q/H test and an auxiliary suction control loop which is connected for the purposes of the NPSH test. The circuit should have suitable instrumentation to obtain the performance data; many pump manufacturers have a fully computerized datalogging system to process the data and display the results.

It makes good sense to start every pump test with a series of *circuit checks*. The act of *checking the circuit* will show that you are adopting a logical, professional approach – and will give you guidance on the likely accuracy of the test results.

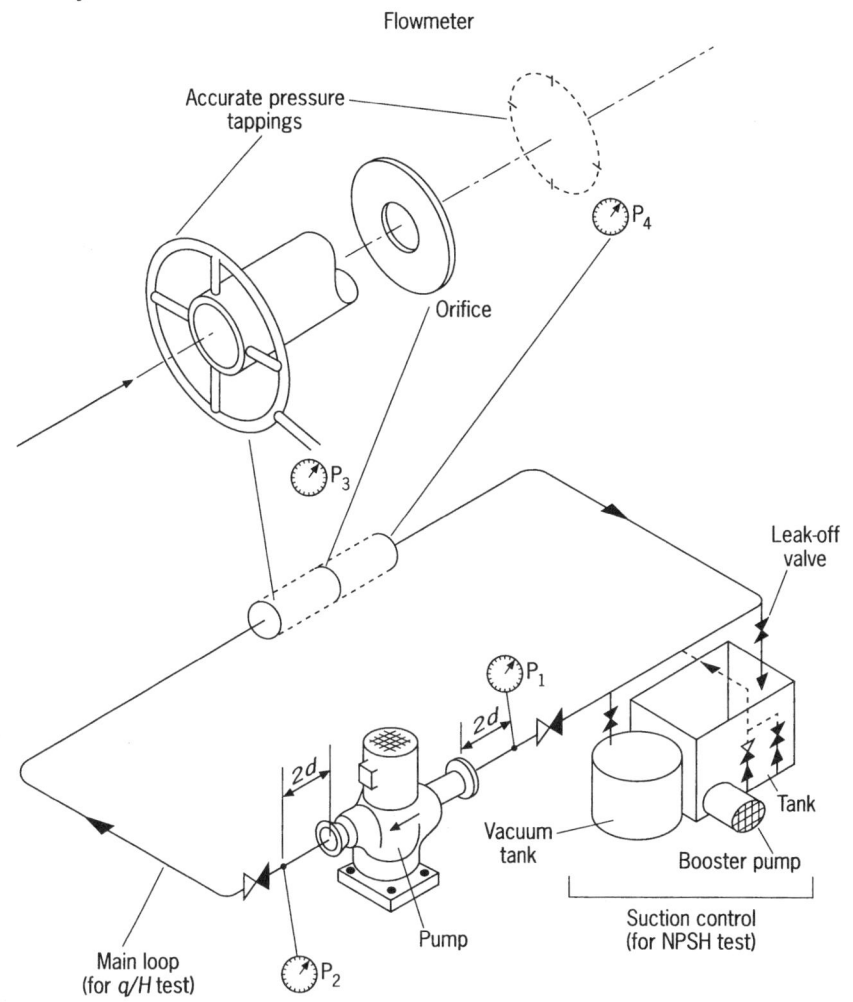

Fig 11.2 The pump test circuit

Some circuit checks

- Normally a shop motor (i.e. not the contract motor) is used to drive the pump. Make sure it has the correct, or higher, power rating.
- Check the pipe arrangements either side of the flowrate measuring

device, there should be sufficient straight run in order not to introduce inaccuracies. Refer to ISO 5167 if in doubt.
- Check the suction and discharge arrangements on either side of the pump. The pressure gauge or manometer connections should be at least two pipe diameters from the pump or readings will be inaccurate.
- Watch for flow straighteners fitted before the pump. These are sometimes fitted to produce the required inlet flow characteristics but they can produce pressure losses and distort the results.
- Ask the manufacturer to explain any variation of vertical levels throughout the circuit. These are particularly relevant to the NPSH test.
- Ensure that the volume of fluid in the circuit is sufficient to avoid temperature rise during the q/H test. If pump input power is high in relation to the volume, then additional cooling may be required.
- *Calibration.* Check calibration records for all measuring and recording equipment. Don't forget the transducers.
- *Empirical factors.* The pump manufacturer may have factors built in to his calculation routines that have an empirical basis. Fluid density corrections and level corrections are two common examples. Check what they are.

These checks will only take a few minutes but are an essential part of the test. Make clear notes of what you have found. The pump test routine will, with a few exceptions, follow a well-defined format. Once started, the steps can follow in quite quick succession.

Step 1

Don't just start, without any preparation. Check the circuit.

Step 2

The pump is started and the circuit allowed to attain steady-state conditions by running for at least 30 minutes. Use this period to make an initial check of the measuring equipment readings to ensure everything is working. Watch for any early indications of vibration or noise.

Step 3: The q/H test

The q/H characteristic is determined as follows. The first set of measurements is taken at duty point (100 percent q). The valve is opened to give a flowrate greater than the duty flow (normally 120 or 130 percent q) and further readings taken. The valve is then closed in a series of steps, progressively decreasing the flow (note that we are moving from right to left on the q/H characteristic). With some pumps, the final reading can be taken with the valve closed, i.e. the $q=0$ or shut-off condition. This is not always the case, however – for high power pumps, or those with a particularly high generated head, it is undesirable to operate with a closed discharge valve. During the test, it is useful to pay particular care to the spacing of readings around the duty point, particularly for Class A pumps where greater accuracy levels will be applied. Close spacing around the duty point will help the accuracy of the results by better defining the shape of the curve in the duty region.

Once the test points are obtained, you can now check against the guarantee requirements. There are several discrete steps required here (refer to Fig. 11.3)

- Draw in the test points on the q/H axes.
- Using the *measurement accuracy* levels given for the class of pump, draw in the q/H measured band as shown.

Steps
1 Draw in the test points
2 Draw in the band using the *measurement* tolerances (from the standard)
3 Add the rectangle, representing acceptable limits (from the specification)
4 This area meets the guarantee

Allowable tolerance (from the specification)

Measurement tolerance 'bandwidth' (from the standard)

Allowable tolerance (from the specification)

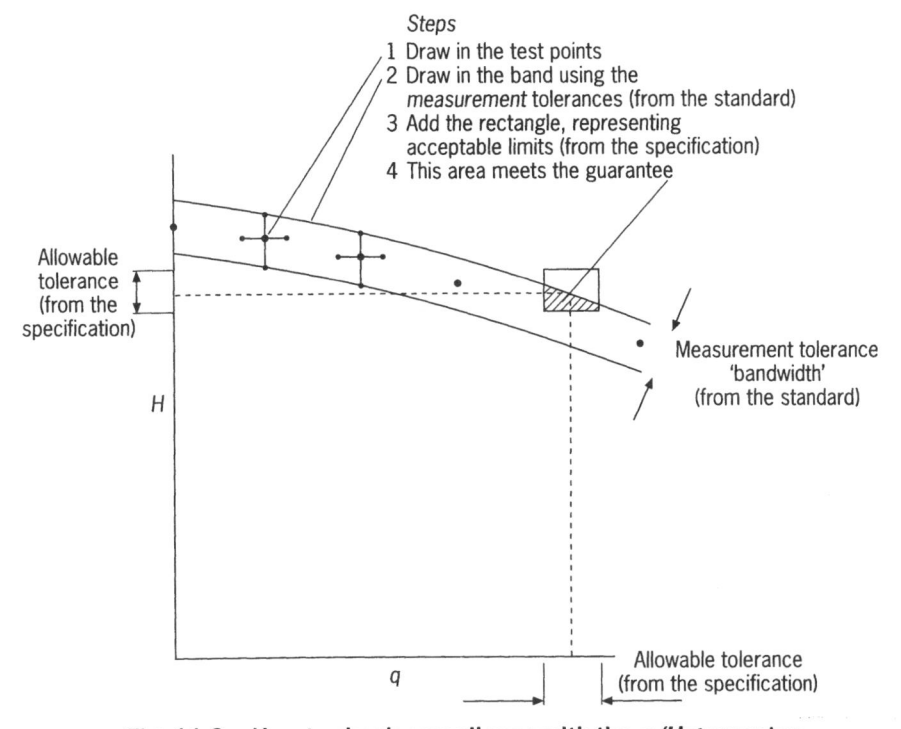

Fig. 11.3 How to check compliance with the q/H guarantee

- Now add the rectangle, which describes the tolerances allowed by the acceptance guarantee on total head (H) and flowrate (q). ISO 3555 indicates tolerances of ± 2 percent H and ± 4 percent q should be applied, if nothing is stated in the specification.
- If the q/H band intersects or touches the rectangle then the guarantee has been met (this is the situation in Fig. 11.3). Note that the rectangle does not have to lie fully within the q/H band to be acceptable.

It is not uncommon to find different interpretations placed on the way in which ISO 3555 specifies acceptance tolerances. The standard clearly specifies *measurement accuracy* levels (± 2 percent q, ± 1.5 percent H) but later incorporates these into a rigorous method of verifying whether the test curve meets the guarantee by using the formula for an ellipse (effectively allowing an elliptical tolerance 'envelope' around each measured point), specifying values of 2 percent H and 4 percent q to be used as the major axis lengths of the ellipse. Strictly, this is the correct way to do it – but I have always found the simplified method shown in Fig. 11.3 easier to use.

Step 4: The efficiency test

The efficiency guarantee is checked using the same set of test measurements as the q/H test. Pump efficiency is shown plotted against q as in Fig. 11.1. In most cases, the efficiency guarantee will be specified at the rated flowrate (q), the same one that you used for checking the head guarantee. The principle of checking compliance is similar, i.e. draw in the characteristic bandwidth using the applicable measurement tolerances, followed by the rectangle representing any tolerances allowed by the acceptance guarantees.

Step 5: Noise and vibration measurements

Vibration levels for pumps are normally specified at the duty (100 percent q) point. The most common method of assessment is to measure the vibration level at the bearing housings using the methodology proposed by VDI 2056. This approximates vibration at multiple frequencies to a single velocity (rms) reading. It is common for pumps to be specified to comply with VDI 2056 group T vibration levels – so a level of up to around 2.8 mm/second is acceptable. Some manufacturers scan individual vibration frequencies, normally multiples of the rotational frequency, to gain a better picture of vibration performance. This does help with diagnosis, if excessive vibration is experienced during a test.

Pump noise is also measured at the duty point. It is commonly specified as an 'A-weighted sound pressure level' measured in dB(A) at the standard distance of 1 metre from the pump surface (the principles are explained in Chapter 7 covering gas turbines). It is sometimes difficult to obtain accurate noise readings during pump tests owing to the considerable background noise which can come from turbulence in the rest of the fluid circuit. Any pump which has noise levels close to its acceptance noise level should be checked *very* carefully for excessive vibration levels, then particular attention paid to bearings and wear rings during the subsequent stripdown.

Step 6: The NPSH test

There are two common ways of doing the NPSH test. The first is simply used for checking that the pump performance is not impaired by cavitation at the specified q/H duty point with the installed NPSH of the test rig. This is a simple go/no-go test applicable only for values of specified NPSH that can be built in to the test rig. It does not give an indication of any NPSH margin that exists, hence is of limited accuracy. The more comprehensive and useful test technique is to explore NPSH performance more fully by varying the NPSH over a range and watching the effects. The most common method is the '3 percent Head Drop' method shown in Fig. 11.4.

Using the test rig as used for the q/H test but with the suction pressure control circuit switched in (see Fig. 11.2), the suction pressure is reduced in a series of steps. For each step, the pump outlet valve is adjusted to keep the flowrate (q) at a constant value. The final reading is taken at the point where the pump head has decayed by at least 3 percent. This shows that a detrimental level of cavitation is occurring and defines the attained NPSH value, as shown in Fig. 11.4. In order to be acceptable, this reading must be less than, or equal to, the maximum guarantee value specified. Strictly, unless specified otherwise, there is no acceptance tolerance on NPSH, although note that ISO 3555 gives a *measurement* tolerance of ± 3 percent or 0.15 metres NPSH. Sometimes you will see this considered as being the acceptance tolerance – I have used this interpretation in the specimen inspection report shown in Chapter 15 of this book.

322 Handbook of Mechanical Works Inspection

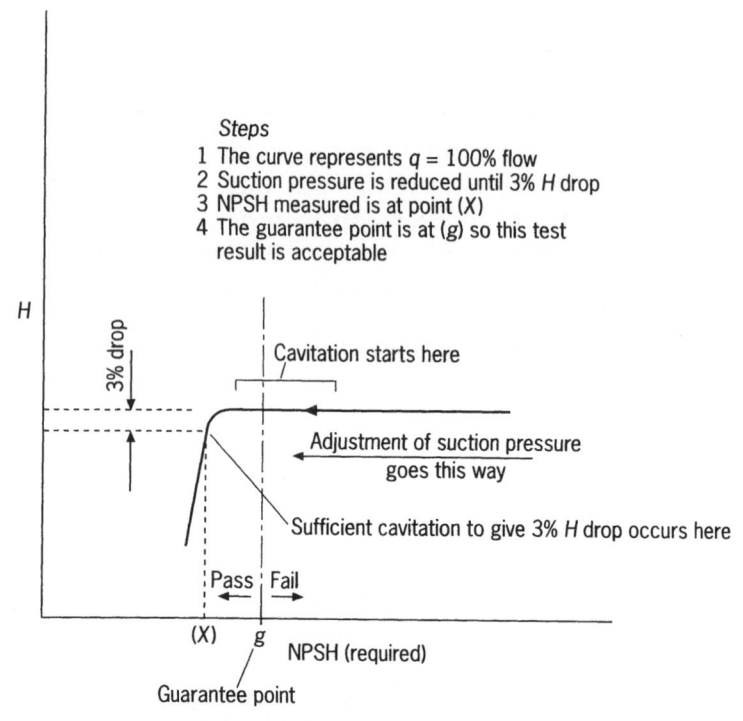

Steps
1 The curve represents q = 100% flow
2 Suction pressure is reduced until 3% H drop
3 NPSH measured is at point (X)
4 The guarantee point is at (g) so this test result is acceptable

Fig 11.4 Measuring NPSH: the '3% head drop' method

Corrections

There are a few commonly used correction factors that you need to use if the test speed of the pump does not match the rated speed. This often happens. The following factors will give sufficiently accurate results. Remember to apply them to *q, H, P* and NPSH.

- Flow q (corrected) = q (measured) $\times (N_{sp}/n)$
- Head H (corrected) = H (measured) $\times (N_{sp}/n)^2$
- Power P (corrected) = P (measured) $\times (N_{sp}/n)^3$
- NPSH (corrected) = NPSH (measured) $\times (N_{sp}/n)^2$
  n = speed during the test
  N_{sp} = rated speed

Step 7: The stripdown inspection

Always try to witness a stripdown inspection after the performance test. This may seem to be an almost incidental part of the test procedure, but it is the inspector's best opportunity to check and report on some important design and manufacturing features of the pump. By inspecting carefully and reporting accurately you add value to what you do – a good example of *effective works inspection*.

Let us look at the best way to do this. First obtain some good background technical information (remember the concept of 'qualifying information' I introduced in Chapter 2?). The pump standard API 610 is useful and will give you clear guidance on desirable mechanical design features, irrespective of how much you know already. Then do some preparation – make a list of the points to check, including any specific requirements of the pump purchase order. Your list should look similar to the following.

Stripdown checks

- Watch the pump run-down – it should be smooth without any undue noise or unbalance.
- Check how the casing sections come apart – they should be a firm press fit but separate without needing excessive force.
- Check the casing joint faces for flatness, there should be no warping.
- Spin the shaft bearings by hand to check for any tightness or radial wear.
- Check the bearing surfaces – there should be no evidence of lubrication breakdown or overheating
- Check the mechanical seals – any chipping or wear indicates incorrect assembly.
- Check the impeller fixing – it should be secured to the shaft with a cap nut so the spindle threads are not exposed to the pumped fluid. The impeller should have an acceptable fixing to the shaft (some specifications require a keyed drive, others do not).
- Check the wear rings for excessive wear (get the limits from the manufacturer's drawings). You can use this opportunity to have a look at the wear ring fixings, they should be locked against rotation by a threaded dowel – not tack welded.
- Surface finish is important. If in any doubt, you can use a comparator gauge – check for a finish of 0.4 µm R_a or better on shaft and seal surfaces. Pump casings should have a finish better than 25 µm R_a on outside surfaces and 12.5 µm R_a on internal surfaces.
- Visually inspect the impeller water passages – smooth surfaces (12.5 µm R_a) indicate good finishing during manufacture. Look also for any evidence of the impeller having been trimmed, or 'underfilled'

on the trailing edges to make it meet its q/H requirements. These are acceptable but only within limits.

Then carefully record the findings for your inspection report.

Common non-conformances and corrective actions : pumps

Non-conformance	Corrective action
The q/H characteristic is above and to the right of the guarantee point (i.e. too high).	For radial and mixed-flow designs, this is rectified by trimming the impeller(s). The q/H curve is moved down and to the left. Watch for resultant changes in dynamic balance. Repeat the test.
The q/H characteristic is 'too low' – the pump does not fulfil its guarantee requirement for q or H.	Often, up to 5 percent head increase can be achieved by fitting a larger diameter impeller. If this does not rectify the situation there is a hydraulic design fault, probably requiring a revised impeller design. Interim solutions can sometimes be achieved by: • installing flow-control or pre-rotation devices • installing upstream throttles.
NPSH is well above the acceptance guarantee requirements	This is most likely a design problem, the only real solution being to redesign. Then repeat the test.
NPSH result is marginal	This can sometimes be a problem of *stability*. The right thing to do is to try the test again and see if you get an exactly reproducible result, paying particular attention to the measurement of the 3 percent head decay (watch and listen for evidence of cavitation). It is sometimes possible to accept marginal NPSH performance under concession – to do this properly you need to check the system NPSH available, to see whether a satisfactory pressure margin (about 1 metre) still exists.

Non-conformance	*Corrective action*
Excessive vibration over the speed range	The pump must be disassembled. First check the impeller dynamic balance (you can use ISO 2373/BS4999 part 142/IEC.42 or ISO 1940 for guidance).
	Next check all the pump comp-onents for 'marring' and burrs – these are a prime cause of inacc-urate assembly. During re-assembly, check concentricities by measuring total indicated runout (TIR) with a dial gauge. Check for compliance with the drawings. Then repeat the test.
Excessive vibration at rated speed	Check the manufacturer's critical speed calculations. The first critical speed should be a minimum 15–20 percent *above the rated speed*. Then do all the checks shown above. It is important to describe carefully the vibration that you see. High vibration levels at discrete, rotational frequency is a cause for concern. A random vibration signature is more likely to be due to the effects of fluid turbulence.
Noise levels above the acceptance guarantee levels	Pump noise is difficult to measure because it is masked by fluid flow noise from the test rig. Take this into account. If high noise levels are accompanied by vibration, a stripdown and retest is necessary.

COMPRESSORS

Compressors work on simple principles but their tests are not always so easy to understand. Compressor designs vary from those providing low pressure delivery of a few bars up to very high pressure applications of 300 bars. For general industrial use the process fluid is frequently air, whilst for some specialized process plant applications it may be gas or vapour. There are several basic compressor types, the main difference being the way in which the fluid is compressed. These are:

- *Reciprocating compressors.* This is the most common positive displacement type for low pressure service air. A special type with oil free delivery is used for instrument air and similar critical applications.
- *Screw compressors.* A high speed precision design used for high volumes and pressures and accurate variable delivery.
- *Rotary compressors.* High volume, lower pressure applications. These are of the dynamic displacement type and consist of rotors with vanes or meshing elements operating in a casing.

Other designs are: lobe-type (Roots blowers), low pressure exhausters, vacuum pumps and various types of low pressure fans. These are covered by different standards and acceptance procedures. Be careful not to confuse the types.

Fitness-for-purpose criteria

The main FFP criterion for a compressor is its ability to deliver a specified flowrate of air or gas at the pressure required by the process system. Secondary FFP criteria are those aspects that make for correct running of the compressor – the most important one, particularly for reciprocating designs, is vibration. It is a feature of compressors that some of the ancillary equipment forms part of the compressor unit. This comprises the unloading system (used to vary the delivery) and the various interlocks and trips to ensure safe operation. These are important functional items.

Basic technical information

To understand compressor testing you need to have a clear view of those definitions which cover the performance aspects you are interested in.

There is a comprehensive set of definitions listed in ISO 1217 **(8)**; here we will restrict ourselves to those key ones that are needed to discuss any 'first order' issues that arise from the acceptance tests. These are:

- *Total pressure (p)*. Total pressure is pressure measured at the stagnation point, i.e. there is a velocity effect added when the gas stream is brought to rest. In a practical test circuit, absolute total pressure is measured at the compressor suction and discharge points for use in the calculations.
- *Volume flowrate (q)*. There are three main ways of expressing this so you need to be precise when discussing flowrates. Refer to Fig. 11.5 to complement the following three definitions.
- *Free air delivery* (FAD) is the volume flowrate measured at compressor discharge and referred to free air (the same as atmospheric conditions). It is the definition nearly always quoted in compressor acceptance guarantees.
- *Actual flowrate* is the volume flowrate, also measured at compressor discharge, but referred specifically to those conditions (remember these are 'total' measurements) existing at the compressor inlet during the test. You really only need this if the suction condition of the compressor is above atmospheric pressure.
- *Standard flowrate* can be slightly confusing as it is nearly the same as FAD. It is the volume flowrate, measured at the discharge, but this time referred to a standard set of inlet conditions. A common set of standard conditions is 1.013 bar and 0°C (273K). A correction factor is needed to convert to FAD.
- *Specific energy requirement* is the shaft input power required per unit of compressor volume flow rate. Power is normally an acceptance guarantee parameter.

Type testing v acceptance testing

Many compressor models, particularly smaller sizes, are manufactured by batch or mass production methods; often the design is well proven and 'type-tested' and has been expanded into a range of catalogue models to meet customers' common requirements. In such situations it is normal practice to carry out a shortened performance test, with specific attention paid only to key parameters and their deviation from the proven and certified type-test results.

The overall objective is to check compliance with the specified performance guarantees, which will look something like this:

- Specified inlet pressure (p_1) and inlet temperature (T_1). Note that these are often left implicit, perhaps being described as 'ambient' conditions.
- Required FAD capacity (q) at delivery pressure (p_2).
- Power consumption (P) at full load
- Vibration – normally specified for compressors as a velocity (using VDI 2056)
- Noise – expressed as an 'A-weighted' measurement in dB(A)

The guarantee schedule should specify the standard to be used for the acceptance tests if it is to have any real meaning. It is not usually stated in the guarantee specifications, but is generally understood, that all acceptance test results are subject to the required corrections. If true accuracy levels are to be maintained, then it is necessary to make explicit statements about the position of the compressor suction and discharge measuring points.

Specifications and standards

Compressors, like pumps, have specific standards devoted to acceptance tests. The most likely one that you will meet is: ISO 1217, *Methods for acceptance testing* (identical to BS 1571 Part 1) **(8)**.

ISO 1217 gives comprehensive testing specifications and arrangements for the major compressor types. The correct time to use this standard is if you are testing an unproven or 'special' compressor design. In practice however, most compressors that you inspect will at some point have been subject to type-testing to verify their overall performance parameters: Part 2 of BS 1571 **(9)** describes simplified methods which can be used in such situations. The core method of proving FFP is essentially the same in both standards.

A simple guideline: unless ISO 1217 is *specifically* imposed by the specification, or there are important safety implications requiring precisely controlled fluid conditions, then BS 1571 Part 2 is generally adequate. It is also simpler to follow.

There are other relevant standards which may be specified in some cases. These are:

API 617 **(10)**: This is a design standard for centrifugal compressors.
API 618 **(11)**: This is a design standard for reciprocating compressors.
ASME PTC 10 **(12)**: PTC stands for performance test code. PTC 10 is

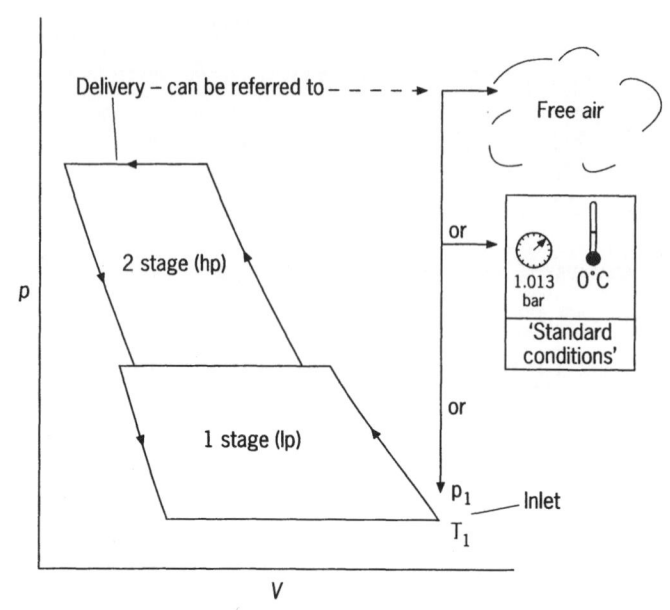

Fig. 11.5 Three ways of expressing compressor flowrate

an American code for compressor testing and is recommended by the API standards.

ISO 5388/BS 6244 **(13)**, *Code of practice for stationary air compressors*. This is mainly devoted to design and safety features.

ISO 2372/BS4675 **(14)**, *Basis for specifying evaluation standards for rotating machines* and VDI 2056 **(7)**. Refer to these if you experience vibration problems.

If you get into trouble understanding fluid systems diagrams, all the symbols are explained in ISO 1219/BS 2917 **(15)**.

Inspection and test plans (ITPs)

ITP formats for the compressor units themselves are quite straightforward, but expect them to be made more complicated by the inclusion of associated pressure circuit components such as suction/discharge vessels and heat exchangers. A specimen is shown below with typical acceptance standards.

Suction/discharge manifolds and dryer/filter vessels

Vessel design review	ASME VIII, BS 5169 or BS 5500
Material certification	BS 5500 or EN 10 204
Ultrasonic testing on plates	BS 5996
Welding procedures	EN 287/288
Radiography on welds	ISO 1106 (see also Chapter 5)
Dye penetrant testing on welds	BS 6443 or ASTM E165
Hydrostatic pressure test	2 × working pressure (or as code)
Painting preparation and thickness check	See Chapter 14
Interstage pipework	ANSI B31.3

Accessories

Safety valves	Manufacturer's certificate of conformity
Limit switches, etc.	Manufacturer's certificate of conformity

Assembly and test

Performance test	ISO 1217 or BS 1571 Part 2
Function check of unloading, trips and safety valves	Manufacturer's procedure
Full documentation review	Manufacturer's procedure

Test procedures and techniques

The best route to understanding compressor testing is to look at an acceptance test carried out to the principles of BS1571 Part 2. This core test of guarantee parameters is the most likely type that an inspector will be called to witness. An essential first step is the validity check.

Validity check

The acceptance test will only have full validity under BS1571 Part 2 if the differences between the planned test conditions and those of the *documented type test* are within the following limits:

- Speed ± 5 percent
- Cooling water temperature ±8 percent
- Pressure ratio ±1 percent
- Intake pressure (absolute) ±5 percent. Note that this is a key

requirement, it is not acceptable to test using, for example, 5 bar suction pressure if the type test was done at atmospheric suction conditions.

The test circuit

There are several possible layouts of test circuit (ISO 1217 shows them). The most common type for air applications is the 'open' circuit, i.e. the suction is open to atmosphere, which is representative of the way that the compressor will operate when in service. If an above-atmospheric suction pressure is required (for instance some natural gas supply compressors) then the test circuit will be a closed loop.

The second main identifying feature is the position at which the flow rate (FAD) of the compressor is measured. It can be measured using an orifice plate flowmeter on either the suction or discharge side of the compressor. Typically, as in the following example, it is measured on the *discharge* side, with a receiver vessel interposed between the orifice and the compressor discharge. The purpose of this is to damp out unacceptable pressure fluctuations which would invalidate the test. The open loop test circuit, with its measured parameters, is shown in Fig. 11.6.

Now we are ready for some preliminary test activities and checks. We will take the steps in order, starting with some *circuit checks*.

Circuit checks

- *Flowrate Measurement*. The orifice nozzle arrangement used for flowrate measurement must comply with the parameters shown in Fig. 11.6. Check the test circuit that is being used against the dimensions given. Check also that a perforated plate is being used in the position shown – the purpose of this is to smooth the flow into the orifice nozzle. If necessary check the size and spacing of the perforations in the plate – ISO 1217 sets out the requirements.
- *Control valve sizing*. This valve should have the same, or smaller, diameter than the diameter of the approach pipe leading into the nozzle. It must not be larger.
- *Receiver sizing*. A receiver is frequently used to damp out pulsations. Check the size as follows: if the pressure drop across the control valve is more than 30 percent of the absolute pressure upstream, the receiver should be larger than that required to contain 50 pulsations at the receiver pressure.

Fluid systems 333

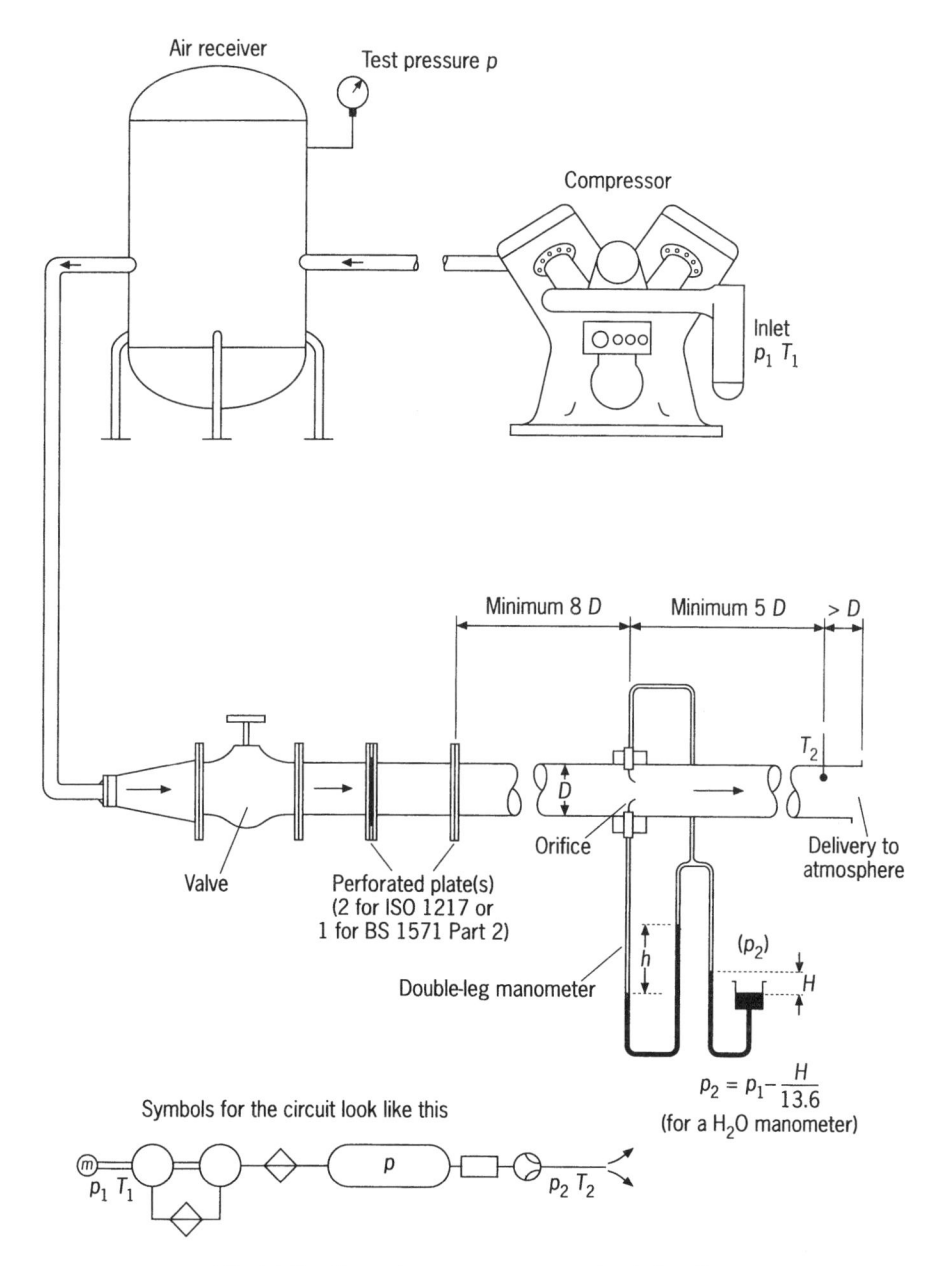

Fig 11.6 Open loop compressor test circuit

Example (for a typical large compressor):

$$\text{swept volume per rev} = \frac{\text{FAD (m}^3\text{/hr)}}{n \text{ (rpm)} \times 60} = \frac{700 \text{ m}^3/\text{hr}}{745 \times 60} = 0.016 \text{ m}^3 \text{ per rev.}$$

So the minimum acceptable receiver size in this case is $0.016 \times 50 = 0.78 \text{ m}^3$

- If the pressure drop across the valve is *lower* than the 30 percent level it is wise to double the required receiver size. If in doubt, check the extent of pressure fluctuation on the manometer or gauge measuring the nozzle inlet pressure. If there is more than 2 percent pressure variation (you can see this level of fluctuation reasonably easy), then the receiver is probably too small.
- *Nozzle characteristics.* You need to check that the nozzle selection is correct. You can do a quick check here from BS 1571 Part 2 – look at Section 9.3 to check that the nozzle coefficient (k) is suitable for the actual test being carried out. Keep a note of this k value – you will need it later for the FAD calculation.

The test

The performance test itself consists of the following steps:
- *Circuit checks* as previously described.
- Run the compressor until the system attains steady state conditions (up to 4 hours).
- Check that the system parameters comply with allowable variations, as in ISO 1217.
- Make minor adjustments as necessary, but only those essential to maintain the planned test conditions.
- Take readings at regular intervals (say 15 minutes) over a period of 2–4 hours with the compressor running at full load.
- Check again for obvious systematic errors in the recorded parameters.
- Carry out functional checks of unloading equipment, relief valves, trips, and interlocks.
- Perform the noise and vibration measurements.
- Do the performance calculations and compare the results with the guarantee requirements. See Fig. 11.7.
- Stop the unit – and perform a stripdown test (if required) for reciprocating designs

STEP 1	Calculate q (FAD) by $q(\text{FAD}) = \dfrac{k\,T_1}{p_1}\sqrt{\dfrac{h\,p_2}{T}}$	Where q = Volume flowrate h = Pressure drop across nozzle (mmH$_2$O) k = Nozzle constant (remember the check described earlier) T_1 = Temperature (absolute K) at compressor inlet T = Temperature (absolute K) downstream of the nozzle p_1 = Pressure (absolute mmHg) compressor inlet p_2 = Pressure (absolute mmHg) downstream of the nozzle
STEP 2	This q (FAD) will be in litres/second. Convert to m^3/hr using m^3/hr = l/sec x 3.6	
NEXT	Do you need any conversion factors?	Absolute pressure = gauge reading + atmospheric pressure (check the barometer)
		REMEMBER
STEP 3	If the test speed is different from the rated speed, correct the q(FAD) by: qFAD (corrected) = qFAD(test) x $\dfrac{\text{rated speed}}{\text{test speed}}$	This simplified correction is normally the only one you will need for a test under BS1571 Part 2.
	THEN	REMEMBER
STEP 4	Compare it with the q(FAD) requirements of the guarantee	There is an allowable tolerance of ± 4–6% at full load depending on the size of compressor – check with BS1571 if in doubt.
STEP 5	Check power consumption kW	Normally measured using two wattmeters
	Apply correction if necessary Power (corrected) = Power(test) x $\dfrac{\text{rated speed}}{\text{test speed}}$	
	If specifically required by the guarantee, express power consumption in 'specific energy' terms by: Specific energy = $\dfrac{\text{Energy consumption}}{q(\text{FAD})}$	Watch the units: a normal unit is kW hr/litre

Fig. 11.7 Evaluation of compressor test results

Common non-conformances and corrective actions: compressors

Non-conformance	*Corrective action*
The compressor does not attain its specified q (FAD) performance, i.e. it produces a lower than required flow rate at rated speed, taking into account the allowable tolerance.	The first step is a double-check of the measured parameters. You must check the test rig until you are satisfied that everything is compliant with the standard. It is a well proven fact that the test results are heavily sensitive to even small errors in the test rig. Have you made any of these common mistakes? • Confusing manometer readings (between mm H_2O and mm Hg) • Using the wrong nozzle coeffecient (k): check against the guidelines in BS 1571 or, the most common mistake: • Are you confusing q (FAD) which is referred to absolute inlet conditions with other calculated flowrates? (See the definitions provided earlier.) • Have you properly assessed the accumulation of errors that will occur? (Check the standard again.) If these checks all prove negative, then there is a problem. The most common reason for reduced flowrate is: • Loss of volumetric efficiency due to internal leakage or wrong design. Corrective actions available are limited – the most common outcome is an application for concession to accept the compressor with its reduced flow rate. First do a check to make sure there have been no assembly mistakes such as wrong piston/liner or valve clearances.

Non-conformance	Corrective action
Stripdown observations Any significant mechanical wear is *unacceptable*. Likely areas are: • irregular wear of valve seats • irregular piston/liner wear on reciprocating types – look particularly for signs of scuffing or oil film breakdown • crankshaft journal and big-end bearing wear on reciprocating designs – check the full periphery of each bearing, not just the area exposed by removal of the bearing cap.	If mechanical wear is found the only correct solution is to rebuild the compressor to original specification, re-test and see if it happens again. Expect most of these problems to be caused by inaccurate assembly, rather than design problems. *But* Keep on looking until you find the source of the problem. Report carefully what is done and what you find (see Chapter 15).

DRAUGHT PLANT: DAMPERS

Draught plant comprises components of air and gas flow systems, either linked to a boiler installation for power generation or for more general process use. Main components are dampers, air heater/exchangers, fans, and associated ductwork and fittings. Of these, experience shows that flue gas dampers are a common source of problems, resulting in enforced plant outages during the commissioning and early life phases. We will therefore look specifically at dampers – note though that many of the general principles of material selection and construction accuracy also hold good for the gas ductwork and other fabricated gas-path components.

Fitness-for-purpose criteria

Gas dampers are large items of equipment of simple construction, but nevertheless play an important part in the operation of a plant. The main FFP criteria are:

- *Sealing*. The efficiency with which the damper prevents gas leakage both within the gas path and to the outside atmosphere.
- *Operation*. The time the damper takes to open or close (or to adopt different modulation positions).
- *Corrosion/erosion resistance*. Gas dampers often operate in harsh process environments. They can experience particulate (dust) impingement, low temperature dewpoint corrosion and other

corrosion mechanisms, depending on their application. Most dampers have a limited life so service lifetime is an important point.

Basic technical information

There are three main types of gas dampers.

Guillotine dampers consist of a single or double sheet steel 'blade' which slides into the duct from outside. They are usually rectangular and can be sized up to 56 m^2 (7 m × 8 m) on large plants. Their only purpose is isolation. Guillotines can form good gas-tight seals.

Louvre dampers have separate, usually horizontal, blades, each mounted on its own spindle inside the gas duct. Some types (double louvres) have two sets of parallel blades. Louvre dampers can be used for gas modulation and for isolation – those intended for efficient isolation are normally of the double-louvre type, with a ventilated interspace. Seals are fitted between the blades and the frame, and between adjacent blades.

Flap dampers have a pivoted flap inside the duct. They normally have a travel of 90 degrees or less and seal at both positions. Their main use is for bypass arrangements, in which the damper directs the gas to either of two discharge paths.

With all of these types, the issue of sealing efficiency is important. Larger size dampers, particularly louvre types, are difficult to seal with 100 percent efficiency even when new. In use, erosion, corrosion, and dust blockage can reduce the efficiency of the seal – hence seals are normally removable, and subdivided into replaceable parts. Seals are patented designs and can be of either the 'finger type' (overlapping shim sheets) or the 'bulb type' which compresses onto its seat. Functional tests are performed to check damper operation and the efficiency of the sealing arrangements. Specific methods are used to calculate the leakage. Because of their inherently simpler seal design, guillotine damper tests are more straightforward than those for louvres and flap types.

A typical acceptance guarantee schedule will be:

- sealing efficiency expressed as a percentage of volume flow (or in m^3/minute referred to a standard temperature and pressure)
- opening/closing time in seconds
- seal air fan power consumption in Watts.

Because dampers are fabricated components, material selection, welding standards, and accuracy of assembly are important.

Specifications and standards

Damper design is not well covered by international standards. Industry practice tends to follow general steelwork and fabrication standards. You can use the following common standards as a basis: these, along with the standards covering welding (see Chapter 5) form the basis of ITPs used for damper manufacture. Apart from the functional test, there are no special test requirements.

Fabricated steel material specifications	BN EN 10 000 Series **(16)**
Corrosion resistant steel for seal strips	ISO 683 or BS 970 **(17)**
General fabrication standards	BS 5135 **(18)**
Overall dimensional tolerance	DIN 8570 **(19)**
Flow measurement	ISO 5167 **(20)**

Test procedures and techniques

For dampers which have an operating temperature of less than about 100°C, functional and leakage tests are generally carried out 'cold' in ambient air – the temperature differential between test and operating temperature is deemed low enough to produce representative results. Seal air is normally also supplied at ambient temperature, hence correction factors are needed when calculating seal air leakage and seal air fan power. Cold tests may also be specified when the damper is too large, or its operating temperature is too high, to be accommodated by normal heating methods. Cold tests do have the disadvantage, however, that expansion and distortion, particularly of the damper frame, are not properly simulated.

In hot tests, the damper is heated to operating temperature in a lagged furnace. Gas is normally used as the heating fuel. The assembly is soaked for a period of several hours, the damper being supported in a way which is representative of 'in-service' support provided by the ductwork and structural steel framework. In both hot and cold tests particular emphasis is placed on the measurement of seal air flow – this is the way that sealing efficiency will be measured. Similar methods are used for louvre and guillotine types with slight differences in measurement techniques depending on whether single or double louvres, or guillotine blades, are fitted.

Unlike some equipment items, functional tests on dampers can never fully reproduce all the stresses and distortions that act on the damper when installed, so it is important to monitor the accuracy of fabrication and assembly during manufacture. This helps to avoid major dimensional inaccuracies in the finished item. We will look at a typical inspection and test programme for a louvre or flap-type damper. The general principles involved can be adapted for individual damper designs as applicable.

Design review

Dampers are not normally subject to formal design review by a third-party organization. Fortunately they are relatively straightforward components so you can carry out a broad design review yourself to get an overall impression of FFP. Do this before manufacture starts or in the *very* early stages. Use the following guidelines.

- Materials – check the material specifications for the frame and blades against the typical grades mentioned in the material specification. Look particularly at carbon content and verify ductility by reviewing the 'percentage elongation' results. This is often relatively low grade material so it may not have full traceability. Follow the general guidelines given in Chapter 4 of this book.
- Seal materials should be of stainless steel or similar with corrosion resistant properties (compare Cr and Ni content with the specification).
- Blade spindles (for louvre type dampers) should have corrosion resistant properties. Check the hardness value, it should be above 400 HV or equivalent to avoid wear and grooving of the shaft where it passes through the seal stuffing-box.
- Blade spindles – it is good practice to have spanner flats on the exposed ends of blade spindles. This allows blades which become twisted or 'stuck' in use to be moved without shutting down the plant.
- Spindle bearings – check they are designed to keep gas and dust out. They should have a shaft seal on the inner frame member.
- Welding – check the proposed welding electrode type to ensure it is compatible with the parent metal. This is also a good time to check the weld procedures for butt and fillet welds (see Chapters 5 and 6).
- Blade seal fixings – check they are corrosion resistant for easy replacement of seal sections. Stainless steel studs, preferably with cap nuts, are good.

Construction

Try to arrange your first works inspection at the stage where the damper frame has been fabricated and the initial set-up of the blades is being made. Check the following points (refer to Fig. 11.8):

- Frame distortion – does the frame lie flat without obvious distortion? The main uprights should not be bowed in either plane more than approximately 1–2 mm per metre length. Make sure the frame is *square*, particularly if any cross-bracing strips have already been welded in. There should be not more than 1–2 mm per metre length of 'lean' to the frame – this can be checked using corner-to-corner measurements. Check the accuracy of the corner fit-ups; you want sound, square joints here.
- Blade clearances – check that the blades fit easily into the frame but are not too loose. End clearances between louvre blades and the frame are typically 2–3 mm but check the drawing.
- Spindle locations – check their centre positions against the drawings, any inaccuracies will cause later problems with the blade alignment and sealing.

In all cases, make sure the frame complies with the tolerances shown on the *manufacturing drawings*, not the general arrangement drawing, which is often not detailed enough.

The hot functional test

The following description is for a hot functional test on a flap-type damper (see Fig. 11.9). It could apply equally to other types. The principles are the same.

Step 1: Check the test arrangement

Seal air flowrate will be measured at inlet and exhaust points as shown in Fig. 11.9. Check the flowmeter constants and calibration records. Pressure measurements need to be taken in several places – because of the relatively low pressures in the plenum area, accurate gauges with calibration certificates are required. Temperature measurements should be taken at three or more locations to give an accurate average reading.

Step 2: Sealing

The three apertures are sealed off using temporary blanking plates. It is important to get a good gas-tight joint otherwise the leakage test results

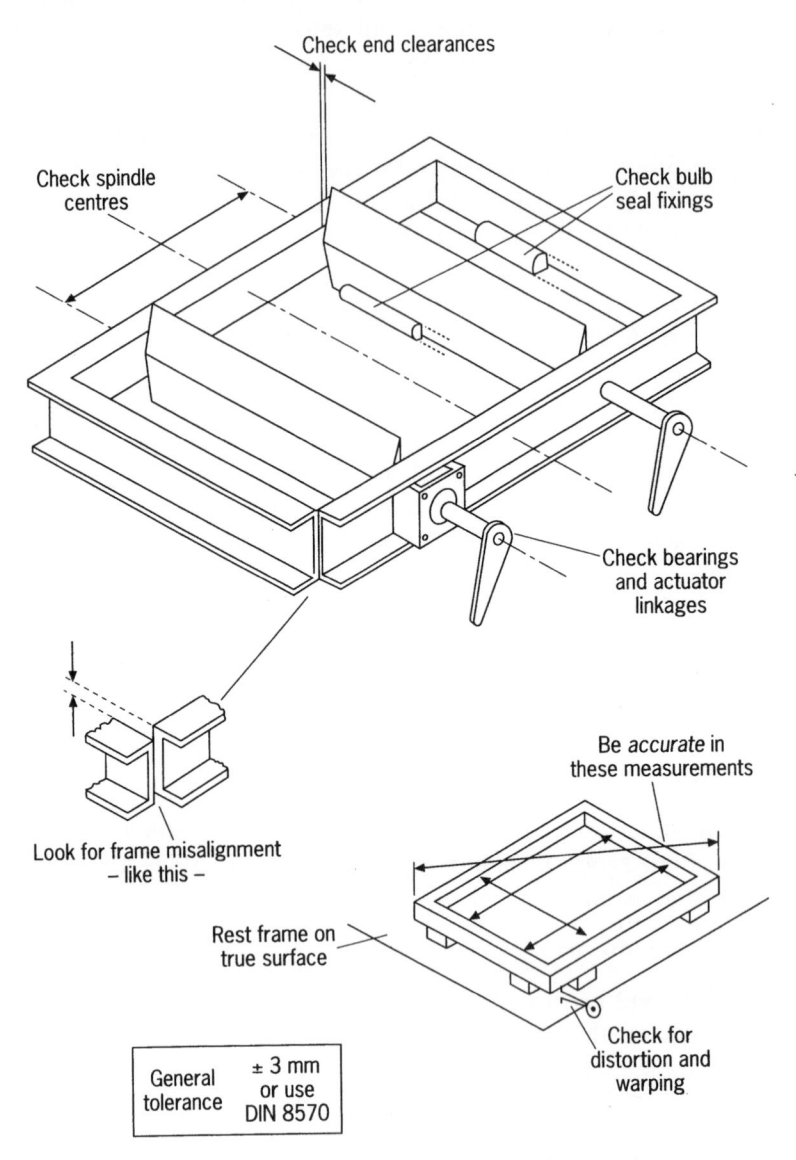

Fig 11.8   The right way to check a louvre damper

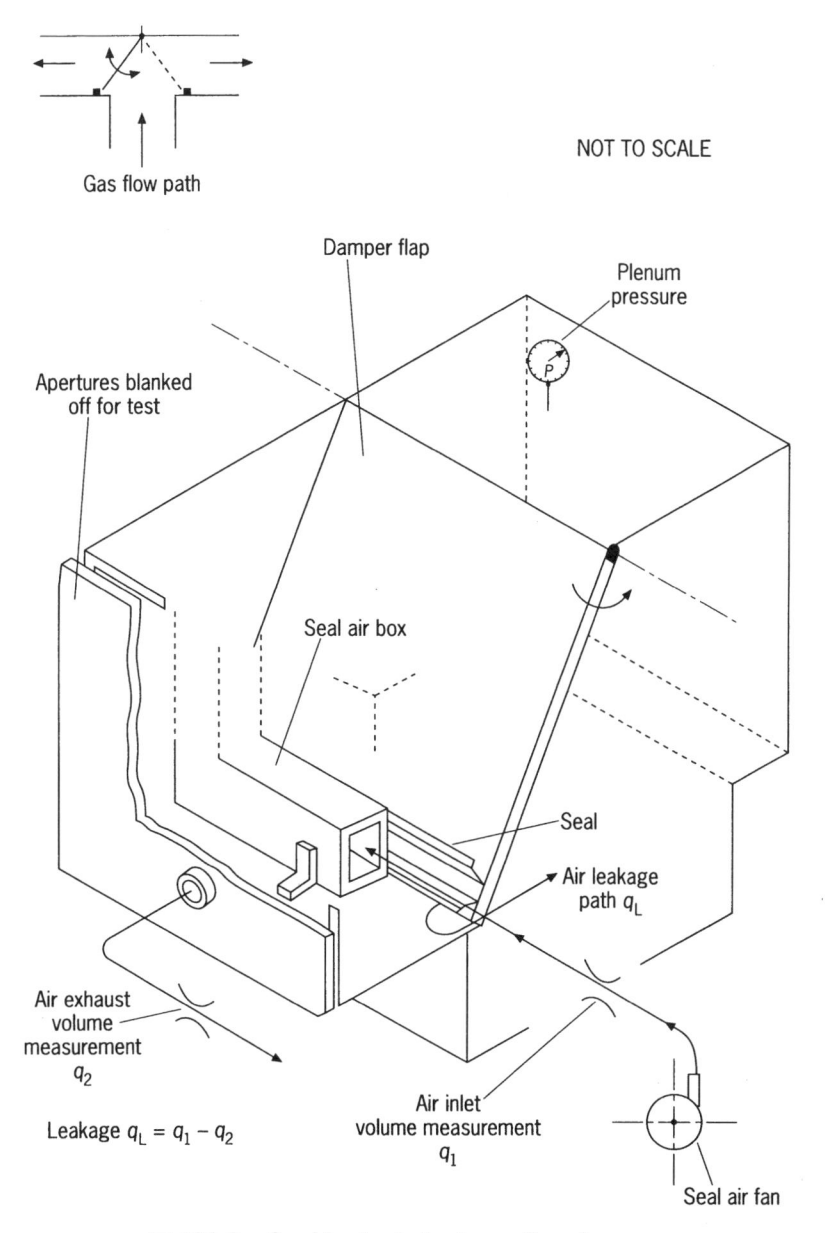

Fig 11.9 Seal leakage test on a flap damper

will be meaningless. Check that temperature resistant gaskets are used and that the flange bolts are correctly tightened in sequence. Insulation should cover all surfaces to prevent heat loss.

Step 3: Heating

The damper assembly is heated to operating temperature; make sure it is 'soaked' at this temperature for *at least* two hours.

Step 4: Measurement

When steady conditions are reached, take all pressure and temperature measurements. Leakage is calculated by:

- leakage (m^3/min) = flowrate at inlet (m^3/min) - flowrate at exhaust (m^3/min)
- then convert to STP (or free air conditions) to enable a proper comparison with the acceptance value.

Calculate the power absorbed by the seal air fan by measuring the current. For flap-type dampers, as shown in Fig. 11.9, all these readings need to be taken at both 'open' and 'bypass' positions of the flap.

Step 5: Operation

Check the operating time of the damper, still in the hot condition. You will not be able to see the flap so you will have to rely on the external actuator and position indicators. The open/close time is normally specified with a tolerance of 10 percent. Check the drive motor current reading, particularly at extremes of flap position, to make sure it is within its limits. Watch the actuator mechanism – it should operate smoothly over the full stroke. There should be a firm, positive action as the flap closes onto its seat, with no evidence of stiffness or juddering.

Step 6: Dimensional check

When the damper has cooled, do the functional checks again, at ambient temperature. After this, you *must* now do another dimensional check to ensure that there is no distortion or permanent set – careful measurement of the corner-to-corner dimensions of the duct apertures will show if it has occurred. Check the damper frame, looking for squareness and truth.

Step 7: Final inspection

If it has not been done already, ask that all major components are match-marked; somebody is going to have to reassemble the unit on site. Later, at the final inspection stage, you should check:

- Painting: normally a minimum 200 micron dry film thickness (see Chapter 14 for further details on painting).
- Linings: some designs can have rubber or GRP linings, or be clad with corrosion-resistant sheets (often Hastelloy C276 or a similar nickel-based alloy).
- Check minor details like position indicators, spare seal components, flange joints, expansion pieces, and steelwork.

Non-conformances and corrective actions: dampers

Non-conformance	Corrective actions
Excessive seal leakage. This tends to be more common with shim-type seals. Given the potential for calculation errors it is prudent to allow a tolerance of 10 percent of the specified value before defining leakage as 'excessive'.	Re-check the seal strips. Are they correctly located and tight? For bulb-type seals, check the 'bulb' dimensions – refix or replace the seals and check again. Check again for distortion of the seating surfaces – use a good straight-edge. If the frame is distorted it can normally be re-jigged but this is a major workshop task.
Erratic function (i.e. juddering/ vibration on operating).	First, eliminate any obvious mechanical or distortion problem. You *can* increase the size of the motor actuator, or you can change the spring in the activator linkage on a trial and error basis – try stronger springs first. On louvre dampers, try and identify which is the sticking blade that is the source of the problem. Most designs enable you to disconnect blades in turn from the actuator linkage so you can test them separately. Look for a bent spindle – it can be removed and replaced. Spindle bearings should be of the self-aligning type.
Most other problems are likely to be with design. Make sure you check the various features mentioned in this chapter at an early stage.	As a broad guideline the following occurrences are unacceptable and probably require disassembly of the damper. Excessive twist or 'lean' of the frame or lack of seal-to-seat contact. There must be *full contact* round the periphery. Any tightness at all in the flap (or blades), actuators, or linkages.

PIPING AND VALVES

Pipes and valves are used in all fluid systems – even a small process plant will contain a large variety of types. For large offshore and chemical plant projects these two sets of components can often account for 10–12 percent of the overall contract value. They are best considered separately for inspection purposes.

Piping

The best advice I can give you is not to delve *too* deeply into this subject. Piping is actually a rather complex discipline involving extensive standardization. There are large numbers of American and European technical standards giving detailed coverage of pipe 'schedule' sizes and materials, as well as related components such as flanges and fittings. This can become a rather confusing picture. Essentially, the fitness-for-purpose criteria for pipes are all related to the *integrity* of the component – so there is a lot of commonality with those aspects discussed in Chapters 4 and 5. Material traceability follows the general rules set out in Chapter 4 – the only difference being that smaller pipes are usually traceable 'by batch' rather than individually. Destructive mechanical tests follow the same general principles as for other material forms, often with the addition of a flaring or 'expanding' test to provide an additional assessment of ductility. Larger diameter pipes may be subjected to a bend test instead. Volumetric and surface NDT techniques are in common use for pipes – follow the general principles shown in Chapter 5.

The standard hydrostatic (pressure) test is a common inspection witness point. Note that not all pipes are necessarily individually tested – it may be done on a batch sampling basis. I will reiterate the point I have made in several places in this book – a hydrostatic test is primarily a test for *leakage* under pressure, it is *not* a substitute for carrying out the correct NDT and properly assessing the results using the correct defect acceptance criteria.

I have summarized the main points of piping and valve inspections together in Fig. 11.10. Treat this as useful 'lead-in' information. I have shown the main technical standards that you will meet – look at them by all means, but don't get too deeply involved. Stick to your FFP assessment.

FITNESS FOR PURPOSE CRITERIA

Piping

The prime criteria for piping are:
- It must be free of defects and
- Material selection must be correct for the application

Valves

Predominantly:
- Integrity and leak tightness and (in the case of large and automated valves), *function*

TEST PROCEDURES AND TECHNIQUES

Piping

Non-destructive testing can be DP/MPI/US/RG depending on application
- Eddy current testing is used for on-line inspection during manufacture
- US examination is particularly useful for finding lamination defects in thick-walled tubes
- RG is used mainly for welded joints

Destructive tests

- Normally tested for % elongation and reduction of area
- For some applications, look for flaring/expanding test, flattening test and proof bend test.

Pressure tests

- Hydrostatic test at 1.5 – 2 × working pressure

Valves

- Hydrostatic test carried out on the valve body
- Seat leakage tests – using water and air. Either zero-leakage, or a maximum value (if specified) is acceptable. This test is done with the valve closed
- For valves with actuators, a functional check is necessary

STANDARDS (for testing)

Piping

ANSI B31.1 to 31.8 : All types of commonly-used pipework
BS 3889 **(21)** BS 6072 **(22)**
BS 3601 **(23)** - a general standard for carbon steel pipes

Valves

ISO 5208 **(24)**. Testing of general purpose industrial valves
BS EN 6053 **(25)**. Control valves
API 598 **(26)**. Inspection and testing of valves

Fig 11.10 (a) Inspection summary – piping and valves

Fluid systems 349

ESSENTIAL ITP CONTENT

Piping

- Material tests and certification (EN 10 204)
- Destructive test results
- NDT tests during manufacture
- Sample NDT tests before assembly
- Marking and identification
- Visual and dimensional checks

Valves

- Material tests and certification
- NDT of any welded connections
- Hydrostatic 'body' test
- Seat leakage test
- Functional test
- Marking and identification
- Visual and dimensional checks

THE INSPECTION ROUTINE

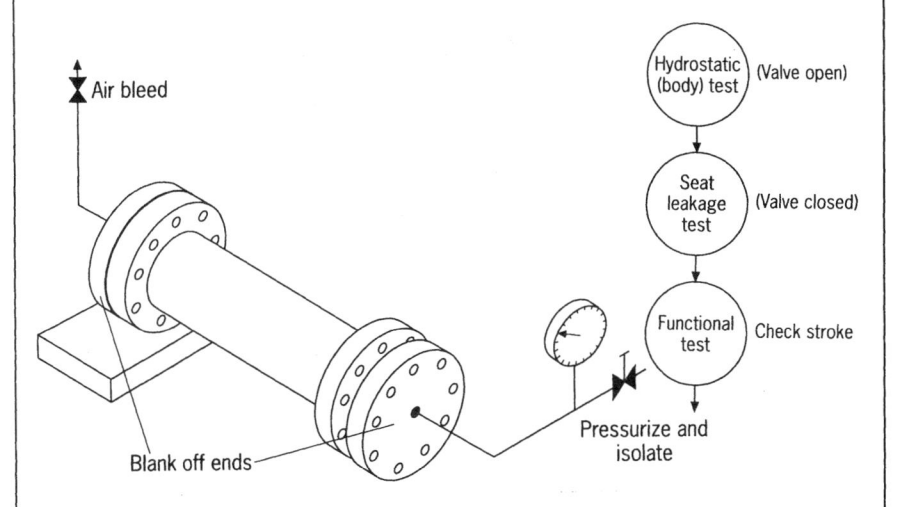

Air bleed

Blank off ends

Hydrostatic (body) test — (Valve open)

Seat leakage test — (Valve closed)

Functional test — Check stroke

Pressurize and isolate

COMMON NCRs AND CAs

Piping

Most NCRs are related to surface defects
- Check acceptability with manufacturing standard. If in doubt: check a further sample.
  Surface cracks are unacceptable: they can cause failure.
  Watch for axial distortion.

Valves

Seat leakage

Check valve seat and disc materials – there should be well defined difference in hardness values.

Check spindle alignment and assembly tolerances against drawings.

Fig 11.10 (b) Inspection summary – piping and valves

Valves

Valve testing is an important, if rather routine, part of works inspection. There are numerous different types of valves, however the principles of inspection are much the same for all types. The main points are summarized in Fig. 11.10 – note the main FFP criteria of integrity and leak tightness. Many valves have cast bodies, often with weld-prepared ends, so the material principles discussed in Chapter 4 are of direct relevance. Valves are also subject to volumetric and surface NDT techniques, RG, US, DP, and MPI are all used, depending on the specific application. The principles and techniques are the same as those discussed in Chapter 5.

Valves are well covered by technical standards. In general, the standards are concise and easy to use – although, as for pipes, they sometimes contain a lot of design-related information that is not directly relevant to works inspection activities. Two of the best standards are ANSI B16.34 **(27)** and ANSI/FCI 70-2 **(28)**.

ANSI B16.34 is a general standard covering all types of flanged, threaded and welded-end valves. It contains useful sections on valve testing and specific tables giving defect acceptance criteria for volumetric and surface NDT. This is good background information to have available, even if the standard is not explicitly mentioned in a contract specification.

ANSI/FCI 70-2 is an American National Standard which specifically covers seat leakage – in this case for control valves. It specifies six classes of allowable seat leakage ranging from Class I (zero leakage) to Class VI, in which a significant amount of leakage is allowed. You should see this standard used quite regularly – so make sure you know what the different leakage classes mean.

A good place to start a valve inspection is to look at the manufacturer's *datasheet* for each valve to be tested. Nearly all valve manufacturers use this system – each valve has a datasheet which is very detailed, containing approximately 100 pieces of information about the valve, including material, NDT, leakage class, and test information. Normally the valve datasheet information exceeds that on the ITP – it makes an excellent checklist.

HEAT EXCHANGERS AND CONDENSERS

There are many types of heat exchangers, ranging from high pressure designs for fuel oil or feed water heating to large multi-compartment condensers for steam turbine installations. Essentially, heat exchangers follow most of the practices associated with fabricated pressure vessels – the materials and extent of fabrication may differ but the main test principles and activities are much the same for all types. I have summarized them in fig. 11.11 for easy reference.

The prime FFP criteria, as we have seen for other fluid system equipment items, are mechanical integrity and leak tightness. Owing to the heat transfer function of a heat exchanger, however, it is also important to place emphasis on the assessment of the materials that comprise the heat transfer surfaces – problems here can reduce significantly the service life of a heat exchanger. Don't neglect your checks on manufacturing conformity just because an exchanger may be operating at low temperatures and pressures.

Standards

There is no single technical standard or set of rules covering all of the different types of heat exchanger that you are likely to encounter. Of those standards that do exist most seem to concentrate on rather complex areas of thermal and mechanical design and reproduce parts of various relevant material standards. For works inspection purposes I think you need to pick and choose carefully only those pieces of information that you need. The external shells of heat exchangers are commonly designed to one of the pressure vessel codes; BS 5500, ASME VIII, TRD, or similar – so for topics such as material testing, welding, NDT, and defect acceptance criteria you can follow the simplified information that I have provided in Chapters 4, 5, and 6. Note particularly the points made in Chapter 6 about code intent (see Figs 6.16 and 6.17). It is quite common for heat exchangers to be designed and manufactured under this 'code intent philosophy' whereby the main stress calculations use a pressure vessel code methodology and assumptions, but the rest of the design – and generally the manufacturing and testing activities – do not. Let me remind you again that you will find this a common occurrence – and that it is not always easy to deal with.

Another standards-related difficulty is that of trying to gain familiarity with non-ferrous material standards. Many heat exchangers use copper or nickel alloys for the heat transfer surfaces. It is likely that you will be less familiar with these than with the common steel standards. The difficulty is compounded by the fact that some non-ferrous alloys are known by their manufacturers' trade names. A little research should pay good dividends here. Surprisingly (perhaps) these material standards, once located, are straightforward – they follow the same general principles of chemical analysis and mechanical testing as steel. By all means obtain the relevant non-ferrous material standards and use them during the inspection. Be careful not to go *too* far though, works inspection is about material *verification*, not material selection. Figure 11.11 lists some commonly used standards. If you wish to read them I would suggest that BS 3274 **(29)** is the easiest place to start – it provides good broad (and simple) coverage.

Inspection and test plans (ITPs)

The format and content of an ITP for a heat exchanger is similar to that used for a pressure vessel, except that the level of detail may be lower in some areas. You will often find that material traceability is limited to EN 10 204 level 2.2 (rather than 3.1B) certificates for the shell components. The tube and tubeplates generally have 'batch' traceability – but there should be a separate ITP entry for the mechanical tests on the tubes.

Welding and NDT of the heat exchanger shell parts should be separately identified in the ITP whether the design is 'full code' or 'code intent', for reasons of good engineering practice. You can use the model 'welding ITP' format shown in Chapter 5 for guidance. Large heat exchangers utilizing sea-water cooling (such as condensers) will often be rubber lined on the sea-water side. A good ITP will have separate entries for the inspection and testing of the lining – I have shown a typical example in chapter 13 (and explained some of the content of the ITP which is inferred, rather than written down).

Test procedures

Material tests and NDT activities for heat exchangers follow general pressure vessel practice. The main witnessed point is the hydrostatic tests on the shell side and tube side of the exchanger. The shell side test is particularly important as it is used to check for leakage from the tube-to-tubeplate joints. These may be of either expanded or expanded and seal welded design – both types can suffer from leaks if the joints are incorrectly made. For a simple expanded tube-to-tubeplate joint it is worthwhile making an interim inspection of the tubeplate holes before the tubes are expanded into them. Make a close check against the drawing: hole diameters and internal surface finish are generally closely specified in order to ensure an accurate fit with the tube.

This section only covers testing of LP fluid heat exchanger vessels. Vessels which operate under vacuum, e.g. condensers, are generally the most demanding type of LP heat exchangers.

FITNESS FOR PURPOSE CRITERIA

- The prime criterion is *integrity* i.e. leak tightness
- Service life - this is mainly a function of the resistance to corrosion (achieved by materials selection, corrosion allowances and corrosion-resistant linings)
- Manufacturing conformity

TEST PROCEDURES AND TECHNIQUES

Hydrostatic (pressure) testing

Normally carried out for shell side and tube side at 1.25 to 1.5 × design pressure for 30–60 minutes

Key points

- Bleed off all air before the test
- Use twin pressure gauges (calibrated)
- Beware of temperature rises – they can distort readings
- A white paint wash helps identify leaks
- Liquids such as kerosine are more effective than water, where their use is practical and safe. Water should include a wetting anti-corrosion agent

Leak testing. Often used for exchangers operating under vacuum. Also for large steam exchangers and condensers. Sophisticated methods include 'black light' and 'search-gas' tests

- Small components can be immersed in a tank (bubble-test)
- Low pressure air or low viscosity gas is used

$$\text{Leakage rate} = \frac{\text{Pressure drop} \times \text{volume vessel}}{\text{time}}$$

Measured in $\frac{Nm}{sec}$ = Watts. Compare with guarantee

STANDARDS

BS 5500: Section on pressure testing
BS 3636 **(30)**: Leak detection
T.E.M.A. **(32)**: Includes pressure and leak testing specifically for heat exchangers

ASTM E432: **(31)**: Leak-testing methodology

ASME VIII

Fig 11.11 (a) Inspection summary – heat exchangers

ESSENTIAL ITP CONTENT

For condensers

- Material test certificates for tubes and tubesheets
- C of Cs for sheets, waterboxes and nozzles
- Welding WPSs, PQRs
- NDT as required by code (see Chapter 6)
- Tube/tubesheet expansion : test pieces
- Hydrostatic test of shell (in works)
- Vacuum test of shell (on site)
- Rubber lining examination (thickness, adhesion and spark test)
- Hydrostatic test of tube side (leak check at tubeplate)
- Final visual/dimensional check
- Final documentation review
- Painting and packing check

HYDROSTATIC TEST (see also Fig 11.12)

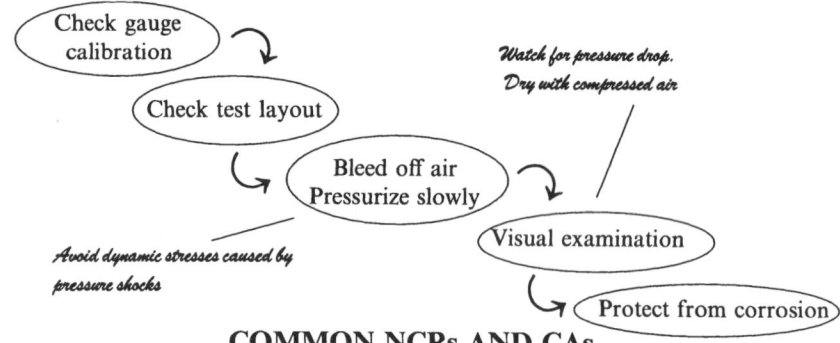

Check gauge calibration → Check test layout → Bleed off air, Pressurize slowly → Visual examination → Protect from corrosion

Watch for pressure drop. Dry with compressed air

Avoid dynamic stresses caused by pressure shocks

COMMON NCRs AND CAs

NCRs	CAs
Mechanical distortion of the shell (detected visually or by measurement).	If yielding has occurred, causing permanent set – *No real solution*.
Tube/tubeplate leaks.	If less than 5% of tubes leak they can be removed and re-expanded or seal welded. If more than 5% leak, check for more fundamental manufacturing or welding problems. Look for rough tubeplate hole finish, or work-hardening of tube ends.
Directional leaks. These leak in one direction but not when the other side is pressurized.	Remove paint or coating and check again.
Poor rubber lining adhesion: most common on corners and changes of section.	Check application procedure – minor rectification can be made but if problem is widespread re-preparation and relining is the best solution.

Fig 11.11 (b) Inspection summary – heat exchangers

Large condensers and exchangers designed for vacuum conditions are frequently subjected to leak-tests using low pressure air or inert gas (see Chapter 6). This is a much more stringent test for leakage than is the hydrostatic test – it will show small leaks that would not be found using a standard hydrostatic test. For this reason, it is essential to make sure that the test procedure and the leakage *acceptance level* are well understood and agreed before the test commences. If you do not do this you may find yourself involved in a spiral of post-test discussions about whether the results were acceptable or not.

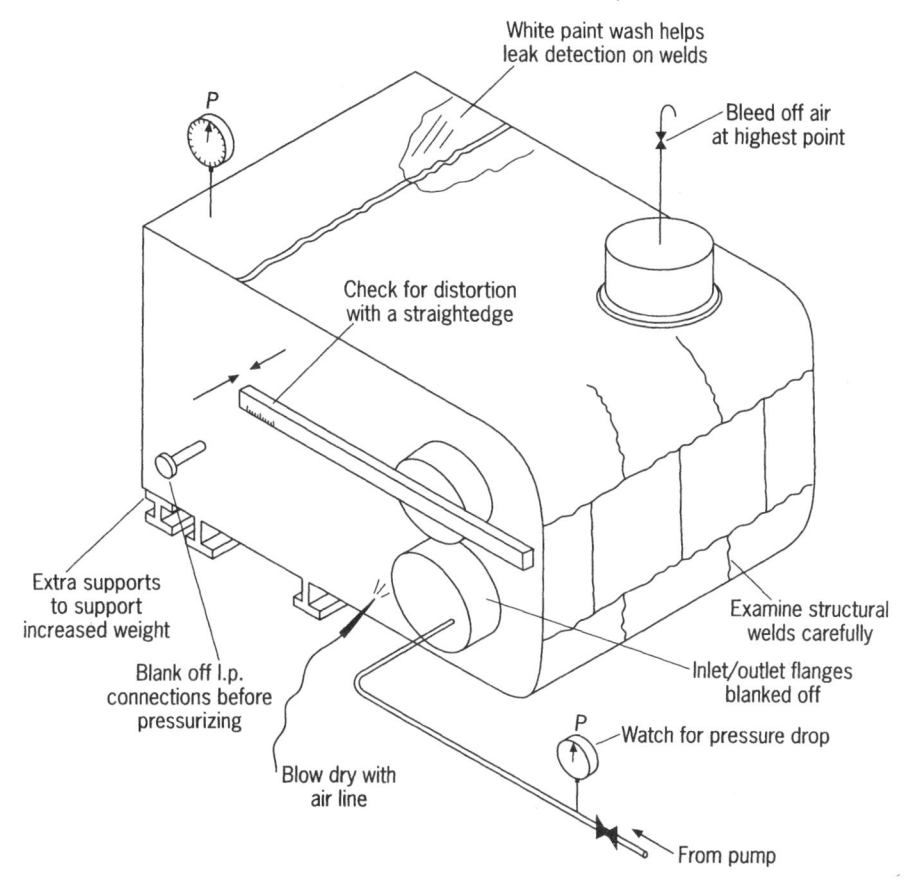

Fig 11.12　Testing a condenser or LP heat exchanger

KEY POINT SUMMARY : FLUID SYSTEMS

Pumps
1. Inspectors are frequently required to witness acceptance tests on centrifugal pumps.
2. Make sure you understand the q/H test *and* the NPSH test (which can be a little confusing). Remember that the q/H curve has a tolerance band.
3. It is important to make an *effective inspection* during the pump stripdown. Prepare a checklist in your notebook and keep revising it as you gain experience.

Compressors
4. The delivery of a compressor is normally referred to free air delivery (FAD) conditions.
5. ISO 1217 is a good testing standard. BS1571 Part 2 is more commonly used for proven compressor designs.

Dampers
6. Functional and seal leakage tests may be carried out hot.
7. Manufacturing accuracy is important. Check dimensions, squareness and accuracy of joints at an early stage – pay attention to *detail*.

Heat exchangers
8. Make a careful examination during the hydrostatic test. Yielding or distortion is unacceptable. Remember that some leaks are *directional*.

Pipes and valves
9. Check material selection. Valves have a hydrostatic (body) test, seat leakage test, and functional test.

References

1. ISO 2548: 1973 is identical to BS 5316 Part 1: 1976. *Specification for acceptance tests for centrifugal mixed flow and axial pumps – Class C tests.*
2. ISO 3555: 1977 is identical to BS 5316 Part 2 : 1977 *Class B tests.*
3. ISO 5198: 1987 is identical to BS 5316 Part 3: 1988. *Precision class tests.*
4. DIN 1944: *Acceptance tests for centrifugal pumps* (VDI rules for centrifugal pumps). Verein Deutscher Ingenieure
5. API 610: 8th ed 1995. *Centrifugal pumps for general refinery service.* American Petroleum Institute.
6. ISO 1940/1: 1986 is identical to BS 6861 Part 1: 1987. *Method for determination of permissible residual unbalance.*
7. VDI 2056: 1964 *Criteria for assessing mechanical vibration of machines.*
8. ISO 1217 : 1986, identical to BS 1571 Part 1: 1987 testing of positive displacement compressors and exhausters – methods for acceptance testing.
9. BS 1571 Part 2: 1984. *Methods for simplified acceptance testing for air compressors and exhausters.*
10. API 617: 5th ed 1988. *Centrifugal compressors for general refinery services.* American Petroleum Institute.
11. API 618: 3rd Ed. 1986. *Reciprocating compressors for general refinery services.*
12. ASME PTC 10 for compressors and exhausters: 1984.
13. ISO 5388. This is identical to BS 6244: 1982. *Code of practice for stationary air compressors.*
14. ISO 2372. This is identical to BS 4675 Part 1: 1986. *Basis for specifying evaluation standards for rotating machines with operating speeds from 10 to 200 revolutions per second.*
15. ISO 1219-1: 1991. Identical to BS 2917 Part 1: 1993. *Specification for graphic symbols.*
16. BS EN 10 000 Series. BS EN 10 001 to BS EN 10 242 provide extensive coverage of the general subject of metallic materials.
17. BS 970. *Specification for wrought steels for mechanical and allied engineering purposes.* Parts 1, 2, 3, and 4.
18. BS 5135: 1984. *Specification for arc welding of carbon and carbon manganese steels.*
19. DIN 8570 *General tolerances for welded structures.*
20. ISO 5167: 1980. This is identical to BS 1042 Part 1 Section 1.1: 1992.

Specification for square-edged orifice plates, nozzles and Venturi tubes inserted in circular cross-section conduits running full.
21 BS 3889 Part 1: 1990. *Methods of automatic ultrasonic testing for the detection of imperfections in wrought steel tubes.*
22 BS 6072: 1986. *Method for magnetic particle flaw detection.*
23 BS 3601: 1993. *Specification for carbon steel pipes and tubes with specified room temperature properties for pressure purposes.*
24 ISO 5208. *Testing of general purpose industrial valves.* This covers the same subject matter as BS 6755 Part 1: 1991. *Testing of valves – specification for production pressure testing requirements.*
25 BS EN 60 534/2/3 1993. *Industrial process control valves - test procedures.*
26 API 598 6th Ed. 1990. *Valve inspection and testing*
27 ASME/ANSI B16.34 1988. *Valves, flanged, threaded and welding end.* The American Society of Mechanical Engineers.
28 ANSI/FCI 70-2 1982. *American National Standard for control valve seat leakage.* The American National Standards Institute.
29 BS 3274: 1960. *Specification for tubular heat exchangers for general purposes.*
30 BS 3636: 1985. *Methods for proving the gas tightness of vacuum or pressurised plant.*
31 ASTM E432. *Guide for selection of a leak testing method.*
32 TEMA (Tubular Exchangers Manufacturers Association): 1985. *Standards for design and construction of heat exchangers.*

Chapter 12

Cranes

Overhead cranes are the main items of lifting equipment that you will be called to inspect within a process engineering or power generation contract. True, there are other items such as hoists, passenger and stores lifts, and mobile cranes but they do not normally warrant as much attention as the main overhead cranes. Some contractors consider cranes as almost a proprietary item – on the basis that they appear to be of relatively simple design. They are, however, subject to very tight legislation in most countries, which means that they have well-defined certification requirements. Largely because of this, cranes are served by well-developed and empirically based technical standards.

Fitness-for-purpose criteria

It is necessary to view FFP criteria for cranes in the context of their role relating to statutory equipment. Statutory controls can help you to focus on what FFP actually is. Statutory rules and requirements have developed iteratively over a long period and incorporate a large body of empirical feedback and technical knowledge. Figure 12.1 is a pictorial representation of the three main FFP criteria which are also described below. As a general rule for statutory equipment, the criteria are perhaps not quite as discrete as I have shown them. There is always a degree of commonality of technical requirements between, for instance, the statutory requirement criterion and that of function and safety, so don't perceive them too rigidly.

Design classification

Overhead cranes have a well-developed system of design classification, which is accepted by ISO and most other standards organizations. Its purpose is to define accurately the anticipated duty of the crane and

360 Handbook of Mechanical Works Inspection

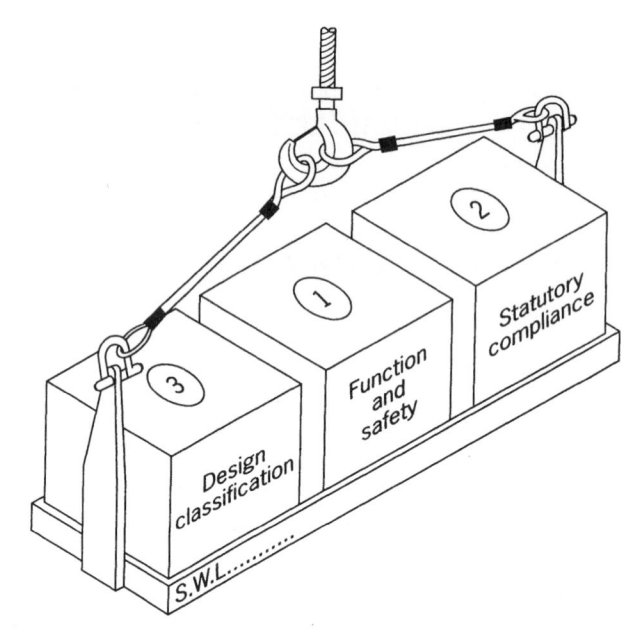

Fig 12.1 Cranes: fitness for purpose criteria

provide a framework for clear technical agreement between the customer and manufacturer. This design classification becomes a valuable tool for the works inspector. It distils a large number of technical criteria into a few design categories, so it becomes possible for an inspector to make a quick, but effective, design check very simply. Always check a crane's design classification as the first part of your FFP assessment. I will show you later in this chapter how the classification system works.

Function and safety

The easiest way to think of these is as criteria that have to be met *after* the crane has passed all its design checks. Function is concerned with whether the crane will lift its design loads, at the speeds specified, without excessive deflection or plastic distortion – then the various safety features incorporated into nearly all cranes must be proven to operate correctly. The main ones are the brakes and motion/load limiting devices. Note that not all of these tests can be completed in the manufacturer's works, some may have to be carried out after installation.

Statutory compliance

In most countries, legislation requires that cranes are *certificated* by a third-party body or classification society. The statutory instruments which impose this requirement would seem quite general if you were to review them – their strength is in the way that they can impose, rigidly, a set of design standards. These standards are more prescriptive for cranes than for some other types of equipment. To obtain certification it is also necessary to subject the crane to prescribed tests, and to demonstrate high standards of material traceability and manufacturing control. It is often not easy to keep a clear focus on how statutory requirements relate to fitness for purpose. For cranes though, you can be excused for making the *assumption* that there is a comfortably close relationship between the two, rather than statutory requirements being 'extra'. This makes things a little easier. Electrical aspects also form a part of a crane's certification requirements. The main areas are:

- isolation arrangements
- emergency stop and control devices
- motor rating and tests
- insulation rating and condition.

Basic technical information

There are many different types and designs of overhead cranes (they are shown in ISO 4301-1 / BS 466) **(1)**. Fortunately they exhibit very similar engineering principles and features, even between types which are visually quite different. Figures 12.2 and 12.3 show the main technical features that you will find on an overhead crane – in this case I have shown a 'top-running' double girder type. We can look in a little detail at some of these features.

The structure

The largest stressed member is the bridge, which is typically of double-girder construction. The girders are fabricated stiffened box sections, constructed with full penetration welds. Note the stress regime on these members – it is a straightforward bending case because of the 'simple' supports provided by the crab and bridge wheels. It should be clear that the maximum tensile stress will occur on the lower flange of the girders. Use this as guidance to where to perform witnessed non-destructive tests. A maximum vertical deflection of 1/750 bridge span is allowed when the crane is lifting its safe working load (SWL). Design calculations use the SWL as a reference point. Bridge end-carriages tend to be designed with a relatively small wheelbase (less than one-seventh of the crane span) so they are stiffer and deflection is less significant. The crab is also relatively stiff but it is subject to significant additional inertia loads. Construction therefore incorporates a lot of cross bracing and deep sections to resist the superimposed stress regime.

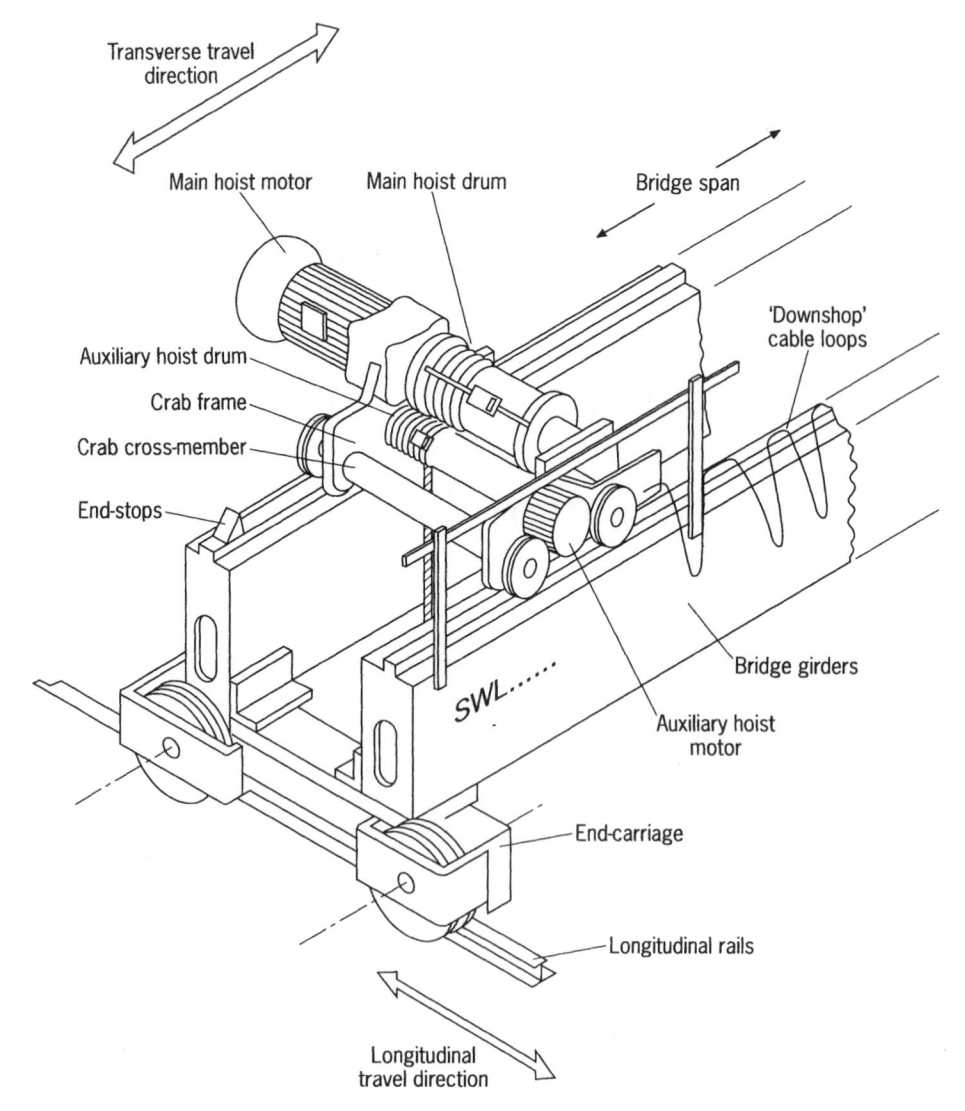

Fig 12.2 An overhead crane – technical features

This component can contain a lot of load-bearing fillet welds, in addition to the full penetration welds used for the fabricated base-frame.

The mechanisms

Crane mechanisms, although not in continuous use, are subject to high static loadings *and* an unpredictable set of dynamic conditions due mainly to inertia loads. Figure 12.3 shows the main components. The most important mechanism is the winding arrangement. The winding drum is normally manufactured to a pressure vessel standard (ASME, TRD, BS, etc.) because of the high compressive stress imposed by the rope turns under load. Rope grooves are machined to a carefully designed profile and spacing. Note the rope guide/sheave arrangement and the two 'dead turns' left on the drum when the load is in the fully lowered or raised position. A load cell or mechanical cantilever arrangement prevents excessive weights being lifted. There is also a 'hook approach' mechanism to stop the hook being wound too near the drum – it works by limiting the axial position of the rope guide and is adjustable.

Separate electric motors drive the main hoist, auxiliary hoist, (a smaller capacity hoist often fitted on the same crab), cross traverse, and longitudinal travel motors. There is sometimes an additional 'inching' motor on the main hoist to move the load very slowly when lifting, for instance, heavy turbine rotors. All these motors have friction brakes, which are a key safety feature of the crane – those for the hoisting motors will be of the centrifugal type. All brakes work on the 'fail-on' principle and are designed to stop the relevant movement in a well-defined short time. A variety of electrical protection devices are installed on the various electric motors and mechanisms.

Crane design classification (and how to check it)

Despite some of its inherent technical weaknesses, I have no hesitation in recommending that you consider crane design classification as one of the major fitness-for-purpose criteria. If there is a better 'quick design check' system I have not yet found it. Its stated purpose is to provide a technical frame of reference between the purchaser and the manufacturer, in a way that encompasses the important aspects of crane design. This helps match the design to its predicted service conditions. A key aspect is that this takes into account fatigue analysis – often neglected (in spite of its importance) in the design appraisal of many other types of equipment.

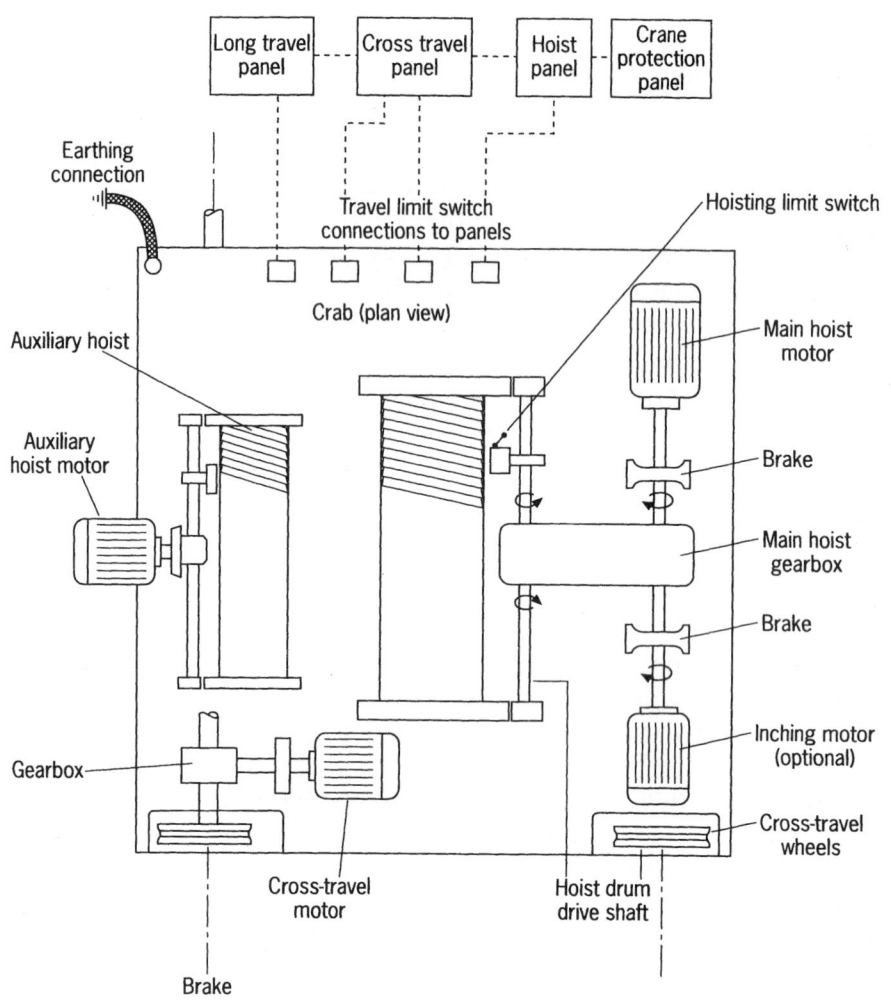

Fig 12.3 The crane mechanisms

The principle

The common system of classification is set out in detail in ISO 4301-1. It is also recognized in the more detailed design stress standard BS 2573 Parts 1 and 2 **(2)**. Expect to see this classification system in common use in contract specifications for power generation, chemical process, and steel-related industries. The main principle is that the crane *structure*, comprising the fabricated bridge girders and crab frame, has a different set of classes to the crane *mechanisms* – the rotating parts and connected components. This means that you can specify the structure class separate to the mechanism class. There is no imposed direct dependency between the two.

Structure class

Structure class (numbered A1 to A8) is determined by the combination of two design factors, the 'utilization' (U1 to U9) and the 'state of loading' (Q1 to Q4). The utilization factor relates to the projected number of operational lifting cycles of the crane structure and ranges in a series of preferred numbers from 3200 cycles (U1) up to 4×10^6 cycles and above (U9). The state-of-loading factor refers to the frequency with which the structure in use will actually experience the SWL – Q1 is where it lifts the SWL infrequently and Q4 is where it lifts it very regularly. Figure 12.4 shows these factors expressed pictorially as a matrix. I have shown the typical classification positions of common crane structures that you are likely to meet.

Mechanism class

This uses a similar principle but different letter-designations (see Fig. 12.4) Mechanism class (M3 to M8) is determined by the 'utilization' factor, this time termed T1 to T9, and the 'state of loading' (L1 to L4). For mechanisms, the utilization factor is based on projected lifetime operating hours (400 hours for T1 up to 50 000 hours for T9), rather than operating cycles. The state-of-loading factor is on the same basis as that used for structures. Owing to this simplified system of classification it becomes possible to perform an effective design *check* on cranes. Use the following guidelines to help you *before* you visit the factory.

- Check the manufacturer's classification against the structure and mechanism class requirement in the contract specification. You can make a quick estimate of utilization and loading state by finding out the planned application of the crane – but don't get too involved in the detailed stress information in BS 2573 Part 1.
- Try and decide which mechanism represents the 'weakest link' and pay just a little more attention to its design classification. Think what the manufacturer would have to do to prove to you that a mechanism is designated M3 or M4 (hint: look at the concept of mechanism duty factor 'G' in BS 2573, Part 2). Make a list of questions to ask before you witness the crane tests in the works. Take Fig. 12.4 with you.
- Be careful not to go too far. Effective works inspection is about verifying FFP – not designing bigger or better cranes.

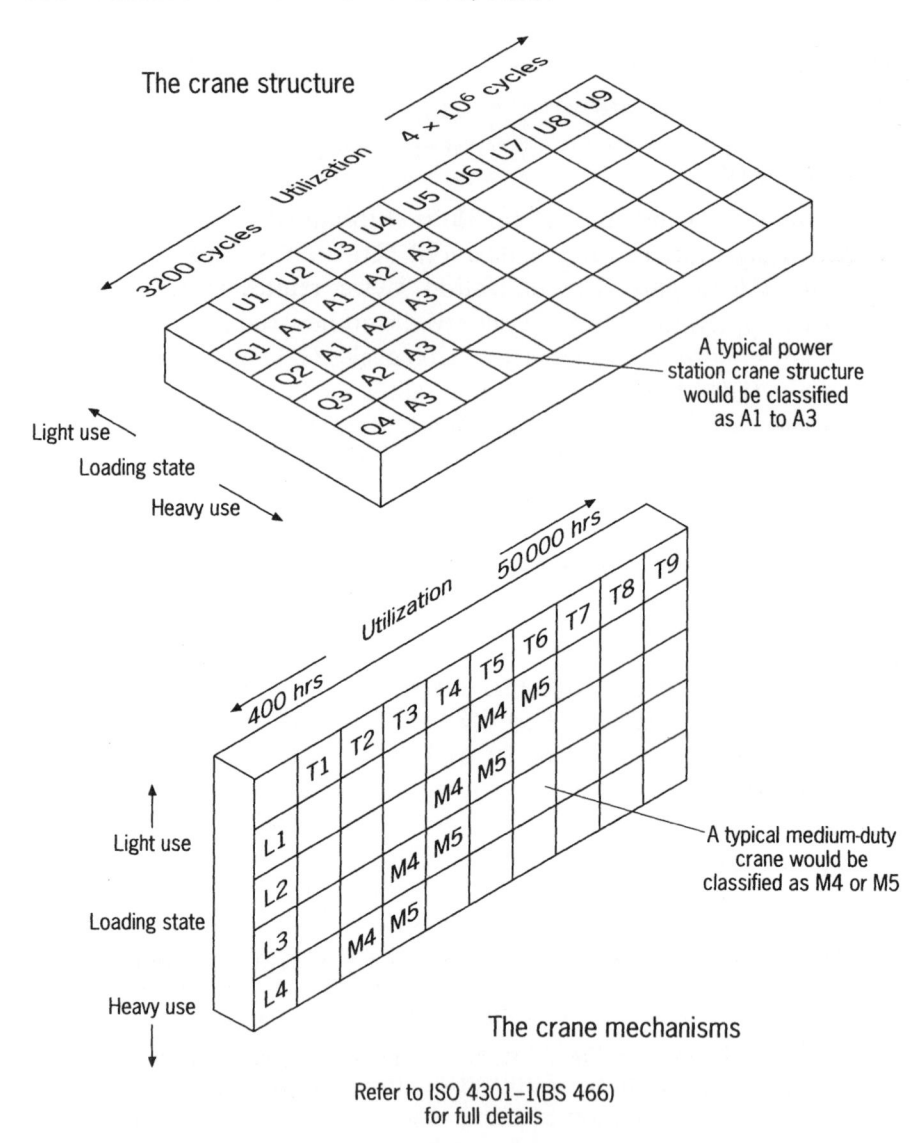

Fig 12.4 How to classify a crane

Acceptance guarantees

The acceptance guarantees format should follow the typical pattern used in most contract specifications. The general acceptance standard is normally taken as ISO 4301-1, qualified by contract-specific performance data. I have shown indicative speeds and data for a 130 tonne turbine house crane so you can see what to expect.

- Crane structure to meet the requirements of group classification A1.
- Mechanisms to meet the requirements of group classification M3 (for hoists, cross-traverse and longitudinal travel).
- Safe working load 130 t. Crane to be capable of lifting 125 percent SWL at mid-span of bridge girders in accordance with ISO 4301-1.
- Speeds of operation ($\pm$ 10 percent tolerance):
  - main hoist = 1.25 m/minute
  - crab traverse = 20 m/minute
  - longitudinal travel = 30 m/minute.
- Minimum hook approach distance = 1.5 m
- Vertical movement of the hook required = 17 m
- Minimum rope factor of safety (fos.) > 6.
- Maximum braking distance:
  - travel = 40 mm
  - traverse = 40 mm
  - hoist = 15 mm.

Note that, for a power station turbine-house crane, this is a relatively light-use example. Some will have higher design classifications than those shown.

Specifications and standards

We have already encountered the main technical standards in the previous few pages. Crane technology does not develop rapidly, so the technical standards that are available fit neatly (almost) into the requirements of purchasers and manufacturers alike.

ISO 4301-1 reflects the content of BS 466 *Specification for power driven overhead travelling cranes*. This is the best general industry standard and will tell you nearly all that you need to know. It covers some design and classification aspects but concentrates on testing. It makes direct reference to the detailed design standard, BS 2573.

BS 2573 *Rules for the design of cranes* is divided into two parts. Part 1 is a thick document with detailed information on stress calculations and design criteria for the structural members of the crane. It repeats the information about the classification of structures that is given in ISO 4301-1. The rest of the information is quite specialized and is difficult to apply to cranes if you are not familiar with structural design. I am sure that this is a good standard but it is not one which is essential for works inspection. Part 2 covers mechanisms. This is simpler, but again not essential for most works inspection situations. Use it if a design question arises, it has been developed empirically and has a sound basis.

Fédération Europiene de la Manutention *Rules for the design of hoisting appliances* **(3)**. This is a French standard, known more generally as the 'FEM regulations'. The part most relevant to overhead cranes is Part 1 *Heavy lifting equipment*. The FEM document is commonly used in European countries; it is predominantly a set of design rules (not unlike BS 2573 Part 1), concentrating on structural aspects.

BS 1757 *Specification for power-driven mobile cranes* **(4)**. This, as the title suggests, deals only with mobile cranes.

BS 2853 *Specification for the design and testing of steel overhead running beams* **(5)**. This applies only to the crane rails, which are made of rolled steel. Some works inspections may involve testing of the rails and you should find the standard is easy to follow, albeit having a limited application.

Two standards which are applicable to most cranes are BS 2903 *Specification for high tensile steel (crane) hooks* **(6)** and BS 302 **(7)** which covers the steel ropes. They contain simple but useful information on material tests and hook testing – you may not often witness these but the records should form part of the documentation requirements.

Inspection and test plans (ITPs)

ITPs for cranes are generally more detailed and precise than is the case for general fabricated equipment. The need for statutory certification imposes a significant pressure on all the involved parties to design and implement the ITP in a positive way. From an inspector's viewpoint the ITP is an excellent tool, so it is worth paying particular attention to some of the finer detail. Some of the points which influence inspection activities are:

- The ITP is normally subdivided by *component parts* of the crane. Structural members, drum and electrical equipment should each have their own section.
- Witness points shown on the ITP must include those relevant to the third-party certification body. Make sure the design appraisal activity is shown, if it is a legislative requirement.
- There will be a number of proprietary (bought-in) items such as the hook, wire ropes and brake assemblies. These are included in the

statutory certification requirements, so provision should be made to specify the correct certification requirements to sub-suppliers.
- Material traceability requirements are high. This will be reflected in the ITP entries showing material identification and marking activities. Some ITPs may specify a separate traceability mechanism for 'batch' materials which are drawn from the manufacturer's stock. Figure 12.5 shows a typical crane ITP.

Component	Check	Reference document	Acceptance standards	Witness M C TP
Crane bridge material	100 percent ultrasonic laminations scan. Visual inspection	BS 5996	BS 5996	X X
Fabrications (girders, end-carriages, and hoist drum)	Review of weld procedures.	EN 287/288	EN 287/288	X X X
	Ultrasonic test 100 percent	BS 5500 (typically)	BS 5500 Table 5.7(2)	X X X
	MPI 100 percent	BS 5500 (typically)	BS 5500 Table 5.7(3)	X X X
	Visual/dimensional examination	Drawing	Drawing tolerances	X X
Wire ropes	Proof test	BS 302	BS 302	X
Hoist components (load bearing members)	Full document review	Material specification	Specification	X X
Control panels	Functional test			X
Motors	Insulation test	ISO 4301-1	ISO 4301-1	X X
Proprietary items: wheels, rails, shafts, and gears	Full documentation review	Specification and ISO 4301-1	Specification and ISO 4301-1	X X
Hook	Proof load test	BS 970	BS 970 (or type test certificate)	X X
	Material tests	BS 970		
	Dimensional check	BS 970		
Crane assembly	Visual/dimensional inspection	GA drawings	Checklist	X X X
	Interim document review			X X X
Crane assembly	Light run check	ISO 4301-1		X X X
	Functional check	ISO 4301-1		X X X
	Overload test	ISO 4301-1	Checklist	X X X
	Insulation checks	ISO 4301-1		X X X
	Final certification review	ISO 4301-1		X X X
Crane assembly	Painting and packing check	Purchase specification	Referenced paint standard	X X

M – Manufacturer
C – Contractor/purchaser
TP – Third Party

Fig 12.5   A typical crane ITP

Test procedures and techniques

With the exception of very large designs, most cranes can be tested effectively in the manufacturer's works. There are a few limitations (again based on size) which can necessitate some simulation of a number of factors, but these have little effect on the validity of the fitness-for-purpose assessment. Because cranes are subject to statutory certification it is normal for works tests to be witnessed by several inspection parties (I provided some general advice on this situation in Chapter 2 of this book, which you may find useful to review in the context of crane tests). As a preliminary to the works tests, I have mentioned that you should carry out a design classification review. It is wise to ensure that the third party inspector is also doing this. Treat the design check as a useful familiarization exercise for the design of the crane to be tested – a benefit which will show through in your technical report.

Visual and dimensional examination

The visual and dimensional examination can be carried out before or after load testing of the crane. In practice, the major measurements are best taken afterwards, as part of the checks for post-test distortion. It is always useful, however, to try and start the works test procedures with some visual and dimensional checks. This will help you become familiar with the features of the crane. The minimum checks you should make are as follows.

Span

The bridge wheel span dimension is critical. It must comply accurately with the drawing. The important dimension is the centre-to-centre distance of the bridge end-carriage wheels – it should be accurate to better than ± 3 mm to match up with the equivalent tolerances used for the longitudinal rail alignment.

Roof clearance

An overhead crane, by definition, operates near the roof of a building and is designed so it does not foul the structure or fittings such as ventilation ducts and lights. Check the roof clearance 'envelope' dimensions shown in the general arrangement (GA) drawing and make sure that there are no protuberances on the crane itself that exceed these dimensions.

Cranes 371

Pendant length

Once installed, the crane will be controlled from the floor using a suspended pendant control. Check the installed length of pendant cable – it should be long enough to meet the specified pendant height, which will be shown on the GA drawing.

Manufacturing accuracy

Check the accuracy of the main bridge girders, particularly if they are of the fabricated box-girder type. Follow the fabrication drawing tolerances on straightness and squareness of the assembly. You can do a preliminary check by eye, sighting along the girders – the minimum distortion (bend) that you can *reliably* detect using this method, however, is probably no better than about 30–40 mm per 10 m girder length. This is well outside acceptable tolerances. The best way is to use a steel tape, measuring the distance between the inside faces of the two girders at several points along the bridge length. This gives a better estimate of truth in the horizontal plane. Vertical truth is easier to check – you can take measurements from the lower face of each girder to the floor (make sure the bridge is in the unloaded condition when you do this).

Paintwork

Crane paintwork falls into two categories. The painting specification for the large components (crab frame, end-carriage frames, and bridge girders) will normally be customer-specified in the purchase order. A standard of 200–250 microns is the norm. The top coat is nearly always bright yellow. Use the guidelines given in Chapter 14 when checking the paintwork on these parts. Often, smaller components such as motors and parts of the hoist assembly will be 'bought-in' from sub-suppliers. Check that these comply with the specified paint requirements and do not have a significantly lower dry film thickness (dft) or different paint type. Pay particular attention to painting on cranes to be installed in buildings housing chemical process plants – environmental conditions near the roof can sometimes be highly corrosive. This will soon expose the weaknesses in an unsuitable paintwork system, or one which has been poorly applied.

Insulation test

The technical standards, and certification organizations, require that an insulation test be carried out on all the electrical components fitted to the crane. In practice this is normally performed on the motors and electrical panels after assembly. The test is quite straightforward, a dc voltage of twice the rated voltage is applied and the resistance to earth measured. The minimum allowable resistance is 0.5 MΩ. Make sure the test is carried out in dry conditions. You can also make a useful visual examination of the insulation to check for mechanical damage – pay particular attention to the conductors which run the length of the bridge, and to the loops of the pendant control cables which move with the crab.

The light run test

This is a no-load test. The purpose is to check the function of the various operating and safety systems before operating the crane in the loaded condition. Note that this is essentially a safety procedure, as the function of the systems could be checked, if required, with the crane under load. Some manufacturers do it this way. The test consists of the following elements.

Speed checks

The hoist motor, crab traverse motor, and longitudinal travel motor are operated (separately) to check that they are correctly geared and provide the correct driving speeds required by the contract specification. If speeds are not specified, check ISO 4301-1 for acceptable values. Specific points to watch are:

- Check the rotation of motors in both directions, including up and down and at 'inching' speed for the hoist. Use a tape measure and stop-watch to measure all the travel and hoisting speeds.
- There is a normal tolerance of $\pm$ 10 percent on allowable speeds. The preferred speed categories given in ISO 4301-1 are quite broad, so small differences are unlikely to be critical. It is preferable for speeds to be too slow, rather than too fast – this is particularly applicable to the long travel motion, where a too-high speed can increase significantly the inertia forces experienced as the crab or bridge is stopped when carrying full SWL.
- Most large cranes are tested with the end carriages mounted on a short section of test rail, or with the carriages resting on blocks and the wheels suspended, so it is not possible to check the longitudinal travel speed. Under these circumstances, simply measure the rotational speed of the wheels over 30 seconds or so and calculate the corresponding longitudinal speed.

Hook approach test

The factory test rig should have enough vertical clearance underneath it to enable the hook approach test to be carried out. The hoist is wound up at slow speed until the hoist limit trip operates, isolating the power to the motor. Note that this is an adjustable limit (it works using an adjustable axial position switch on the hoist rope-guide) so after the test, make sure the setting is sealed to prevent accidental movement. The minimum acceptable hook approach distance will be stated in the contract specification or the technical schedules/data sheets.

Hoisting height

If the crane manufacturer is using a low level test rig it will not be possible to measure directly the hoisting height. It is acceptable to calculate this from the number of rope turns on the hoist drum. Note the following points:

- Use the effective winding diameter of the hoist drum (defined by the position of the rope centre), not the drum external or flange diameter.
- All crane design standards require that two full rope turns on the drum be classed as 'dead turns'. This means there should be two empty grooves (sheaves) when the hook is at minimum approach distance and two full rope turns left on the drum when the hook is fully lowered. You should not include these 'dead turns' in your hoisting height calculation.

Brake operation

Brakes are fitted on the hoisting, cross traverse and longitudinal travel mechanisms. Brake efficiency is only proven properly during the loaded test, however the light run test is a good opportunity to test the function of the brake mechanisms. Check for:

- *Fail-safe mode.* The brakes should *fail on* when the power supply is cut off.
- *Free operation.* There should be no binding of the brakes when they are in the 'off' position. You can detect binding by excessive noise or heating of the brake housing, accompanied by high current reading from the relevant drive motor. Most drum brakes can be adjusted.

SWL performance test

This is the main proving test of the operation of the crane. The load is a concrete block in a steel lifting frame – expect the manufacturer to have a 'lifting schedule' which shows the weight of the individual components of the lift (frame, shackles, etc.) to demonstrate the correct overall SWL is being used. A separate smaller weight will be required for the auxiliary hoist. The test consists of the following three elements.

Hoist up and down

The load is hoisted up and down several times, as near to the specified vertical limits of travel as is possible on the test rig. Important points to watch are:

- Use slow hoisting speed first and do not let the load swing too much.
- Check brake operation. Stopping should be smooth in both directions – there should be absolutely no juddering as the load comes to rest. Estimate the stopping distance using a tape measure or a chalk mark on the hoist drum. Pay particular attention to the 'worst case' condition, i.e. when the load is being lowered, and watch for any slipping of the brake. As a guide, maximum braking slip should be not more than 7–8 mm per m/minute of hoisting speed.
- Check brake release. The brakes should release cleanly when the power is restored. Check the motor current again and listen for any undue noise.

The hoisting test is the best time for checking for any obvious problems with the crab – use the same checks as those shown for the overload test. Note that it is necessary to do the checks *twice*, first during the SWL test and then again during the overload test. It is not good practice only to do them during the overload test – if there is a problem with cracking or yielding it will be useful to know whether it occurred during the SWL or the overload test.

Traverse and longitudinal travel

The crane traverse and travel motions need to be tested under SWL conditions. Longitudinal travel cannot usually be tested fully, however, as the test set-up will consist, at best, of a short longitudinal rail section. This is normal. The main points to check are:

- Make sure that the crab is tested over several complete traverses of the bridge length. A single traverse is not always sufficient to show any problems.
- Check the *tracking* accuracy of the crab and the bridge rails (see Fig. 12.6). You can do this using an accurate measuring tape or feeler gauges. It is important to make sure that the clearances are the same

Cranes 375

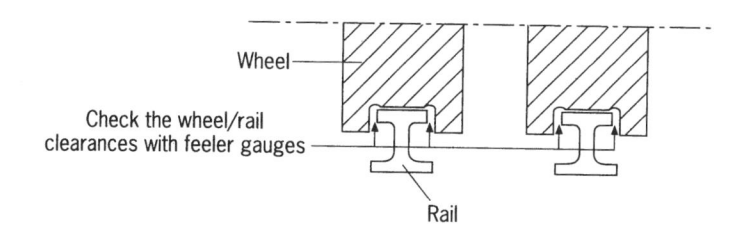

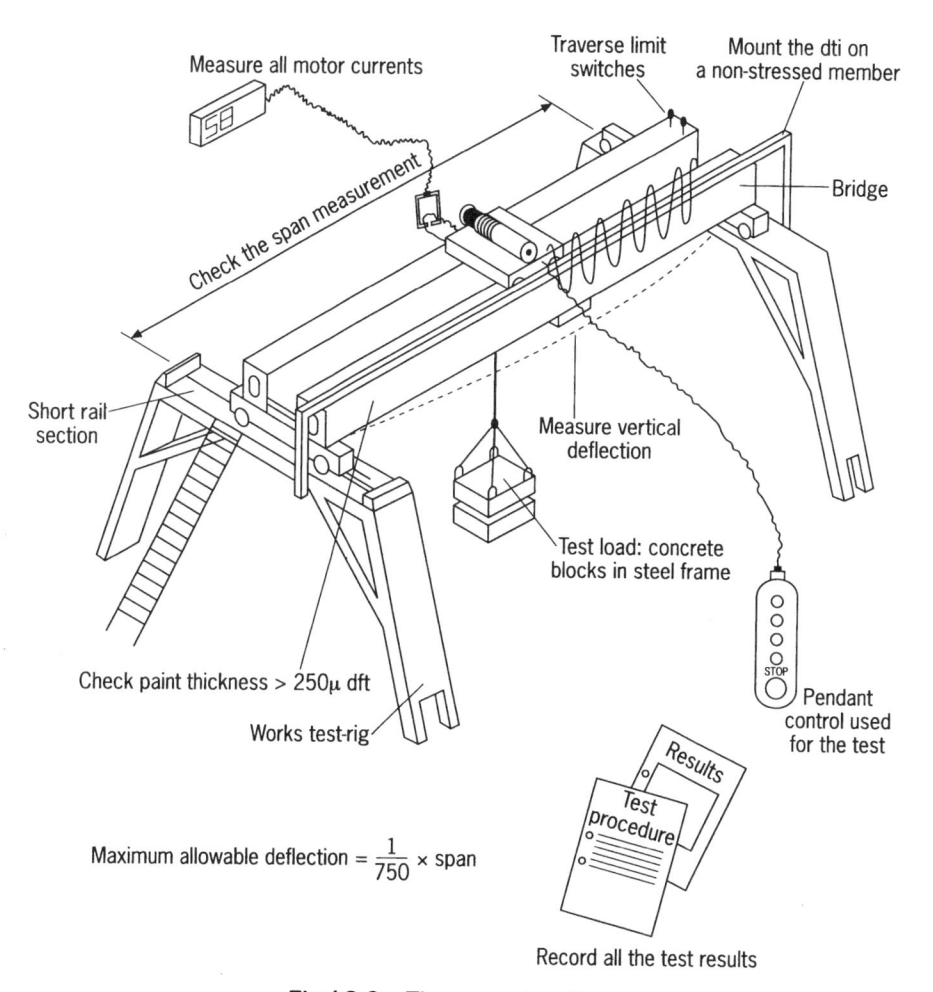

Fig 12.6 The crane load test

- Wheel
- Check the wheel/rail clearances with feeler gauges
- Rail
- Measure all motor currents
- Traverse limit switches
- Mount the dti on a non-stressed member
- Check the span measurement
- Bridge
- Short rail section
- Measure vertical deflection
- Test load: concrete blocks in steel frame
- Check paint thickness > 250μ dft
- Works test-rig
- Pendant control used for the test
- Maximum allowable deflection = $\frac{1}{750}$ × span
- Test procedure
- Results
- Record all the test results

at several points along the rail – this will show whether the rail is straight. Acceptable tolerances on rail distortion are quite small (I have shown some approximate values in Fig. 12.7) because misalignment will cause quite rapid wear in use, causing the crab to become unstable (it will 'wobble') in some positions, most often at the end of the bridge.
- Check the operation of the crab traverse limit-stops. These are normally a simple micro-switch. There should also be physical stops or buffers to stop the crab traverse if the limit switch becomes defective – they should be positioned accurately so the crab will contact both stops at the same time, hence spreading the impact load.
- Watch the crab closely as it stops after a full traverse. You will need to climb into the access gantry for this – it is of little use viewing it from the ground. Check what effect the inertia force has on the crab assembly. There should be no 'jumping' of the crab on the rails as the load swings, or visible movement of the hoist mechanisms bolted to the crab. Everything should be securely dowelled and bolted.

Deflection measurement

It is normal to measure vertical deflection of the bridge under SWL, as required by the design standard. Sometimes it is measured at overload conditions as well. Measure deflection at mid-span on the bridge using a dial test indicator (dti). Figure 12.6 shows the general arrangement. A good test should incorporate the following points:

- The 'undeflected' reading should be taken with the crab positioned at one end of the bridge.
- Don't take the deflection reading immediately after lifting the load, move the load up and down a few times and then wait a minute or so for the structure to settle.
- It is essential to have an accurate fixed datum point on which to mount the dti. The mounting point should not be connected in any way to the stressed structure. If it is, the deflection reading taken will be meaningless (you will get an under-measurement). Usually, it is best to try and mount the dti on the access gantry or similar. A more difficult way is to measure the distance vertically from the underside of the bridge to some reference point on the floor. If the manufacturer insists on doing it this way, think very carefully about whether you are happy with the accuracy of the measurements taken.
- Double-check that the crab is at the exact mid-span of the bridge

Cranes 377

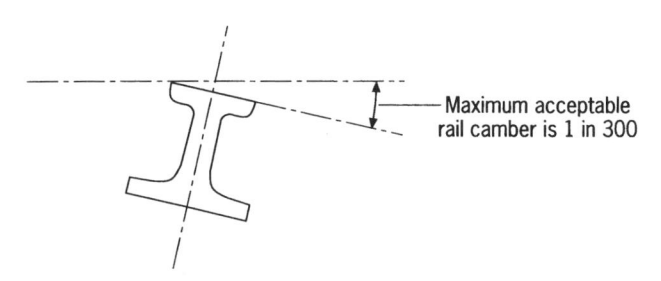

Fig 12.7 How to measure crane rail alignment

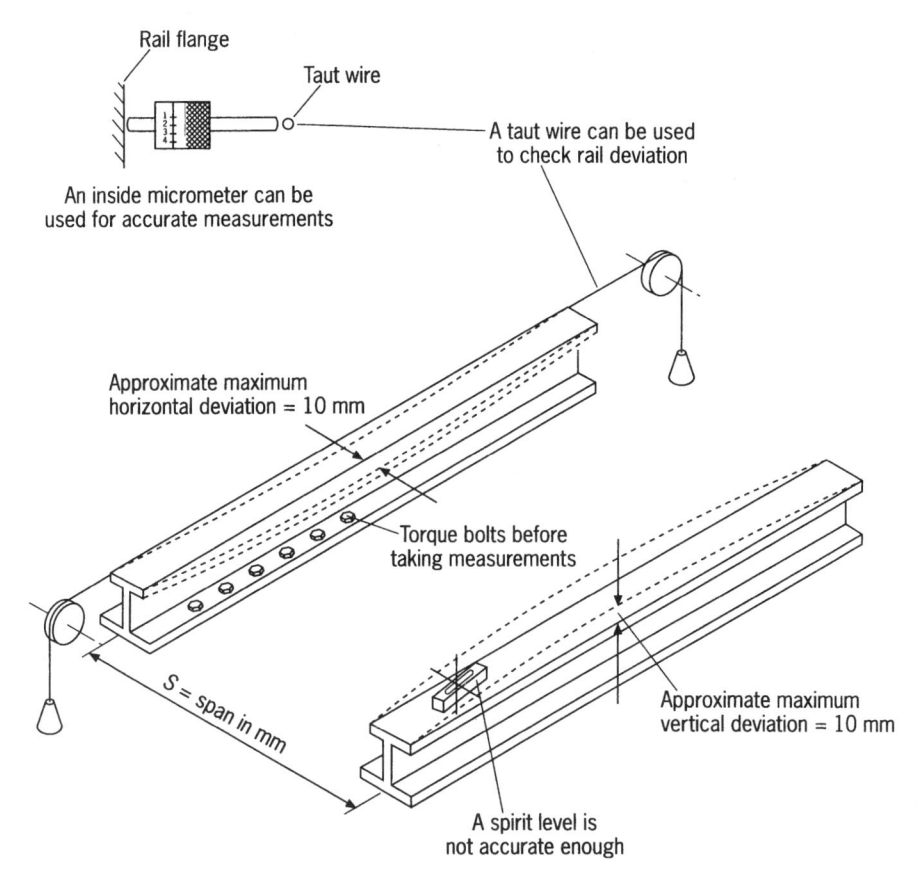

Labels in figure:
- Rail flange
- Taut wire
- An inside micrometer can be used for accurate measurements
- A taut wire can be used to check rail deviation
- Approximate maximum horizontal deviation = 10 mm
- Torque bolts before taking measurements
- Approximate maximum vertical deviation = 10 mm
- A spirit level is not accurate enough
- S = span in mm
- Maximum acceptable rail camber is 1 in 300

Span tolerance: $S \pm 3$ mm for $S \leq 15$ metres
$S \pm (3 + 0.25 [5–15])$ for $S \geq 15$ metres

when measuring the bridge deflection. There should be a hard-stamped reference mark. If there isn't, make one before the works test so that the test can be repeated accurately after installation on site.
- Vertical deflection under SWL should be an absolute maximum of 1/750 of the span (longitudinal rail centre-to-centre distance). Anything more is a major non-conformance point.

The overload test

Crane design standards and most good contract specifications specify an overload test at 125 percent of SWL. This acts as a proving test for the factors of safety incorporated into the mechanical design, and of the manufactured integrity of the structure and mechanisms. Your task as an inspector during the overload test is to look for three things: yielding (by this I mean *plastic* distortion), cracking, and breakage. Surprisingly, none of these three will be easy to find, they will only show themselves as a result of a structured approach to the test. Once again, this is not an area in which you can use a 'broad-brush' approach. Consider two preliminary points.

- *Checklists.* You cannot inspect thoroughly, and then report effectively, without your own checklist listing those areas of the crane to be examined during an overload test. There is absolutely no reason why the manufacturer has to provide you with one that is comprehensive enough. A good checklist will cover all the main stressed components of the fabricated structures and the rotating mechanisms.
- *Stripdown.* Unlike some other equipment, cranes do not normally require an extensive stripdown examination after the overload test. There are two areas that you should inspect, however: the drive gear trains for the hoisting mechanism (be prepared to ask for removal of inspection covers to check for gear tooth breakage) and the internal stress-bearing welds of bridge girders (it may be necessary to remove inspection hatches or cover plates on the bridge box sections).

The overload test consists of lifting a 125 percent SWL weight to impose *static* stresses on the structure. Note that crane standards do not specify directly that the overload test should test for the structure's ability to withstand dynamic (inertia) loads resulting from movement in the overload condition. Dynamic stress concentration factors can be very large so beware of inertia loads – it is acceptable to use slow speeds for hoisting and traverse, and to wait until the load stops swinging before starting the next movement. You can measure vertical deflection during the overload test but this is generally not subject to an acceptance level.

The test should consist of a complete set of hoisting, crab traverse and, if applicable to the test rig, longitudinal travel movements. During and after the test, go through your checklist item by item, making a full visual inspection of all the stressed components. A brief analysis of the stress case will quickly indicate the areas of highest tensile and shear stresses. For the structural sections of the crab and bridge, areas of *stress concentration* such as fillet weld toes (particularly on stiffeners which are subject to a shear stress, transverse to a web axis), changes of section from thick to thin material, and sharp corners are the most likely locations in which to find cracks. Fillet and gusset-pieces that do not have 'mouse-holes' at the enclosed corner are also common areas (see Fig. 12.8). Some useful points to note are:

- Make sure fillet welds are clean before the test so you can see any subsequent cracking. Very thick paintwork which obliterates the weld is not a good idea; it can and will hide cracks.
- You should ask for a dye penetrant test of critical areas if you have evidence that there is any cracking at all. If this is not available, a useful technique is to clean the weld and apply a light 'dust' coating of white cellulose spray primer (a proprietary aerosol can is fine and dries in seconds). The paint will make small cracks much more visible. It is also useful to help detect any movement of the main hoist on its mountings. This is only a stand-by technique, however, and does not replace a proper dye penetrant or magnetic particle examination.
- Yielding (plastic deformation) is more difficult to detect, particularly if it does not progress to the extent where it causes cracking or breakage. The *only* quantitative way to detect this type of yielding is by direct measurement before and after loading the particular component. In practice, because of the size of overhead crane components compared to the accuracy of measurement (using a tape or straightedge rule) the only yielding that you can identify with any certainty is that related to the bridge girders. The crab is much smaller and stiffer so the resulting small deflections are difficult to measure. For the bridge girders make sure, during the deflection check, that the dti reading returns to zero when the load is removed.

This gives some comfort that the bridge has only been loaded within its elastic limit and that yielding has not occurred.
- Drive gear teeth do not have such a high factor of safety as some of the structural parts of a crane. This is particularly the case if the mechanism design classification is in the M3 or M4 category. You should inspect all gear teeth for the main drives. The objective is to look for tooth breakage or obvious visible cracks. Normally, if there is a problem with gear teeth you can expect to find teeth completely broken, or chipped due to overstressing or shock load – you are not looking for scuffing or any of the failure mechanisms associated with rotational wear.
- Load display: if this feature is fitted, it should be correctly indicating the overload during the test. Check this carefully, it is an important safety consideration.
- Inspect the rope grooves on the hoisting drum after the test is finished. Look for any obvious abrasion of the groove lips which would indicate the rope guide mechanism is not adjusted correctly.
- Double-check for flaking paint on all components – this is a sure sign of excessive stress. The only possible exception to this is on the bridge

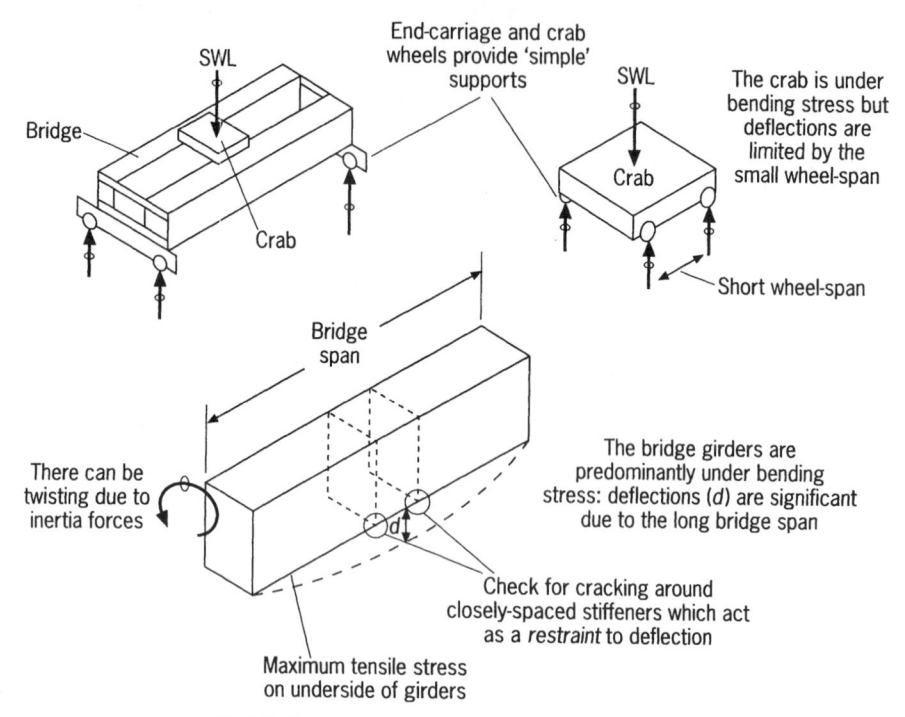

Fig 12.8 Stresses on crane components

girder, particularly if it is a single box section girder or offset crab type sometimes used on lightweight cranes. These may twist within their elastic limit, causing some slight paint-flaking on the box section corners around mid-span. This can be acceptable – but check the vertical deflection accurately to make sure that yielding has not occurred. Designs with double box section girder bridges (the type shown in Fig. 12.2) should not twist or deflect enough to cause the paint to flake.

It is worth mentioning again the importance of completing your checklist. I have purposefully not given you an example because it is better that you develop your own. The best type will encourage you to make specific comments about what you found rather than reduce it to a simple list of 'ticks in boxes'. The main benefit of a more detailed report is in the event of a crane failure after installation – there will be a positive record of what was actually checked during the works inspection. Try to make your crane checklist as comprehensive and incisive as you can. A poor checklist (or no checklist) sometimes looks like an attempt to avoid responsibility for making important, but difficult, observations about fitness for purpose.

It is difficult to think of many really valid reasons why a manufactuer cannot perform all the tests (except for the longitudinal travel, as I have mentioned previously) in the works, rather than to defer any problems to the construction site. The governing factors are generally with length of the bridge and the size of test load required – some manufacturers will just not have large enough test facilities.

Common non-conformances and corrective actions: cranes

Non-conformance	Corrective action
Incorrect speeds	Remember there is a ± 10 percent tolerance on speeds. Marginal deviations don't matter *that* much, as long as they are not outside the broad ranges given in BS 2573 Part 2. Minor differences are an acceptable basis for application for a concession. Support the manufacturer in this. Make it clear in your report that you feel, prima facie, the concession should be approved.
Malfunction of limit switches or other small electrical items.	This is probably caused by minor faults in the wiring arrangement, or it is not uncommon for overcurrent trips to be incorrectly adjusted. You should issue an NCR but ensure that you refer to a rectification list that you have agreed with the manufacturer. Rectifications can be done on site during final installation assembly. Persuade the third-party certification organization that this is not the type of defect that should have any effect on the issue of statutory certification.
The crane design cannot be verified.	This is a fundamental FFP criterion. It warrants an NCR. Write it so that you place the action on the manufacturer to demonstrate the design classification – *then* do what you can in discussions with the manufacturer's designers while you are still in the works. Clarify as many points as possible and produce a detailed set of minutes of your meeting. A good designer will suggest that you separate carefully points relating to structure and mechanisms so you don't get confused. You can refer to the clauses in ISO 4301-1/BS 466 and BS 2573 Parts 1 and 2 as applicable.
Incomplete documentation	This is important because a crane is a statutory item. A third-party certification organization is acting correctly by withholding certification under these circumstances. Firstly, identify *exactly* which documents are missing, using the ITP listings as a reference. Then prepare your NCR in conjunction with the third-party inspector, making sure that you agree (so you can state in writing) which documents are missing. Give the manufacturers a timescale for completing the documents and tell them, in writing, who to send the documents to. Continue with the crane tests and record the results as you would have done if there had been no documentation problems. You may find a strategic approach will help you – briefly review Chapters 2 and 3 of this book to remind yourself what it's all about.

Cranes 383

Common non-conformances and corrective actions : cranes—continued

Non-conformance	Corrective action
Poor braking	This normally just requires adjustment so don't issue an NCR prematurely. Witness the adjustment and then retest without further delays. Check the 'lift-off voltage' of the brake solenoids and note it in your report.
Bridge rail misalignment	If you observe clear mechanical *wear* on the rails then the misalignment is serious. You should: • issue an NCR • inspect and report on the condition of the wheel flanges as well as the rails • try and agree with the manufacturer 'on the spot' whether the rail or wheels (or both) need replacement. If the misalignment is slight, i.e. you have only detected it by close *measurement*, the correct action is: • Issue an NCR, placing the action on the manufacturers to produce a report comparing the actual rail alignment measurements with the limits in ISO 4301-1. Witness the 'actuals' measurements while you are in the works, make a sketch of the results, sign it and then pass it to the manufacturers to do the comparison. Encourage them to do this while you wait. • Don't withdraw *any* NCR until you are satisfied. Be difficult to satisfy, but not unreasonable.
Excessive bridge deflection (> 1/750th span)	Repeat the test, paying careful attention to the accuracy of the measuring technique. If the retest confirms the excessive deflection then: • Look for yielding, broken webs or stiffeners as an obvious cause. • If there is no obvious yielding or breakage it is likely to be a serious design problem. There are no quick solutions to this. Issue an NCR describing the failure and asking for a design report demonstrating compliance with BS 2573. Involve the third party in this. • Don't apply for a concession.
Yielding, cracking or breaking of the structure or mechanisms during SWL or overload test.	Step 1: *Quantification*. Agree with all parties the extent and nature of the damage. Don't draw instant conclusions on the possible cause. Strive for technical accuracy in your description of the failure (read chapter 15 of this book). Step 2: *Design check*. Make a quick check on the design using the guidelines I have introduced in this chapter. Step 3: *Material check*. The earlier you can eliminate the possibility of incorrect material use the better. Check the traceability records – the third-party inspector will *probably* be quite experienced at this.

Common non-conformances and corrective actions : cranes–continued

Non-conformance	Corrective action
	Note: if material records prove elusive, look at the section in Chapter 2 I have called 'Asking and listening'. You can try these techniques here. Step 4: *Ask the manufacturer for a design report.* Then, using the information from the above steps, write an accurate and concise NCR.

KEY POINT SUMMARY: CRANES

1. The main FFP criteria for overhead cranes are:
    - Design suitability (classification)
    - Function and safety
    - Statutory compliance.

2. Lifting equipment is influenced by statutory requirements, which means that Third Party *certification* is required. Expect an overhead crane to be subject to:
    - Formal design review
    - Third Party inspection during manufacture
    - Comprehensive documentation requirements.

3. Cranes are well covered in standards ISO 4301-1 (BS 466) and BS 2573 Parts 1 and 2. Most contract specifications will reference these.

4. There is a simple system of design classification that you can learn to use. Crane structures are classified by 'A' numbers and mechanisms by 'M' numbers.

5. Works testing includes an insulation test, light-run test (to check function), performance test at 100 percent SWL, bridge deflection test and overload test at 125 percent SWL.

6. You need to develop your own detailed checklist for the overload test – to give a structured approach to your observations.

7. Don't support a concession application if there is *any* breakage, cracking or yielding. Yielding can sometimes be difficult to identify.

References

1 ISO 4301-1: 1984. This is technically equivalent to BS 466: 1984. *Specification for power driven overhead travelling cranes, semi-goliath and goliath cranes for general use.*
2 BS 2573: *Rules for the design of cranes* Part 1: 1983. *Specification for classification, stress calculations and design criteria for structures.*
BS 2573 Part 2: 1980. *Specification for classification, stress calcuations and design mechanisms.*
3 Fédération Europiene de la Manutention. Design rules sections I and IX.
4 BS 1757: 1986. *Specification for power-driven mobile cranes.*
5 BS 2853: 1957. *Specification for the design and testing of steel overhead runway beams.*
6 BS 2903: 1980. *Specification for higher tensile steel, hooks for chains, slings, blocks and general engineering purposes.*
7 BS 302 Parts 1 to 8. *Stranded steel wire ropes.*

Chapter 13

Linings

RUBBER LININGS

A disproportionate number of problems on power generation, desalination and general process plants are caused by the failure of rubber linings. It seems that these failures tend to happen at the most awkward time, perhaps 6–12 months after plant commissioning. Because linings are used as an alternative to expensive materials (the purpose of linings is to resist corrosive and/or erosive effects of the process fluid) they find use on large vessels and equipment items. This means that downtime and repair costs are inevitably high.

Fitness-for-purpose criteria (FFP)

More than 90 percent of all lining failures can be attributed to poor application. It needs only a single application error to break the seal between an aggressive process fluid and a vulnerable base material. General corrosion, erosion, and failure then follow in quite rapid succession – with some process fluids it may take only a few days. For this reason the overriding FFP criterion for linings is the way in which the lining is *applied* to the base material. You will find it useful to apply this simplified view during works inspections to help your focus. It is easily within the capability of a works inspector to find and eliminate nearly *all* the application errors that cause linings to fail – techniques are straightforward and there is a set of well developed and proven tests. Consistency is also on your side – remember the point I made in Chapter 2 about engineering problems tending to repeat themselves in a rather regular way – about them being *predictable*? Linings are one of the best examples of this. We will work through the technical issues, to see what these application errors are, and where to look for them.

Basic technical information

The purpose of a rubber lining is to protect vulnerable materials against corrosive and erosive attack. Rubber is commonly used to protect components in sea-water cooling, condenser cleaning, and chemical dosing systems, and to resist aggressive process liquors in specialized chemical plants. There are two distinct types of rubber compound used. Natural rubbers are used for general low temperature sea-water or slurry system applications. Synthetic compounds such as nitryl, butyl or neoprene are used for operating temperatures up to 120 °C or when oil is present. Both natural and synthetic rubber formulations can be classified broadly into either hard or soft types. Hard rubbers have a higher sulphur content, which in some cases forms hard compounds commonly called ebonites. You will sometimes see hard rubber actually termed ebonite. It is mainly used for temperatures up to 100 °C. The terms 'hard' and 'soft' rubber are generally accepted as referring to specific hardness levels.

The most widespread method of application of both natural and synthetic rubber linings is to apply the rubber in sheet form, bonding it to the vessel or component surface with a suitable adhesive such as isocyanate. The rubber sheet is normally 3, 4, 5, or 6 mm thick. This process is done manually – in large vessels the sheets are laid down in overlapping courses while for smaller components such as valves and pipes, smaller pieces are stretched over the component profile, stuck down, and the excess trimmed off. As a final step the lining is vulcanized by heating to approximately 120 °C in a steam-heated autoclave or oven for several hours. This develops the final physical and chemical properties of the rubber.

From this description, two points stand out. Firstly, rubber lining is a highly *labour intensive* activity – it is practically impossible to automate the sheet-lining of engineering components, apart from perhaps small, mass produced fittings. This means that the whole activity is skill dependent – and subject to human error. Secondly, rubber lining is an activity which requires careful *process control* – small variations in the way in which application, bonding, and vulcanizing are done will significantly affect the final result. These two points have direct links to FFP and so have important implications for the way in which you should approach the inspection of rubber linings.

Acceptance guarantees

The process of rubber lining is analogous to welding, in that acceptance guarantees in contract specifications normally reference technical standards, rather than specifying directly a separate list of tests and pass/fail criteria. Expect to see only a rubber type, thickness and hardness value stated. It is the case, however, that one of the main issues governing FFP remains largely implicit – you will rarely see it written in a guarantee schedule. This is the issue of *operator skill*. The assessment of the operators' skill in applying linings, with all the subjectivity that implies, is the responsibility of the works inspector – and it forms a major part of the *implicit* requirements of a contract specification.

Special design features

Those parts of a process system that are lined as a means of protection normally incorporate a number of 'special' design features. These are by no means major design changes (for instance amendments to the popular vessel or valve codes). They are small design *features*, incorporated in the recognition of the specific types of problems that arise when a lining has to be applied to a component (and has to stay on). Figure 13.1 shows those that I consider most important. It is good practice to review these features as one of the first stages of your works inspection of a rubber lined vessel or component (there are several technical areas in this book where I have recommended this type of simple design *check*). You do not need to be a designer to do it. You may wish to look at the technical standards that can help you specifically with this activity. The references are included at the bottom of Fig. 13.1.

Specifications and standards

Although the chemical analyses and mechanical properties of rubbers are the subject of quite comprehensive technical standards, you do not need large volumes of information with you for the purposes of works inspection. Many of the standards concentrate on the chemistry of rubbers and are predominantly of 'laboratory interest' and are therefore not of direct use during a works inspection. There are a small number of standards, however, which *are* of use to an inspector:

390 Handbook of Mechanical Works Inspection

Pipework

Nominal pipe dia	Max. length between flanges
32 mm	2500 mm
40 mm	3000 mm
50 mm	3500 mm
65 mm	4000 mm
80 – 600 mm	6000 mm

All pipes < 450 mm dia. should be seamless

Additional flanges needed to give access to reducer and bend internals

All connections are flanged, not screwed

Flanges

Lining wrapped round to form a seal

Radii on external corners

Corners and edges

Internal weld-fillet ground to radius

External radii ≥ t

Weld located clear of corner if possible

Weld-caps ground flush

Lining thickness (t)

Internal radii ≥ 2t

Rubber lining

Butt welds

Smooth convex   Smooth flat   Smooth concave

GOOD
BAD

Undercut   Badly ground   Excess penetration

Refer to BS 6374 part 5 or DIN 28051–5 for further details

Fig 13.1 Rubber lined components – design features

BS 903 *Physical testing of rubber* (**1**) is one of the biggest technical standards that you will meet – there are more than 60 parts to it, of which you will probably need to refer to only three:

BS 903 Part A9 covers determination of abrasion resistance (equivalent in parts to ISO 4649 and ISO 5470). This is really a standard used to assist in the *choice* of rubber compound rather than during inspections, but it contains useful technical background, particularly if you are doing a design check.

BS 903 Part A36 (equivalent to ISO 4661/1) is about the preparation of samples and test pieces for use in physical testing. Again this provides useful background, but note that works inspection of rubber lined components does not normally involve test pieces. This is done earlier, at the design or type-testing stage.

BS 903 Part A57 (equivalent to ISO 7619) explains methods for the the determination of indentation hardness using hand-held hardness meters. This is the type of test that *is* commonly used during a works inspection. In practice, the test is so straightforward that you don't need step-by-step instructions from the standard, but you may at some stage need the information on calibration and hardness scales that it provides. Interestingly, BS 903 does not give much useful information on spark testing, which is one of the important tests for lining integrity.

BS 6374 *Lining of equipment with polymeric materials for the process industries* (**2**) is a good standard. It has real practical use and is divided into five sections, each dealing with a particular category of lining materials. They are:

- BS 6374 Part 1: This covers the application of sheet thermoplastics. The most common linings that you will meet in this category are those with a high resistance to acid. These are commonly used to line process vessels in chemical plant applications.
- BS 6374 Parts 2, 3, and 4 cover the application of non-sheet thermoplastics, stoved thermosetting resins, and cold-cured thermo-setting resins respectively. You may also meet these lining materials in chemical plants.
- BS 6374 Part 5 is a specification for lining with rubbers. It shows desirable design features (complementing the more general ones I have shown in Fig. 13.1). Some care is needed in implementing these, though, as the standard makes clear that it contains guidelines rather than mandatory requirements. There are useful sections on fabrication of vessels intended for lining, common lining defects, and relevant rubber properties. It describes tests for rubber hardness

and continuity – which are two of the common tests that you will witness frequently during works inspections. This is a simple and well-written standard which you should find of real practical use.

The German standards DIN 28 051-5 **(3)** have similar coverage to BS 6374 Part 5 in the areas of design features and testing. There is little technical contradiction between the two but DIN is a little more definitive in its definitions of desirable design features. It references DIN 53 505 for hardness testing.

ASME VIII and BS 5500 cover the issue of rubber lined vessels in slightly different ways, but the principles are the same. There are no common technical standards relating specifically to the lining of pumps, piping, or valves.

Inspection and test plans (ITPs)

At best, the ITP for a rubber lined vessel component will contain only broad statements of inspection requirements. Often it merits merely a single line entitled 'lining inspection' towards the end of the ITP – this seems particularly common for all vessel types, even large ones. It is also a poor and rather misleading interpretation of what is actually required. The main reason for this is because of the key roles played by operator skill and process control that I described earlier. Small operator or process errors made during the application process have an absolutely key impact on the final integrity and longevity of the lining. Many of these errors are so small as to be very difficult to detect at the final inspection stage using the practical test techniques that are available. The message therefore is that you must inspect *during* the lining process – armed with a knowledge of the types of errors that can occur. You need to get really close to the process – it is not sufficient only to inspect the lining at the final inspection stage. Such interim inspections are rarely shown on the ITP. Be prepared therefore to apply a little interpretation to what is written – in the knowledge that the body of technical evidence about the root causes of lining failures is largely on your side.

If, as part of your early input to a contract specification, you have the opportunity to influence ITP content, raise rubber linings as an important issue. Try as a minimum to get BS 6374 or DIN 28 051-5 quoted as a specific requirement. Specify interim inspections of the lining preparation and application process in the ITP if you can. Figure 13.2 shows an example of what to include – this gives a much more comprehensive approach and decreases the probability of subsequent lining failures.

Inspection step	Comments
1. Check design feature	Check for features which will cause lining problems (BS 6374 part 5).
2. Fabrication check	Check weld profiles, edge radii and general finish for smoothness.
3. Surface preparation	A minimum grade of SIS 05 5900 Sa 2½ is required.
4. Materials check	Check the shelf-life of the unvulcanized rubber sheet and the bonding adhesive – also make a visual examination of the rubber sheet for defects (pores, blisters, or tears).
5. Witness lining application	Monitor ambient conditions (at least 3°C above dewpoint), adhesive, and methods of sheet jointing.
6. Pre-vulcanization inspection	Check the workmanship of the completed lining. Do a continuity (spark) test before fitting cover-straps to seams. Review the vulcanization procedure – time, temperature (125–160°C) and humidity requirements.
7. Post-vulcanization inspection	Visual inspection, hardness test, full continuity (spark) test, and adhesion (rapping) test.
8. Repairs	Check of local repair procedure and witness of re-tests after further vulcanization.
9. Pre-shipping inspection	Final check for correct packing to avoid damage to the lining in transit.

Fig 13.2 Rubber linings – ITP steps

Test procedures and techniques

There is an accepted set of straightforward testing techniques which is used during final works inspections of natural and synthetic rubber linings. Do not think of these as perfect test techniques – they have a few weak points. They are, however, the best tests available for use in a practical 'shop floor' situation.

The visual inspection

It is essential to perform a close and thorough visual inspection, preferably before and after vulcanization. A general cursory examination is not sufficient, your objective should be to inspect each full – or part-sheet – that has been used in the lining, and every joint and seam. Use the following guidelines (refer also to Fig. 13.3).

- Use adequate lighting when making a visual inspection of an internally lined vessel. You need an ac lead lamp for this – a torch is not bright enough. In large vessels you should use a wooden ladder to reach the upper walls and top surfaces. Remember that you are making a *close* visual inspection – not just observing from a distance.
- Adopt a methodical approach. Start your examination at one end of the vessel and inspect each lining sheet 'course' in turn, working along each seam to the other end. Check each sheet for:

    - Physical damage such as tears, cuts, and punctures.
    - Adhesion to the surface (just do this visually for the moment). Look for obvious bulges, blisters, or ripples, especially on concave surfaces, that indicate where the sheet is not properly glued down. Check carefully at the location of any set-through nozzles, you will sometimes find air bubbles where the sheet has not been properly smoothed over the internal nozzle-to-shell fillet weld. Pay particular attention to the internal lining of small diameter pipes where application of the rubber sheet is difficult. One of the best areas to check is on a tight outside radius – for instance where the rubber sheet is lapped over the face of a pipe flange. If you 'push' the lining with the heel of your hand you should be able to feel whether or not it has adhered properly to the surface. Unfortunately, there is no in-situ quantitive test for adhesive bond strength, the only way is to use test pieces.
    - *Smoothness*. Sight along the sheet with the light behind you and look for evidence of any weld spatter or small foreign bodies left underneath the sheet.
    - *Seams*. Check the scarf joints or overlaps where the lining sheets have been joined. For a typical 4 mm thick rubber lining there should be at least 16 mm (four times the sheet thickness) of surface contact between adjoining sheets but not normally more than 32 mm. Figure 13.3 shows the three main types of scarf joint that you will see. Run your hand along each seam to make sure the edge is firmly stuck down. You should not be able to feel

a loose edge when running your fingers against the lap directions and the scarf edge should be sharply feathered to give a good tight joint. A raised 'butt' on the scarf edge is bad practice.
- *Nozzles and flanges.* The lining sheet should be correctly wrapped round all nozzles and flanges to provide a complete seal against the process fluid. Pay particular attention to small fittings of less than 30 mm diameter – it is more difficult to get the lining to adhere to their tight internal radii.
- *Internal fittings.* Fittings such as baffles, separators, and internal tubeplates are often designed as permanent fixtures in a lined vessel to try and avoid practical lining difficulties. If you encounter removable fittings, make sure that the lining sheets have been applied in a way that enables a full seal to be made. If separate cover strips of lining have been used, make sure they are well jointed and stuck down. Do not forget miscellaneous fittings such as handholds, brackets, and access door pivots.

As with other types of visual inspection, it is useful to use a checklist. This will ensure that nothing is missed, and form a useful addition to your inspection report. The best way to develop this checklist is to go through BS 6374 – listing the features it mentions – then sort them into a logical order for use during your visual inspection.

Rubber hardness check

The hardness of a rubber lining is a good measure of desirable properties such as abrasion resistance, and an indicator of whether the vulcanization process has been completed correctly. Post-vulcanization hardness is therefore one of the main specified properties. Hardness is measured using a simple hand-held indentation meter. The tip is pushed into the rubber and the hardness reading is shown on a spring loaded dial. There are two main hardness scales in common use. Soft rubbers use the International Rubber Hardness Degree (IRHD) scale – this goes from 0 to 100 'degrees' and is roughly proportional to the Young's modulus of the rubber. A soft rubber is generally considered as having a hardness of between 40–80 degrees IRHD (below 40 degrees is a very soft rubber and these are used less often). Hard rubbers range from 80–100 degrees IRHD *or* may be referred to the Shore D scale (see BS 903 part A57). For most engineering applications, hard rubbers are considered as those in the range 60–80 Shore D degrees. There is a generally accepted tolerance of ± 5 degrees allowed for both IRHD and Shore D measurements. It is important to check hardness readings at

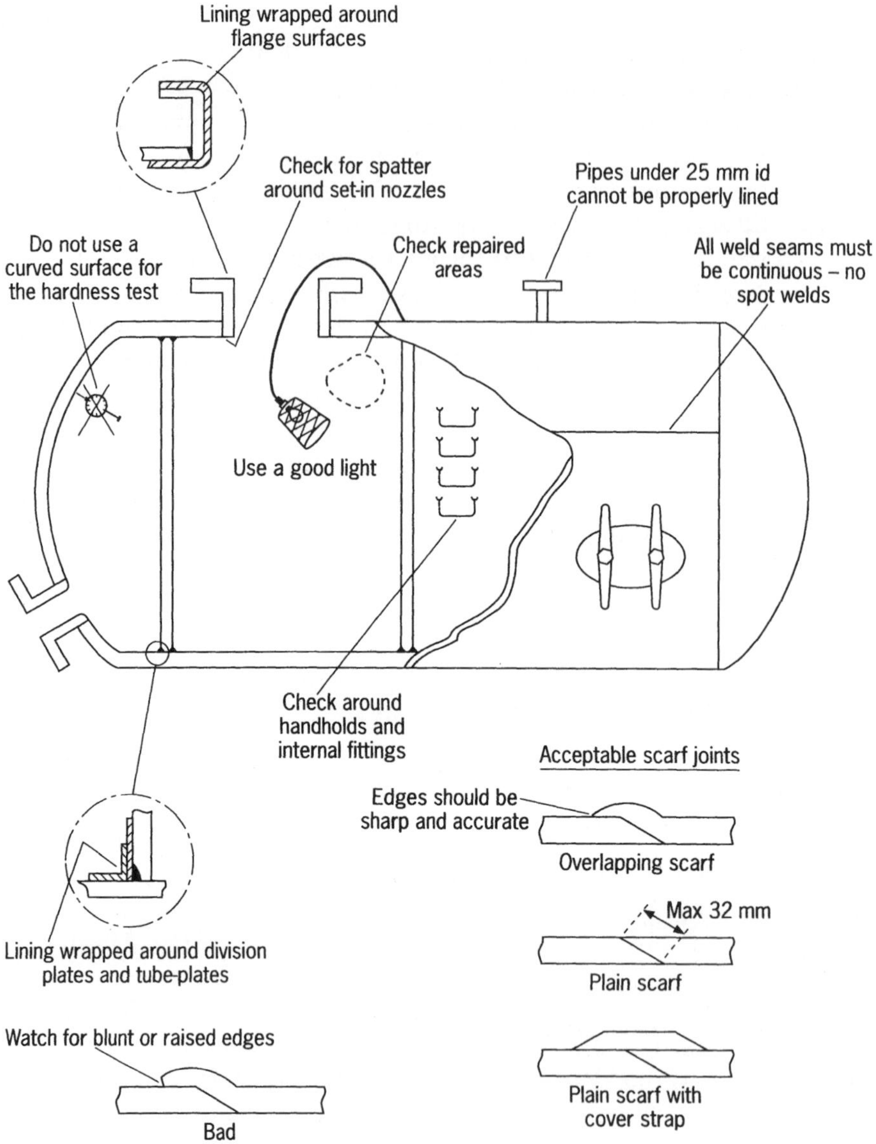

396 Handbook of Mechanical Works Inspection

Fig 13.3 Visual inspection of rubber lined vessels

several points around the vessel lining to obtain the average reading. Use the following general guidelines:

- Always take readings on a flat surface. Measurements from curved surfaces can introduce errors of perhaps 10–20 degrees IRHD.

- Take the readings at the centre of the lining sheets or at about one metre apart. To get representative results, it is best to take at least three readings per individual rubber sheet.
- Concentrate on those regions of the vessel which were in the *lower* position when the vessel was placed in the autoclave for vulcanization (obtain this information from the lining contractor). You will find that these lower areas are sometimes softer. This is due to condensation accumulating from the steam heating, preventing the vulcanization from being completed properly. You may see some evidence of this during your visual checks – poorly vulcanized areas will be a noticeably different shade (normally lighter) than the rest of the lining.
- The ambient temperature will have some effect on the hardness readings that are taken. Try to ensure that the local temperature inside the vessel is between 15 °C and 25 °C. If the vessel is placed in strong sunlight the internal temperature will rise and give misleading readings.
- Check the calibration of the hardness tester before use. Calibration test pieces should be kept with each tester.

Spark testing

The purpose of spark testing is to check for *continuity* of the lining. This is quite a 'searching' technique – an electrical spark will locate a pinhole or small discontinuity in a lining sheet that would not be detected visually. It will not detect any errors in adhesion. A high frequency, high voltage ac supply (it is common to use a minimum of 20 kV) is applied between the parent metal of the vessel and a hand-held probe. The probe is then passed just above the surface of the rubber. A strong spark will jump the gap when there is a conducting path caused by a pinhole or discontinuity – there are a few points to note:

- It is essential that the voltage be sufficient to enable the spark to jump the maximum air path that it is like to encounter. This would be the path through a faulty scarf joint. For a typical 4 mm thick lining it is approximately 32 mm. You can do a test by simply checking that the spark will jump this length of gap from the probe to an unlined area such as a bolt hole.
- Don't confuse stray 'air path sparking' from the probe with the large blue spark that will jump to a real leakage path. There will always be some small 'streamer' sparks into the air, particularly in damp conditions.

- Dim lighting makes it easier to see the sparks.
- Again, be methodical in your approach to spark testing. It is worth working round all scarfed seams, nozzle joints and any cover straps over fittings. Check around access holes and handholds. Test the bottom surface of the vessel where physical damage is more likely – and in difficult locations where accurate seams are more difficult to make.

For large rubber lined items such as vessels and ductwork, it is common to perform a spark test both before and after vulcanization. This is also good practice for vessels which are lined and vulcanized at the construction site, under non-ideal working conditions. For seams which are fitted with cover straps a preliminary spark test should be done before the cover straps are glued on.

Adhesion tests

Poor adhesion is a common cause of failure of rubber linings; the lining sheet can peel off the base material, often starting from an air bubble or loose radius, until a seam is reached. The process fluid gains access underneath the lining and general corrosion and failure happen within a short time.

Under laboratory conditions it is possible to obtain adhesive bonding between rubber and metal that is stronger than the rubber itself (BS 903 part A19/ISO 188 covers this). In practice, variables such as the adhesive mix, ambient conditions and operator skill make this difficult to achieve, particularly around internal radii, sharp corners, and fillet weld profiles. A crude but reasonably effective works check on adhesion can be carried out by doing a *rapping test*. The rubber is rapped using a special ball-headed hammer. Areas of good adhesion will give a firm ringing sound. If there is an air gap between the lining sheet and base material, the sound will be 'duller'. This is by no means an exact test, it is easy to disagree on the comparison of sound obtained. Small variations of hardness of the lining can also cause differences. The test will also only identify areas where there is significant lack of adhesion leading to an air gap. It will not detect low strength or partial adhesion, both of which are root causes of many lining failures, nor will it detect small patches of poor adhesion on scarfed seams. The rapping test is a useful and practical site test, as long as you recognize its limitations. It is certainly not a substitute for witnessing the lining activity (see the recommended ITP steps in Fig. 13.2), and making careful checks of surface preparation, adhesive mix, and application technique.

Common non-conformances and corrective actions: rubber linings

Non-conformance	Corrective action
Poor surface preparation	Don't accept poor surface preparation – you should ask for it to be done again. Check that: • Weld caps are properly smoothed off (they don't necessarily have to be ground totally flat). • Shot blasting is to the level SIS 055900 Sa 2½. There should be no significant scale or rust and the surface should be a light, speckled grey.
Wrong rubber lining thickness	There is a general allowable tolerance of ±10 percent on the (unvulcanized) thickness of the rubber sheet. If the sheet is too thick, this is unlikely to cause any harm. Accept it. If the sheet is more than 1 mm too thin, reject it. The vessel needs stripping, shot blasting, and relining.
Defects found before vulcanization	Local repairs are possible. The solution is to cut out the defective area and insert a repair piece. Use scarf joints with a 10–15 mm overlap. Make sure the new seams are very well stuck down and all air bubbles are excluded.
Incorrect hardness results	First check the allowable tolerances in BS 903. A tolerance of ±5 percent is acceptable on any individual reading. Hardness is usually specified as a *minimum* value required. The most likely cause of low hardness is incorrect vulcanization – it is acceptable to revulcanize the component. If this does not improve the situation, it is almost certainly the wrong grade of rubber. Discrepancies of less than 10 degrees (Shore D or IRHD) can often be accepted under concession – more than this and the lining cannot really be considered fit for purpose.
Defects found after vulcanization (by spark tests)	Again, repairs are possible. After repair, the area should be locally vulcanized using special heated 'irons' or similar. Repeat the spark test, hardness measurements and rapping test on the repaired area.
Poor seam overlaps or short scarf lengths	Poor scarf joints can be overlain with cover strips – separate strips of lining approximately 70–100 mm wide. This is acceptable to a limited extent but is not a replacement for neatly scarfed seams. Cover straps may peel off sharp corners or tight inside radii in use. It is best to use them sparingly.
Lack of (or poor) adhesion	Patches of poor adhesion are usually symptomatic of a more general problem with the adhesive or surface preparation. Always assume that the situation will get quickly worse in use. Except for very small areas near the centre of lining sheets, poor adhesion is unacceptable. Issue an NCR and ask for it to be done again.

METALLIC LININGS

The term 'metallic lining' is often rather loosely used to encompass a large number of processes. These range from cladding base material by welding on sheets of corrosion resistant alloy to more complex processes where a very thin layer of material is added to a metal surface by deposition, spraying, or electrolytic action.

Fitness-for-purpose criteria

As with rubber, the main purpose of most metallic linings is to prevent corrosion and erosion of the base material by an aggressive process fluid or environment. The main FFP criterion therefore is *integrity* – the lining must provide a perfect seal if it is to be effective. This seal is provided by welding in the case of loose-clad components and a metal-to-metal bonding in the case of sprayed, dipped, or electrolytically applied metallic coatings. For some coatings such as chrome plating, the surface finish of the plating itself may also be an important FFP criterion – it depends on the application.

It is safe to say that, *in general*, metallic linings involve less works inspection activities than do rubber linings. One reason is that the processes themselves tend to be more technologically complex and need to be quite closely controlled if they are to work at all. Metal spraying and electrolytic plating are good examples of this. A second reason is that the inherently simpler processes, such as cladding, come under the control of accepted welding procedures and practices. This is a reasonably well controlled regime, as we saw in Chapter 5.

Procedures and techniques

We can look briefly at the three most common metallic lining techniques that you will meet in a works inspection situation.

Loose cladding

Loose cladding is most commonly used in fluid and process systems, in particular where there are aggressive process conditions, such as in chemical or desalination plants. The cladding material is usually either a high-nickel alloy such as Inconel or Monel, or stainless steel. The term 'loose cladding' means that the lining sheets are not bonded to the base material (usually low carbon steel) over their complete surface area – attachment is achieved by welding around the periphery of the lining sheets. This may be supplemented by plug welding – a MIG welding technique in which spot welds penetrate through the cladding sheets into the base material at regular intervals. ASME VIII has a section dealing specifically with clad vessels.

Some particular guidelines relevant to works inspections are:

- *Design features.* ASME VIII contains guidelines on desirable design features for clad vessels. You can do a very broad design check using the construction drawings if you are inspecting conventional vessels, but other equipment is more difficult.
- *Welding.* The welding between the cladding sheets and base material should be covered by the system of WPSs, PQRs and welder qualifications described in Chapter 5. It is good practice to carry out surface crack detection on these welds – apply the same standards as you would for other dissimilar material welds.
- *Pneumatic testing.* Some clad components are subject to a pneumatic test.
- *Surface preparation.* It is good practice to shotblast the inside surface of the base material before cladding. This removes any active corrosion products and minimizes subsequent deterioration. The inside surface of the lining sheet is normally mechanically cleaned before fixing.
- *Final inspection.* It is worth making a final inspection of loose-clad components. Check that all the seal welding is properly completed. Small fittings and fasteners should be of the same material as the lining, to prevent galvanic corrosion in use.

Figure 13.4 shows some specific checks on common loose-clad components.

Galvanizing

Galvanizing is often used as a generic term for the coating of iron and steel components with zinc. Its main use is to protect the base material against attack from a corrosive atmosphere, or from water. It is predominantly used externally on engineering components rather than to protect internal surfaces. It can be used instead of painting or occasionally in conjunction with it. Much of the benefit in using zinc lies with its position in the periodic table – it is anodic to steel and will therefore protect it, even if the zinc layer suffers scratching or damage.

Galvanizing is unlike many types of cladding or coating because the zinc forms a *chemical* bond with the iron in the surface layer of the substrate. Purer zinc then bonds with the iron/zinc compound until the surface layer is almost pure zinc. Several different processes are often misleading termed 'galvanizing'. Processes such as sheradizing (coating items using a zinc-rich dust), zinc plating (actually an electro-deposition process), and zinc spraying tend to be used for small, mass-produced items such as fasteners, fittings, and precision components. The most relevant process for works inspection is the technique of *hot dip* galvanizing. This is widely used for structural steelwork and large fabrications that are exposed to the atmosphere. Hot dip galvanizing is a relatively low technology process – the component is cleaned and then dipped in a bath of molten zinc. The zinc compounds form on the surface and remain when the component is removed from the bath.

The most widely used technical standard for hot dip galvanizing is BS 729 **(4)**. This is a good general guide which contains most of the information that you need during a works inspection. The nearest comparable standard in the ASTM range is ASTM A90 **(5)**. Practical checks that are carried out are limited to visual examination and specific tests to determine coating weight and uniformity.

The visual inspection

Although hot dip galvanizing can be considered a proven and reliable process, errors can still be caused by incorrect composition of the zinc bath, or by poor preparation of the parent metal. A good visual inspection is therefore important. Key points to check are:

- *Surface appearance.* If the base metal has an even surface, then the surface of the zinc coating should also have a smooth and even finish. Uneven features such as weld laps and seams will show clearly through the coating – normally this is a purely cosmetic problem.
- *Colour.* It should be bright. The only common exception is for some types of high silicon steels in which the alloying elements cause the coating to be slightly dull. Check the steel type if you see this, to make sure that this is the reason and that it is not a problem with the purity of the molten zinc mix.
- *Staining.* Technical specifications (and good practice) call for the galvanized surface to be free of staining. The most common type of staining is caused by storing the galvanized components in wet conditions for long periods – this produces an effect which looks rather like white rust. If necessary the components can be subjected to a phosphating or chromating surface treatment after galvanizing to stop this happening.

Linings 403

Carbon steel pipe spool clad with stainless steel

- Welds between dissimilar materials should have WPSs and PQRs
- Check for a good sound seal-weld

Cladding

Backing

Any welds in the cladding sheets should be completed (with NDT) before assembly

The cladding should be tightly fitted to the backing material (no large air-gap)

Check for cracking or work-hardening around tight radii

Carbon steel gas damper louvre clad with C276/'Hastelloy' nickel alloy

Cladding sheets should overlap and have sound seam welds

DP or MPI test of the welds

C276 cladding sheet

MIG plug weld in drilled hole

Sa $2\frac{1}{2}$ surface

- Check the plug-weld spacing

Carbon steel backing

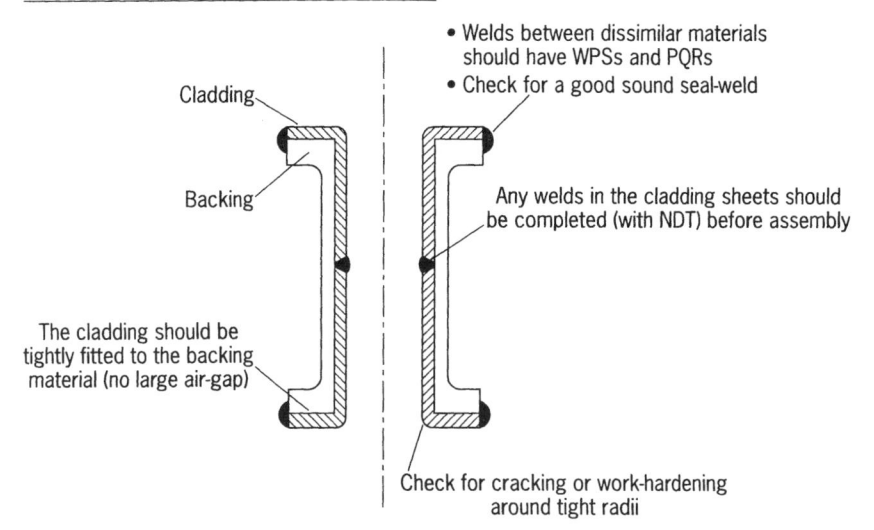

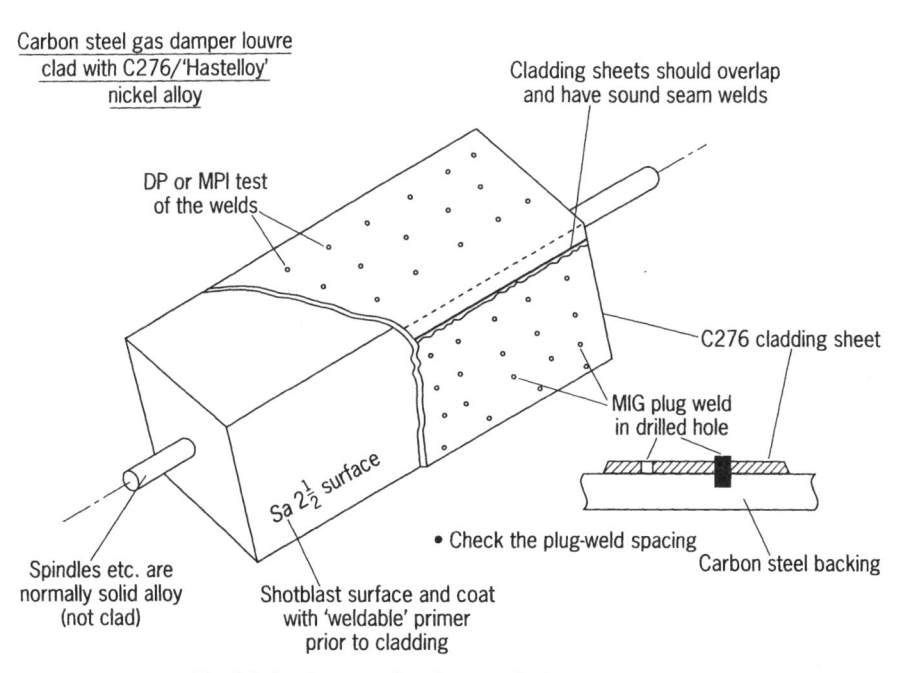

Spindles etc. are normally solid alloy (not clad)

Shotblast surface and coat with 'weldable' primer prior to cladding

Fig 13.4 Inspecting loose clad components

- *Coating discontinuities.* These are more often the result of mechanical damage rather than problems with the chemistry of the coating process. Look carefully, therefore, around exposed edges and corners. Small areas up to 30–40 mm^2 in size can be repaired (using a low melting point zinc filler rod, applied rather like solder) but this is not good practice for large areas.

Coating weight test

It is convention for galvanizing specifications to specify the coating *weight*, expressed in g/m^2, that has to be applied. Strictly, the best way of checking this is by a 'stripping test' in which a small test-piece is immersed in an acid solution and the weight of coating removed (by dissolving) measured. Practically, the most common method used during works inspection is a simpler weight difference method – the component is weighed before and after galvanizing and the difference divided by the surface area to give a g/m^2 coverage figure. This is an acceptable method. Figure 13.5 shows typical coating weights suitable for various components. Note that a rough (shotblasted) surface of the parent metal will accept a thicker coating than a smooth one.

Coating uniformity test

The uniformity test is used as a complement to the weight test – it is not sufficiently accurate to use alone as a test of coating thickness. The purpose is to check the 'evenness' of the coating – you should see this written as a desirable criterion in most galvanizing contract specifications. The test is commonly called a 'Preece test' and consists of dipping a test piece into a copper sulphate solution. The solution will expose the base metal in any areas of thin coating and deposit a red-brown layer of metallic copper. This is an indication that the coating has weak areas and is therefore unacceptable. Figure 13.5 shows the general details.

In practice, most works inspections of galvanized components are performed on a small sample chosen from what may be a large batch of similar items. Accept this as convention, but don't neglect a general visual inspection of the whole batch – just to make sure.

Chrome plating

It is unlikely that you will often have to inspect chrome-plated items during a works inspection, as chrome plating is rarely used on items large enough to warrant a specific inspection visit. You will find it, however, on large hydraulic rams such as those used for water gates, cranes, crushing equipment, and some very high technology applications where a very fine surface finish is important. It is worth having a little knowledge of what to look for. There are several technical standards applicable. The most relevant one is BS 1224 (**6**) which covers electroplated coatings of nickel and chromium.

The purpose of chrome plating is not only to provide a fine surface finish but also to provide a coating that will resist the environmental conditions it will see in service. The severity of these service conditions is graded on a scale of 1 to 4 – condition 4 is the most severe and refers to outdoor conditions in a very corrosive atmosphere, whilst condition 1 would be indoors in a warm, dry atmosphere. A system of plating classification has been developed to describe electroplate type and thickness. These are several ways of expressing the classification, a common example being as follows:

Fe/Cu 20 Ni 25(p) Cr (mc)

Note how to interpret this:

- Fe denotes the iron or steel parent material.
- Cu 20 denotes a minimum 20 µm of copper plated onto the steel.
- Ni 25(p) denotes a minimum 25 µm of nickel plated onto the copper. The 'p' denotes that the nickel finish is 'semi-bright' ('b' would represent nickel deposited in the 'fully-bright' condition).
- Cr (mc) denotes that the top layer is chromium plate with a 'microcracked' structure. I have shown the various chromium structure designations and their respective minimum plate thicknesses in Fig. 13.6. Note that although the Cr thickness is not stated explicitly, it is specified *implicitly* by the designation of the chromium plate structure that is used.

When inspecting plated components the best approach I can recommend is to learn to rely heavily on a good close visual examination. Almost all of the quantitive test techniques for plating thickness, adhesion, ductility, and corrosion resistance are 'laboratory-type' tests requiring the use of pre-prepared test specimens representative of the plated components being inspected. Practically these will not usually be available to you during normal works inspections. A sensible course of action is to perform a detailed visual examination first – then you can take action to deepen the investigation (by reviewing the actual plating process and asking for plated test pieces) if you feel it is necessary.

> **COATING WEIGHT**
>
> | Parent material | Minimum galvanized coating weight (g/m^2) |
> |---|---|
> | Steel 1–2 mm thick | 335 |
> | Steel 2–5 mm thick | 460 |
> | Steel > 5 mm thick | 610 |
> | Castings | 610 |
>
> An approximate conversion from coating weight to coating thickness is:
>
> $$1 \text{ g/m2} \simeq 0.14 \text{ μm}$$
>
> Coating weight can be determined by either:
>
> - An acid 'stripping' test (the most accurate 'laboratory' method).
> - Direct measurement of the weight difference of the component before and after galvanizing.
> - Sometimes a simple magnetic or electronic thickness measuring device is used to calculate an approximation of the coating weight.
>
> **COATING UNIFORMITY** (Cu SO_4 or 'Preece' test)
>
> The uniformity test is complementary to the coating weight check. The steps are:
>
> - Prepare test specimens of the galvanized material.
> - Dip the specimens four times successively into a Cu SO_4 reagent solution – each dip should last 60 seconds and the specimens must be rinsed between dips.
> - If a permanent film of red-brown metallic copper appears on the surface of the metal, the coating has failed the uniformity test.
> - For further details, look at BS 729.
>
> **Fig 13.5** Hot dip galvanizing: weight and uniformity tests

The visual examination

Try to adopt a structured approach to the visual examination, based around the following steps (see also Fig. 13.6):

- *Plating classification.* First, find out the classification of plating that is specified (refer to BS 1224 if you need a more detailed explanation than I have provided). The main information that you need is the service condition 'grade' and the electroplating composition and thickness.
- *Dimensions.* Electroplating adds thickness to a component. Although the plating is relatively thin (if it is applied correctly) some hydraulic components do have closely specified dimensional tolerances, particularly for sealing faces. Check plated cylinder dimensions at several points on their length and circumference to make sure that tolerances have not been exceeded. You should also check for any undesirable 'unevenness' or 'wavyness' by sighting along the plated surface.
- *Visual defects.* Check for mechanical damage such as chips, deep scratches and any obvious discontinuities in the plating. Then look for process-related defects; the main ones are visible cracks (crazing), visible pores, and blisters. All these are unacceptable. If you see minor score marks along the axis of a hydraulic ram, check whether this could be due to a tight or brittle hydraulic seal ring rather than being an actual plating defect.
- *Surface finish.* There are three main categories of plated surface finish; bright (which should be mirror-like), satin (with a semi-matt appearance), and dull. They can normally be easily differentiated visually but a comparator gauge can be used for marginal cases. Specialist comparator gauges are available for a more quantitive measurement in μm (R_a). It is important to look for *uniformity* of surface finish – there should be no visible variation over the 'significant area' (this is the terminology used for the critical plated area – it should be clearly indicated on the drawings). Areas such as seal grooves and radii that are not within the significant area often have a poorer surface finish and are generally non-critical. For chromium surfaces which are specified as microcracked (mc) or with microporosity (mp), examine the surface with a microscope using a magnification of × 100. Note that these features should *not* be visible with the naked eye. If they are, they as classified are defects and the plating is unacceptable. The cylinder liners of some high speed diesel engines use (mc) or (mp) chromium plating.

Accurate and careful observation is the watchword when doing a visual examination of plated components. Try to use a detailed checklist and be concientious in its use, reporting accurately what you find.

Chromium plated hydraulic ram

A 'typical' plating specification is:
- 0.3–0.8 μm Cr
- 25 μm Ni
- 20–30 μm Cu
- Backing material

Visible porosity is unacceptable

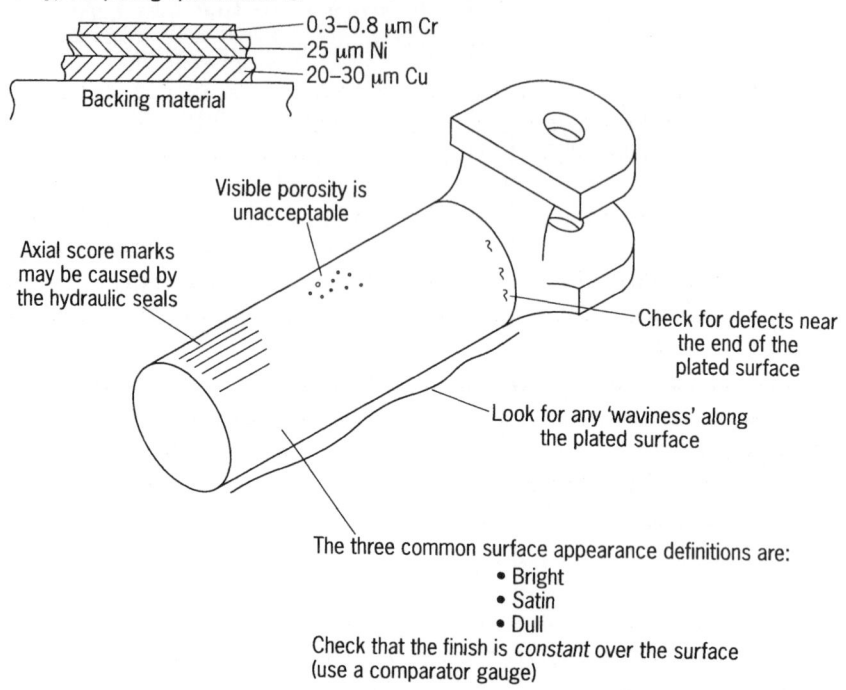

Axial score marks may be caused by the hydraulic seals

Check for defects near the end of the plated surface

Look for any 'waviness' along the plated surface

The three common surface appearance definitions are:
- Bright
- Satin
- Dull

Check that the finish is *constant* over the surface (use a comparator gauge)

Check for the following defects in the Cr layer:

- visible porosity
- mechanical damage
- plating discontinuities
- 'blooming' (dull patches)

The chromium plate 'structure' designations are:
Cr(r): A 'regular' finish – minimum thickness 0.3 μm
Cr(f): 'Free' from cracks – minimum thickness 0.8 μm
Cr(mc): 'Micro-cracked' – minimum thickness 0.8 μm
Cr(mp): 'Microporous' – minimum thickness 0.3 μm

Fig 13.6 Inspection points – chrome plating

Remember that the objective of the examination is to form a clear view as to whether further investigation is required and whether test pieces are needed for further tests. There will often be some subjectivity in these decisions, for the simple reason that the acceptance criteria are not *that* well defined. Think carefully about FFP – then make a judgement.

Further tests

If you do find evidence of problems, there are a number of further tests that can be done to help diagnose the problem. They are all laboratory tests and need test specimens. The most common ones are:

- *Plating thickness check*. The thicknesses of the copper and nickel backing materials are measured by mounting a specimen (macro) section in a plastic block and making a visual measurement of the plating thickness using a microscope with a calibrated graticule. The chromium plating is much thinner and can only be measured accurately using a coulometric method. This involves dissolving the chromium anodically and measuring the time taken.
- *Adhesion test*. A rough assessment of plating adhesion can be obtained by using a coarse hand-file obliquely against the edge of the plated test specimen to see how easily the plating 'lifts off'. It is also possible to do a quench test but this also gives only an indicative measurement.
- *Ductility test*. A plated specimen is bent over a former and then examined for cracking. The generally accepted minimum elongation value is 8 percent. This is not a particularly useful test.
- *Corrosion tests*. There are several of these. The most common one is known as the CAASS (copper accelerated acetic acid salt spray) test and involves spraying the specimen with acid to see how readily it corrodes. It has limited use in diagnosing problems that you will encounter during works inspections.

Failure of chrome plating in normal use is actually quite rare. When it does occur it is nearly always the result of inadequate process conditions during application of the final chromium layer, or of simple mechanical damage. A close visual and dimensional inspection will normally reveal these defects at the final inspection stage. Interim inspections at the preparation or 'backing layer' stage are unusual, and would be unlikely to prevent the majority of failures.

Surface finish

The assessment of fine surface finishes is important, not only for electroplated components but also for accurate machined and ground finishes used on precision rotating machinery components such as turbine rotors, gear wheels, and shafts. It is also specified closely on highly stressed items such as turbine blades, where poor surface finish can cause significant reduction in the fatigue strength. Do not confuse the assessment of such fine surface finish with the grading assessment of shotblasted surfaces covered in Chapter 14 – a completely different set of standards apply.

There are two accepted methods of determining a grade of surface finish. The surface can be subject to *direct measurement* using an accurate stylus device – which gives a true quantitative reading. This method is usually reserved for laboratory work. The technique that is most often used during works inspections is a *comparative assessment*. Here, the surface is compared with a set of prepared reference samples (generally imprinted on a sheet steel comparator gauge) and a qualitative assessment is made. This is a quick and practical test and, with a little experience, is sufficiently accurate for most inspection purposes.

The classification of surface finishes is well described in technical standards. This is a good example of a technical discipline in which the standards *lead* industry practice. There are some slight differences of approach between standards but the underlying principles are much the same. Surface roughness is described by the parameter R_a. This is a measure of the vertical deviation of the peaks and troughs of the surface about its average (do not confuse this with R_z, a lesser used parameter which refers to the absolute vertical distance between the highest peaks and lowest troughs). Note that the R_a parameter has replaced the previously used centre line area (CLA) measurement – the difference is only in terminology, the measurement is exactly the same.

You will find that contract specifications use this R_a parameter in order to specify a certain grade of fine surface finish. Although not always stated, this refers, by inference, to the use of one of the following technical standards (look at Fig. 13.7, which summarizes the important pieces of information that you will need):

DIN ISO 1302 **(7)** and BS 1134 **(8)** are well accepted standards. The preferred R_a values (in microns) are classified by a series of 'N-numbers'. A similar standard is ANSI B46.1 which expresses the preferred R_a values in imperial units (micro-inches) – Fig. 13.7 shows the equivalence to metric units.

Your surface finish examination should follow the same principles as the other visual examinations discussed in this book – close and careful examination of the entire surface under a good light – with accurate and descriptive reporting of the results. There are a few specific points to bear in mind when using the comparator gauge:

- The way in which the surface has been formed is important. The surface texture resulting from lathe turning, cylindrical grinding, milling and honing are very different. This has been taken into account in the technical standards and a good comparator gauge will have specimen surfaces produced by each of these processes that you can use as comparisons.
- Using the comparator gauges needs *touch* as well as sight. You should first rub your fingernail in several directions over the surface to be graded, then find the comparator specimen that you feel is the best match.
- As a guide you should be able to make an assessment accurate to one N-number for fine finishes of grade N1 to N5 (equivalent to 0.025 μm to 0.4 μm R_a) and considerably better than that for grades N6 (0.8 μm R_a) and above, where the difference between preferred grades is much wider. If you want a *rule of thumb* to work to, then a rough turned surface, with visible tool chatter marks, is about grade N10 (6.3 μm R_a) and a reasonably smooth machined surface is likely to be about grade N8 (3.2 μm R_a). Surfaces which mate with other static surfaces, or provide a datum for locating other surfaces, you will usually find specified as grade N7 (1.6 μm R_a) or better. Surfaces which incorporate a relative movement or bearing function vary from grade N6 (0.8 μm R_a) down to the finest 'normal' surface grade N1 (0.025 μm R_a). Finer finishes may be used for special applications or plated components. These may need to be measured using a stylus instrument.

412 Handbook of Mechanical Works Inspection

	FINE FINISH →							← ROUGH FINISH →				
R_a (μm) BS1134	0.025	0.05	0.1	0.2	0.4	0.8	1.6	3.2	6.3	12.5	25	50
R_a (μ inch) ANSI B46.1		1	2	4	8	16	32	63	125	250	500	1000	2000
N-grade DIN ISO 1302		N1	N2	N3	N4	N5	N6	N7	N8	N9	N10	N11	N12

Ground finishes

Seal-faces and running surfaces

Smooth turned

Medium turned

Rough turned finish

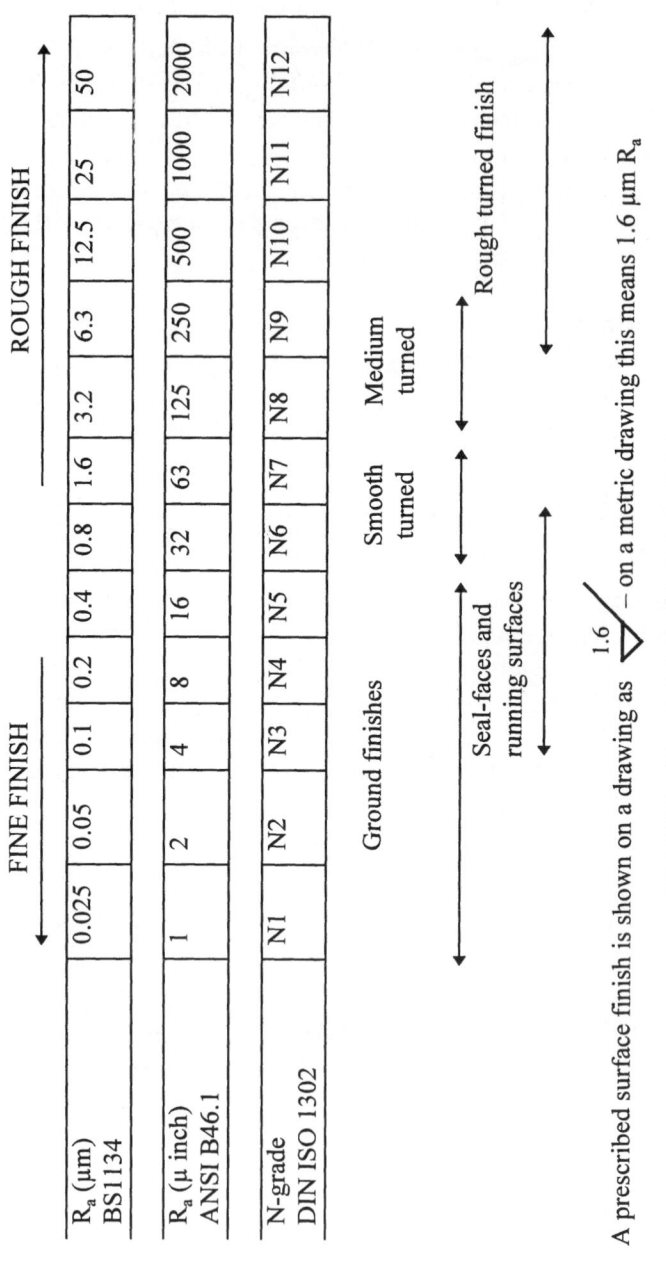

A prescribed surface finish is shown on a drawing as $\overset{1.6}{\triangledown}$ — on a metric drawing this means 1.6 μm R_a

Fig 13.7 How to interpret surface finishes

KEY POINT SUMMARY: LININGS

1. The root cause of most rubber lining failures is poor application – you can help the situation by making *interim* inspections.

2. There are a number of small, but important, design features that are desirable for rubber lined vessels and components. It is worth checking these. Use BS 6374 Part 5 and DIN 28 051-5 to help you

3. Common techniques are used to test for:
   - hardness
   - continuity (the spark test)
   - adhesion (the rapping test)

   Always do a close visual examination – there are several specific things to look for.

4. Defects in rubber lining *can* be repaired – but you should monitor the procedure to ensure it is controlled closely.

5. For metallic linings, the main FFP criterion is *integrity*, so that the base material is not exposed.

6. Loose cladding is primarily a welding-based process. You can use some of the information in Chapter 5 for guidance.

7. Galvanizing (coating steel by dipping in molten zinc) is a common process. BS 729 and ASTM A90 have useful inspection-related information. One of the main tests is to check the weight of the zinc coating.

8. Try to learn the basics of assessing grades of *surface finish* using a comparator gauge. The same principles are used for machined surfaces and plated surfaces. Grades can be expressed as N-numbers (ISO DIN 1302) or using the parameter R_a, which relates to the roughness of a surface expressed in microns.

References

1. BS 903. *Physical testing of rubber* (various parts).
2. BS 6374 Parts 1 to 5. *Lining of equipment with polymeric materials for the process industries* various parts.
3. DIN 28 051: 1990. *Chemical apparatus; design of metal components to be protected by organic coatings or linings.*
4. BS 729: 1994. *Specification for hot dip galvanised coatings on iron and steel articles.*
5. ASTM A90/A90M 1993. *Test methods for weight (mass) coating on iron and steel articles with zinc or zinc-alloy coatings.*
6. BS 1224: 1996. *Specification for electroplated coatings of nickel and chromium.*
7. DIN ISO 1302: 1992. *Technical drawings – methods of indicating surface texture.*
8. BS 1134 Part 1: 1988. *Assessment of surface texture – methods and instrumentation.*
   BS 1134 Part 2: 1990. *Assessment of surface texture – guidance and general information.*

Chapter 14

Painting

You will find that almost everyone likes to comment on painting. It is visible, its purpose is well known and so it *always* attracts attention during works inspections. Despite this attention (often a subtle mixture of the authoritative and the uninformed) you will still hear reports from site engineers and end users about poor painting. Often it is only a cosmetic problem, but sometimes it is more serious and the paint flakes off after a very short time, allowing corrosion to start. Why is this? One reason is poor inspection (hopefully you will not by now even be starting to confuse this with *not enough* inspection). Poor inspection caused by, perhaps, confusion over the relative importance of the FFP criteria relating to painting. Well-intentioned confusion no doubt, but confusion nevertheless.

Fitness-for-purpose (FFP) criteria

One reason for the apparent confusion over FFP criteria for painting is that the relative 'weights' of the criteria are not constant. They vary depending on the item that is being painted, more than with the type of paint 'system' that is being used. For small, robust items of equipment such as pumps, motors, valves, and small vessels the main reason for painting the item is largely cosmetic rather than protective. For large and fabricated items operating in a corrosive environment, however, the priorities change. Structures for offshore and coastal use, fabricated desalination modules, bridges, ships, exterior ductwork, and similar, use very specialized paint systems to reduce the effects of corrosion. Painting has become almost an integral part of these technologies. Different approaches to FFP criteria are therefore required, across the ranges of plant that a works inspector is likely to see. It is dangerous to be too general, but the main FFP criteria are:

- proper preparation
- a suitable paint system
- correct application.

You can see how these are composite criteria – note how they refer to the *steps* of the painting process rather than purely to the assessment of the integrity of the paint after it is applied – this is similar to the approach used for the assessment of welding in Chapter 5. Put simply, effective inspection is about monitoring the *painting process*, not just the end result.

Remember the scenario that I introduced right at the beginning of the first paragraph in this chapter? You can, with a little practice, circumvent this. Firstly, keep a clear focus on the FFP criteria, then try to follow the following principle. It is not absolute, or perfect, but I think it has a certain wisdom.

Principle

When inspecting painting don't waste your time picking at chips and scratches. Instead – look for the big issue.

The big issue is whether any of the three FFP criteria have not been done correctly.

Basic technical information

To be able to inspect paintwork effectively it is necessary to understand the role of the paint coating. This role differs between types of paint but fortunately only to a limited extent. The role is related to the chemistry of the paint system – this is covered in detail in many standard text books so I will not repeat it here. What you *do* need to know is the likelihood of finding problems that will result in failure of the paint film. Figure 14.1 shows the approximate frequency of the various painting 'problems' that you can expect to find over a large number of inspections. Note how I have shown the problems in terms of the role of the paint – the role that is not being fulfilled in each case.

Paint types

There are numerous different types of paint. For the purposes of works inspection, paint type is only important insofar as it influences the methodology of the inspection, and the way in which it affects the weighting of the three FFP criteria. This is a simplified but satisfactory

Painting 417

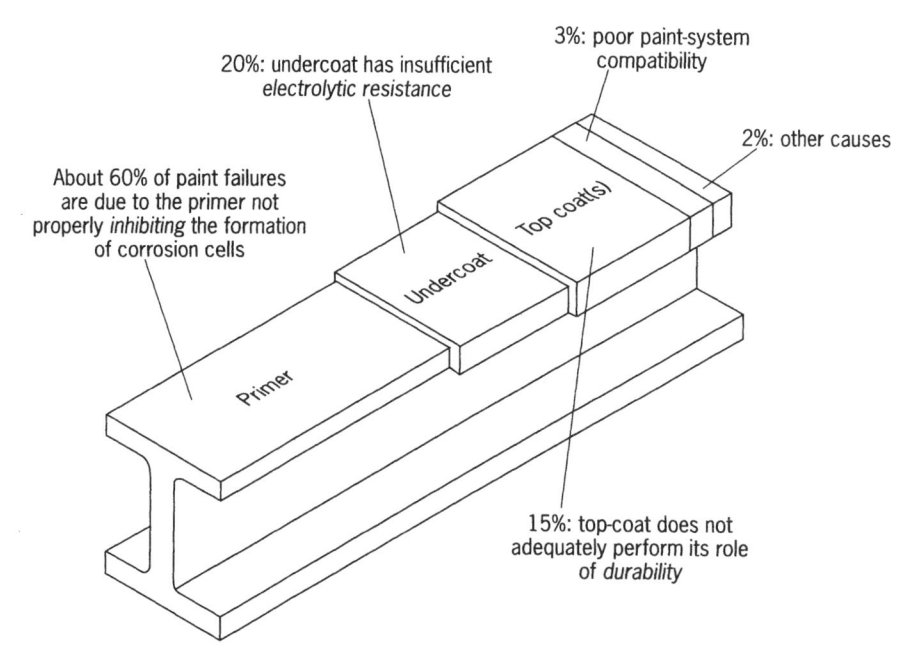

Fig 14.1 Practical reasons for paint failures

Labels on figure:
- About 60% of paint failures are due to the primer not properly *inhibiting* the formation of corrosion cells (Primer)
- 20%: undercoat has insufficient *electrolytic resistance* (Undercoat)
- 3%: poor paint-system compatibility
- 2%: other causes
- 15%: top-coat does not adequately perform its role of *durability* (Top coat(s))

approach – there is no need to have detailed knowledge of the chemistry of the subject. It is useful, though, to try and learn a little about the major classifications to which paints belong, rather than to rely solely on manufacturers' product trade names, which can be confusing. The main classifications are outlined below, along with the main features that have an effect on their inspection.

Air-drying paints

These are the most common type that you will meet in general engineering applications. They account for maybe 70–80 percent of all painting on indoor structures and equipment, or on outdoor installations which are not subject to highly corrosive atmospheres. There are three main types:

- *Alkyd resins* are used in primers and undercoats. Dry film thicknesses (dft) are generally 35–50 µm per coat. Expect to see cosmetic problems caused by the short term wet edge time. Adhesion is not a common problem if preparation is done properly.

- *Epoxy esters*. Expect to find these on structural steelwork and storage tanks. They have better chemical resistance than all alkyd types. The inspection requirements are the same. You may need to guard against problems caused by poor spraying technique with this type of paint.
- *Chlorinated rubbers*. These are used for exterior protection of structural steelwork and fabrications in particularly corrosive environments, such as coastal or offshore locations. The top coat is normally applied over a similar chlorinated rubber alkyd or zinc-rich primer, other types may result in adhesion problems. Coat thicknesses vary, but single coat dft is normally quite thin, about 50–60 μm per coat due to the high level of solvent in the paint. Often, however, specifications will call for a single thick coat of 300–400 μm. For this a thickening agent has to be added to the paint. Chlorinated rubber paints do not often suffer from intercoat adhesion problems, as subsequent coats fuse together quite effectively. They *do* suffer from problems such as sagging and peeling if the coating is applied too thickly.

Two-pack paints

Two-pack paints comprise a pigmented resin and a catalyst (or hardener) which are mixed together. They harden via a chemical reaction rather than by evaporation of the solvent. The most common range is the 'epoxy' type which is highly resistant to water, environmental, and chemical attack. Expect to see it specified anywhere where there are alkaline or acidic conditions, rather than for general service use. Note four specific inspection-related points:

- Epoxy paints are generally 'hi-build' – an average dft is 100 μm per coat.
- They have a very limited pot life once mixed, so application problems *do* occur.
- The ambient temperature during painting is important. Epoxy will only cure successfully at temperatures above 7–8 °C.
- On balance there are more critical process-related factors than for air-drying paints. This means that monitoring of the application process is an important part of the inspection process.

The other type of paint in this category is two-pack polyurethane. This is also used for chemical-resistant applications. Dry film thickness can vary from 40–100 μm per coat, depending on the paint formulation used.

Primers

For metal which has been shotblasted it is common for a thin (30 μm) coat of zinc-based blast primer to be applied. This can be an 'etch' type – containing phosphoric acid to etch the surface – or a two-pack epoxy rich in metallic zinc. Zinc primers have inhibitive properties, the zinc providing cathodic protection to the iron. From an inspection viewpoint, a key issue is to ensure that these primers are applied immediately after shotblasting. The zinc must make intimate contact with the metal surface and not be restricted by corrosion products, which can form very quickly on a freshly blasted surface. For less critical fabrications, you may find zinc chromate or zinc phosphate (or occasionally red lead) primers being used. These are traditionally applied in quite thick coats and tend to be more common for on-site repair work than for new factory-built equipment. Their performance on steel that has not been fully cleaned is quite poor.

Preparation

Incorrect surface preparation is the most common root cause of failure of paint films, particularly those applied over common ferrous materials such as low carbon steel. Poor preparation does not always result in instant failure – a period of two or three years may elapse before the real problems become apparent. By then, however, the breakdown will likely be almost complete – the paint system having effectively given up its protection of the underlying metal.

Proper preparation comprises two objectives. Removal of existing rust cells is one, but the main one is to eliminate any active (or latent) corrosion cells on the surface of the metal by removing the *millscale*. Millscale is hard brittle oxide formed during the steel rolling process. It causes 'mechanical' problems to a paint film because it expands and cracks, causing the paint to flake off. It also causes electrochemical problems because it is cathodic to steel, so it encourages rapid anodic attack on small areas of unprotected surface.

Three key points about millscale:

- Proper preparation means eliminating *all* the millscale before painting.
- Millscale cannot be removed by wire brushing, it is too hard. The

practice of 'weathering' material only reduces the amount of millscale – it does not remove it totally.
- The *only* way to remove millscale properly is by shotblasting with the correct heavy grade of grit.

The degree to which a material is shotblasted is important. Grades of cleanliness of the blasted surface are covered by several well accepted standards (see later). From an FFP viewpoint, however, it is wise to be wary of grades of finish that do not specify complete removal of the millscale. I have shown these grades in Fig. 14.2 along with some key inspection points that you should check.

Application

Most fabrications, vessels, and larger equipment items that you meet during works inspections will have their paint applied by spray. Methods such as dipping (for small components), hand brush or roller application, and electrodeposition are less common. The *technique* of spray painting is one which relies heavily on the skill of the operator – it is perfectly feasible for a properly prepared surface, with a well chosen paint system, to give poor results because of a poor application technique. It is not possible to describe the techniques of paint application here. References **(1)** are available. It is important, however, that as an inspector you have an appreciation of the main variables that have an influence on application. You will find them addressed in the painting procedures and record sheets used by good painting contractors. Use them also as a rough checklist of items to look at when you are *witnessing* paint application. The main variables are:

- *Spraying air quality*. It should be moisture free, by using filter/dryers, in order to avoid contamination of the paint. Some techniques use an airless spray in which the paint atomizes due to pressure drop only as it exits the spray gun.
- *Paint mix and consistency*. This is particularly important for two-pack paint types. Shelf life and pot life restrictions need to be complied with. Note that some paints are heated before being applied.
- *Ambient conditions*. This is a key variable. Ambient temperature and relative humidity must be within prescribed limits (paint specifications clearly state what these are). High humidity (above 80 percent)

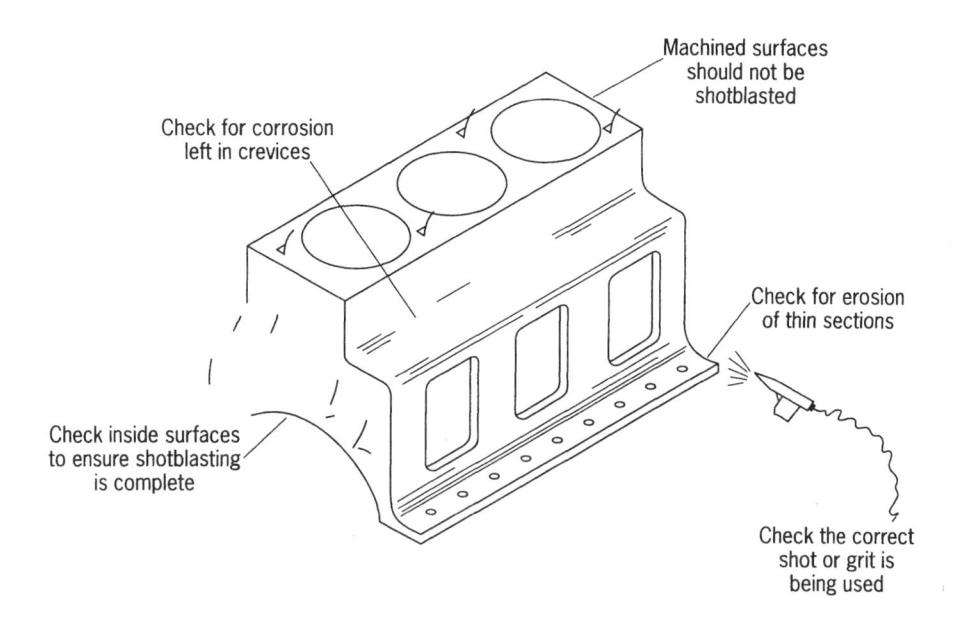

SHOTBLASTING: POINTS TO CHECK

- Ideally, all millscale should be removed
- Check for areas of over-blasting, where critical dimensions have been eroded
- Make sure the blasted surfaces are brushed or vacuumed before priming
- Primer should be applied immediately after blasting. A quick-drying 'blast primer' should be used if there is any delay before the main primer is applied

	Preparation grades (standard SIS 05 5900)
Sa 3	Blast cleaning to pure metal. No surface staining at all
Sa 2½	Very thorough blast cleaning. Only slight surface staining may remain
Sa 2	Thorough blast cleaning. Most millscale and rust is removed then the surface cleaned
Sa 1	Light blast cleaning. Any loose millscale and rust is removed

Other standards that may be used are BS 7079 and ASTM/SSPC

Fig 14.2 Shotblasting – points to check

and low temperatures (below about 4 °C) are cause for concern. Dust-laden or salty conditions are also undesirable.

Specifications and standards

The extent to which technical standards are quoted in contract specifications depends on the purpose of a piece of equipment, and the practices of the industry within which it is used. For power generation, non-specialized chemical plant, and general industrial uses it is *unlikely* that you will need to become familiar with the detail of many of the specific technical standards – effective inspection of painting is more about careful observation and the application of a certain amount of judgement. In short, a commonsense approach works best. The corollary of this, however, is that you cannot rely heavily on technical standards to reinforce your judgements. Experience will help – another reason why it is important to witness painting techniques at the preparation and application stages. It is much easier to understand the techniques of application once you have watched it being done.

There are several sets of standards relevant to surface preparation and paint testing. The most common ones are described below.

BS 7079 (2) covers the generic subject of preparation of steel substrates before painting. It comprises 16 separate documents divided into 4 groups:

- Group A covers visual assessment of surface cleanliness.
- Group B covers more general assessment of surface cleanliness using chemical and pressure - sensitive tape methods. It is unlikely you will see these used during normal works inspections.
- Group C covers assessment of surface roughness (or profile) using comparative and direct instrument measurement methods.
- Group D covers the preparation techniques themselves, including shotblasting and power brushing.

Swedish standard SIS 05 5900 (3) is a similar standard specifying 'grades' of shotblasted surface finish. This standard is accepted and used in most industries in preference to BS 7079. It provides reference photographs for four main grades of blasted finish, termed Sa3, Sa2½, Sa2 and Sa1. You will frequently see these quoted by themselves, without explicit reference to the SIS 05 5900. Figure 14.2 shows the interpretation of these grades. Nearly all contract specifications require that surfaces are prepared to Sa2½. To meet this grade the surface must be blasted to the state where only *traces* of millscale or rust remain. In practice this is normally interpreted as meaning that only a light staining can be seen – it is only slightly short of the Sa3 level which requires that the surface exhibit a pure metal, totally uniform finish.

SIS 05 5900 also makes provision for describing the degree of rusting of hot rolled steel before it is shotblasted. There are four grades A to D. Grade A is where the steel surface is still fully protected by millscale with little, if any, rusting, whereas Grade D represents the worst condition in which the millscale has rusted away and the exposed surface has become heavily pitted. You may see these grades shown in conjunction with the blasted finish designation – for example C Sa2½ represents a moderately rusted surface which has then been shot blasted to Sa2½.

There is a similar American standard from ASTM SSPC (Steel Structures Painting Council). This uses the same principles of describing rusting and blasted finishes but uses different designations. The most commonly used ones, A, B, C or D Sa2½ are designated as grade SSPC-SP10.

BS 5493, on the protective coating of iron and steel structures against corrosion, **(4)** is a very broad technical standard of more than 100 pages. It covers painting requirements for most of the engineering applications you are likely to meet, except those in the offshore industry. There is clear guidance on how to choose and specify a paint system suitable for various environments – it provides some data on the performance of various types of paint but does not go far enough to enable a full FFP assessment to be made from the information provided. Section 2, Table 4 of the standard provides a very useful list of technical questions about the assessment of a paint system. You could develop an excellent checklist from this information, for use during your works inspections – it is a good summary of FFP issues. Section 3 of the standard uses a system of classifying paint types using digit codes. You will see this used in some contract specifications but it is not universally accepted, many contractors (and paint manufacturers) have their own system of paint classification.

The issue of environmental conditions is covered in BS 5493. Various environments are defined ranging from clean indoor conditions to exterior 'polluted coastal atmosphere'. You will see these commonly referred to in the parts of contract specifications dealing with the choice of paint systems.

The issue of paint colour is dealt with by BS 381 **(5)** which cross-references a series of standard colours similar to 'RAL numbers', originally from a German standard. These RAL samples have a four-digit designation and a simple swatch card is available for colour comparison purposes. Most of the colours are not too hard to differentiate from each other, however you may have a little difficulty with the blues and greens (RAL 5000 to RAL 6000) which use a number of quite similar shades. By all means use a swatch card during your works inspections – but be prepared to allow a little leeway regarding the exact shade of paint that is used.

BS 3900 **(6)** is the main British Standard dealing with paint testing. This standard has an extremely wide scope – there are over 100 separate parts, many of these having direct ISO equivalents. The parts are subdivided into ten groups designated A to J, each group containing a number of consecutively numbered parts. Much of the standard is orientated towards the chemistry of paint systems, which is of most use in a laboratory context. Two groups are of direct relevance to works inspection. These are:

- Group E contains 13 parts relating to *mechanical* tests on paint films. Of these, the most useful ones are:
  - Part E2 (equivalent to ISO 1518) describing the scratch test.
  - Part E9 (equivalent to ISO 2815) describing the Bucholz indentation test.
  - BS EN 24 624 (equivalent to ISO 4624) describing the 'pull off test' for a paint adhesion. This was previously part E10 of BS 3900 and still cross refers to other parts of this standard.
- Group H contains six parts relating to the evaluation of paint *defects*. The parts cover the common defects of blistering, rusting, cracking, flaking, and chalking. You will find these standards useful in helping you to describe, objectively, painting defects that you find. They remove some of the subjectivity in this area and so are useful to improve the technical robustness of your reporting of defects. These documents are pictorial standards, so they allow direct comparison with actual paint defects. A grading system is shown in Part H but this stops just short of defining these as acceptance criteria. Interpretation, and a clear focus on FFP, is still required.

The determination of dry film thickness is covered by ISO 2808 (BS 3900 Part C5) and by ASTM D1186 **(7)**.

Test procedures and techniques

There are only a few test techniques that are used practically during works (or site) inspection of paintwork. They are only partially useful in verifying the fitness for purpose of a paint film – there are a multitude of very real paintwork problems that they will *not detect*. This means that effective inspection of paintwork consists initially of verifying the three main FFP criteria (preparation, system, and application) that I introduced at the beginning of this chapter. After this it is heavily dependent on good *visual* examination, assisted by simple tests.

Visual examination

Try and adopt a thorough approach to the visual examination – the objective is to find any major problems that exist. Before you start, make a point of visualizing how the operator would have sprayed the component. It should be reasonably clear which way up the item would stand, and whether the operator would spray by standing at ground level, or whether scaffolding or a mobile gantry crane would be needed (for large fabrications and structures). This will provide insight into where potential problem areas may lie. Check the following areas. They are equally valid at the primer, undercoat, or top coat stage.

Enclosed areas

Check these first. Enclosed areas are those that, although not necessarily totally enclosed, cannot be sprayed directly from 'outside'. Typical examples are the inside of tanks or sumps, box section girders, hollow sections with a lot of internal ribs or stringers, and ductwork – all of these require an operator to reach *inside* to apply the spray. Visibility and access will be difficult so such areas are often the most likely places to find problems with both surface preparation and paintwork application.

Inside corners and radii

Inside radii often have problems with excessive paint application, causing it to sag and peel off. Check also the inside corners of fillet welds, particularly enclosed corners where plate material joins together in three planes (as in the inside corners of a box). Where such areas are some distance from the operator's spray-gun, because of difficult access, look for the opposite problem – these areas are often too thinly coated because the operator is deliberately trying to avoid applying too much paint.

Inside edges

These are edges that face *away* from the operator during spraying – generally because there is no access to the other side. They are not easy areas to spray. Look for too-thin dft on the edges.

Outside edges

Outside edges are those that face *towards* the operator and as a result are slightly more predictable. A good spraying operator will coat the outside edges first, before covering the horizontal or vertical areas behind it – look for areas of incomplete coverage on the edges, to see whether this has been done properly. It is fairly common to see paint ridges perhaps 100–200 mm back from the edge (check by sighting along the edge). This is caused by the operator allowing the wet edge to dry slightly before re-covering it when spraying the nearby areas or by the application of a 'stripe coat' to build up the edge thickness. There is no need to be too concerned about this – it is only a threat to FFP if the edge dft is so excessive that recognized 'over thickness' defects such as paint sagging or runs start to appear.

Large horizontal surfaces

Check that the coating looks even without any excessively thick or thin areas. These can sometimes result from an operator trying to spray from one level, without the correct scaffolding or gantry. Large horizontal planes are a good place to find defects caused by poor ambient conditions. Look for blistering or uneven thickness caused by condensation.

Large vertical surfaces

Large vertical surfaces are normally easier to access for spraying so unevenness is not a common occurrence. The main problem is paint *runs* – these surfaces are a good test of the consistency of the paint mix (particularly with two-pack types) and of the skill of the operator in avoiding over-application.

Around fittings

Most fabrications have integral fittings such as saddles, nozzle stubs, reinforcing rings, hatches, and lifting/support fixtures. Check around them carefully – some have reverse edges, webs, and fillets which can be overlooked by the operator when spraying. You are more likely to find paint defects due to the paint film being too thick in these areas, rather than too thin.

Masked-off areas

It is common for various areas such as flange faces and precision machined mating faces to be masked off during painting. It is worth checking the manufacturing drawing to make sure that the marked areas are indeed the correct ones. Equally, be very wary of any milled, turned or ground surface which has been painted over – this is *almost certainly* a mistake.

A key part of the visual examination is, of course, reporting what you found. Follow the general precept suggested in Chapter 15: *say* what you did but *describe* what you found. For this you need to recognize the common painting defects, and the specific terminology that is used.

Painting defects

The main painting defects that you are likely to see are described below.

Sagging and curtaining

These are easy to identify, consisting of obvious areas of paint overthickness where the wet film has sagged under its own weight. You will most often see it in wide horizontal bands on large vertical surfaces. It is caused by applying too much paint. In extreme cases it will develop into 'runs' down vertical surfaces. Although unsightly, it is not necessarily cause for rejection, unless it is widespread or is accompanied by surface preparation problems.

Orange peel effect

In this case, the paint surface when dry resembles orange peel, with a dappled surface, sometimes accompanied by small blisters. This may be a problem of paint consistency or poor spraying technique. In most cases the adhesion of the film is affected, so significant areas of orange peel effect are a clear cause for rejection.

Wrinkling or lifting

At first glance this appears similar to the orange peel effect with a blistered appearance to the film surface. The quickest way to differentiate is to run your hand over the surface to check whether any significant lack of adhesion is present. If the surface film is clearly 'loose' (especially at the edges of any large blisters) then this is indicative of an *intercoat* problem. The most serious type is caused by incompatibility of the solvents used in successive paint coats, i.e. a paint system error. A similar, but less pronounced, effect can be caused by the operator not allowing the correct time interval between coats.

Rough surface finish

A poor surface finish is normally indicated by the loss of gloss on the top coat. It is easiest to see on large flat surfaces. The main causes are condensation or airborne dust and dirt. Unless dirt contamination is very serious, it should be possible to rub down and recoat, without having to remove the existing paint coats.

Pinholing

This has the appearance of large numbers of small concentrated pinholes in the paint surface. The most likely cause is contamination of the paint by oil or water. Extensive pinholing does require a full repaint, otherwise the FFP of the paint system is likely to be affected.

Thin areas

This is the most common defect that you will encounter – expect to find components with overall too-thin coatings as well as those where only individual areas such as edges have insufficient application. Provided that the primer and undercoat films have been applied, then the top coat thickness can easily be built up by further application of paint. If (as occasionally happens) the primer or undercoat are much too thin, or missing, then the coating will not meet its FFP requirements. Strictly, an additional thickness of top coat is not an acceptable remedy for a deficient primer layer, as primer often has an *inhibitive* role whereas the main function of the top coat is to protect against weathering and mechanical damage.

Dry film thickness (dft)

The dft of a paint coating is an important parameter. Assuming correct preparation and application, the durability and protective properties of a coating are related directly to its dft. Adequate thickness is necessary in order for the film to have sufficient electrolytic resistance to prevent the formation of the local galvanic cells that cause corrosion. Most painting specifications show the dft required for the separate primer, undercoat and top coat, as well as for the three together. You will see that these are shown as *minimum* dft. Sometimes a maximum is also quoted but often there is none – it is inferred that a thicker dft is acceptable, as long as this is not so thick as to result in peeling, sagging, or other defects associated with excessive application. It is only possible to obtain a true dft reading when a paint film is completely dry – this can be up to 24 hours after application of the last coat, depending on the type of paint. During application the painting operator checks wet-film thickness using a 'comb' or 'wheel' gauge. This thickness then reduces to the dft as the solvent evaporates. It is difficult to quote general relationships between wet and dry film thickness as it varies with the type of paint (it will be shown on the paint manufacturer's data sheet if you need to know it when you are witnessing paint being applied).

Simple hand-held meters are used to measure dft. They work on electromagnetic or eddy current principles and provide a direct digital readout. Some have a recording and print-out facility. Here are a few general guidelines to follow when witnessing dft measurements:

- *Calibration.* Do a quick calibration check on the dft meter before use. Small test pieces, incorporating two or three film thicknesses, are normally kept with the meter. You can also do a quick zero calibration test on a convenient exposed steel surface – a machined surface is best.
- *Test areas.* Test a large number of points, two or three are not sufficient. Make sure you include vertical and horizontal areas, corners, radii, and edges. Pay particular attention to enclosed areas. Use your appreciation of which areas would be difficult for the operator to spray easily.
- *Results.* Specified dft requirements are generally accepted as being considered 'nominal' or 'average' values. As a rule, the average dft measured (from multiple readings) over a minimum 1 m^2 area should be greater than or equal to the specified level but, there should be no individual dft readings of less than 75 percent of the specified level. This means that there is always room for some interpretation, unless a purchasing specification is very specific about the number and location of measurements that are to be taken.
- If in doubt, the best standard to look at is ISO 2808 (BS 3900 part C5).

Paintwork repairs

The easiest place to do repairs to paintwork is in the works, before shipping the equipment to the construction site. Some construction sites are well equipped and have the necessary skilled subcontractors for painting large items, but many are not. The site may also have environmental problems such as high humidity, dusty or salty air. Repair techniques depend on whether the faults you have found constitute a major FFP non-conformance (wrong preparation, system or application) or whether they are cosmetic. A situation involving a major non-conformance normally has only one solution – remove the faulty paintwork and repeat the process, taking care to eliminate the previous faults. Given that serious faults have *already* occurred there are several points that you should consider.

- Do not start the repair without a proper diagnosis of the original fault and what caused it.
- The only real way to obtain proper preparation is to shotblast completely to a minimum grade of Sa2½. Witness this activity – repairs are often done to emergency timescales which do not encourage careful and thorough surface preparation.
- You need to check the new paint system. The best way to do this without wasting time is to talk directly to the paint manufacturers – they can give you a very quick response about the suitability of a paint system for a particular purpose, compatibility of coats, and details of preparation and application techniques required.
- Witness the application of all three coats if this is practical. If not, make sure that reliable visual and dft checks are done after each coat.
- Repeat all the final checks again after the repainting is finished. This is one area where site rectification of any subsequent defects would not be good practice. Expect some equipment manufacturers to place you under pressure to release the item before you have checked it properly. Decline such invitations.

Cosmetic painting defects can be treated differently. Unless the substrate material is actually exposed, or you can find evidence of a more serious problem, it is up to you whether or not you feel it necessary to witness or reinspect cosmetic repairs. Cosmetic defects *should* be repaired even if you have concluded that they do not compromise fitness for purpose. Try not to say or do anything that would enable these defects to be misunderstood out of context by a distant client or end user. Do not base a whole inspection report around lists of cosmetic defects – not if you want to be taken seriously.

KEY POINT SUMMARY: PAINTING

1. For painting, the FFP criteria refer to the steps of painting *process*. They are:
   - proper surface preparation
   - a suitable paint system
   - correct application.

2. The main types of paint are air-drying and 'two-pack' (epoxies). They have different properties, thickness and *problems*.

3. Proper surface preparation is essential. This *normally* means shot blasting to the standard SIS 05 5900 grade Sa2½ – a well-blasted surface with all the millscale removed.

4. You will find it an advantage to know the content of the technical standards mentioned covering surface preparation and paint testing. It is not necessary to understand the many chemistry-based standards that exist.

5. *Visual examination* is the core of effective inspection of paintwork. Learn the common types of painting defects and how to describe them properly.

6. The main works tests are for dry film thickness (dft) and adhesion (the 'pull-off' test).

7. Major repairs usually require complete shotblasting and repainting: partial solutions are normally ineffective.

8. *Maintain your focus*. Don't waste time picking out small cosmetic defects. Look for the big issue.

References

1. Dunkley, F.G., *Quality control of painting in the construction industry*. JOCCA, 1976.
2. BS 7079. *Preparation of steel substrates before application of paints and related products*. This document is in 16 separate parts.
3. SIS 05 5900. *Pictorial standards for blast-cleaned steel (and for other methods of cleaning)*. Standardiseringskommissionen I Sverige, Stockholm.
4. BS 5493: 1977. *Code of practice for protective coating of iron and steel structures against corrosion*.
5. BS 381: 1988. *Specification for colours for identification, coding and special purposes*.
6. BS 3900. *Methods for tests for paints*. There are more than 100 parts to this standard.
7. ASTM D1186: 1993. *Test methods for non-destructive measurement of dry film thickness of non-magnetic coatings applied to a ferrous base*.

Chapter 15

Inspection reports

Almost anyone can produce an inspection report, of sorts. Effective reporting, though, is a little different; a tailored technical product pointed directly at your client's need. This is not an easy thing to do.

Your report is your product

Your clients will not attend most of the works inspections and will not see the way in which you act to verify fitness for purpose of their equipment – nor will they be aware of the underlying knowledge that is the basis of your decisions. They *will* see your reports. Consider also the economic factors at play in inspection contracts. Reporting time will be charged to your clients and it is all too easy for these time charges to become too high. If this happens you will price yourself out of the market or, at best, not get as much inspection work as you would like. Reports are your product and a shop window for your future services. They must be good.

Know your objectives

The purpose of an inspection report is not *just* to identify problems. There are three objectives to consider before you start the report, and to keep in mind whilst writing it. They will help to keep you on track. I have shown these objectives in Fig. 15.1 and they are described below.

To communicate with your client

To meet your clients on their own level – give them some confidence in what you are doing. Figure 15.2 shows *what* your clients are interested in.

The objectives of an effective inspection report are:
① To communicate with your client
② To explain FFP requirements
③ To define a single course of action

Fig 15.1   Effective inspection reports – the objectives

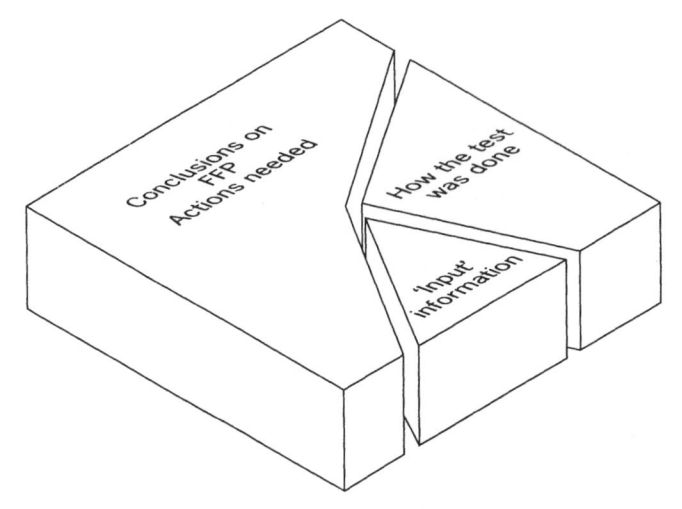

Fig 15.2   What your client is interested in

To explain fitness for purpose

Tell your clients what they want to know about the FFP of the equipment that you have seen tested. The best way to do this is to learn to strike a balance between reporting by description and reporting by exception.

To define actions

More specifically, to define a single course of action. Then to define

Inspection reports

'reference points' from conflicting opinions and positions (remember the points made in Chapter 2 about reaching a *consensus* on FFP?).

One specific point merits further explanation. You will find long-standing differences of viewpoint as to whether inspectors should report by description (explaining what they did and what they found) or by exception (only explaining what they did if a problem was found). Extreme cases of these two approaches result respectively in reports which are expansive (and expensive) or technically superficial. You need to craft quite a fine balance using aspects of *both* these approaches. Figure 15.3 shows a method you can use. For those engineering aspects which impact directly on the main FFP criteria, you should provide a detailed technical description of the tests that you witnessed and the observations that you did (and didn't) make. I should emphasize here that it is *technical* description that is the most important, many procedural details can be just 'padding' and not add a lot to your client's knowledge. For the myriad of other aspects that are not directly related to the main FFP criteria I think it is best to report by exception. This means that you only need to resort to detailed technical descriptions when you find a problem. Exception reporting is a good way of keeping your reporting costs down – just be extremely careful not to let this 'easy' approach spread to the FFP issues.

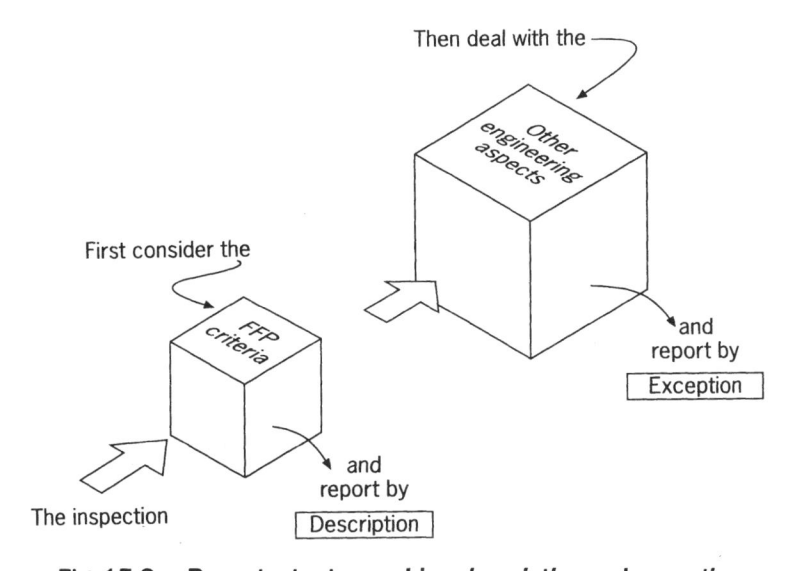

Fig. 15.3. Reports: try to combine *description* and *exception*

Technical presentation

Once you have grasped the concept of achieving a balance of descriptive and exception reporting, we can expand our view and move onto the question of technical presentation. This is about how to construct and present the short technical arguments which are the core areas of an effective inspection report.

Do not be misled into thinking that technical presentation doesn't matter – that manufacturers, contractors and clients alike will not question the statements that you make. They will, most of the time. Remember too that your report has to take its place amongst the huge body of technical information that exists in a large plant construction project. We will look a little later at the breakdown of the contents of an inspection report. First we need to work on the structure of the writing at the technical level – the way to present key technical observations and opinions in a logical and precise way.

Aim for a logical progression

In reporting technical matters you need to achieve a logical progression of the technical argument (I use argument in its loosest form – it does not have to infer disagreement). Each statement must follow on from the last without gaps inbetween. This gives good dense dialogue.

Figure 15.4 shows the steps involved. Use these steps to structure your writing *within* sections of your report where you are making technical descriptions of what you found during your inspection. The structure may spread over several paragraphs but not normally over a total length of more than about 250 words – if it is spread too far your writing will lose its cohesion and gaps will appear in the logic. This will cause your arguments to lose their strength.

Figure 15.5 shows how I structured the technical statement, 'Abstract A', which is given in Fig. 15.6. This is a simple example concerning the discovery, by an inspector, of a visible casting defect on the inside of a steam turbine casing. Note the 'thought stages' in the left hand column of Figure 15.5 which match the logical steps shown in Fig. 15.4. I have then assembled these points into the three paragraphs of abstract A in Fig. 15.6. Note the discrete sentences. These may differentiate technical reports from smooth and erudite writing but they provide the necessary description. This is the object of the exercise.

Abstract B is an example of poor logical progression. Note how I have also disguised the text with a number of 'style points' (use of poor abbreviations, indirect and passive speech, and indecision, among

Inspection reports 437

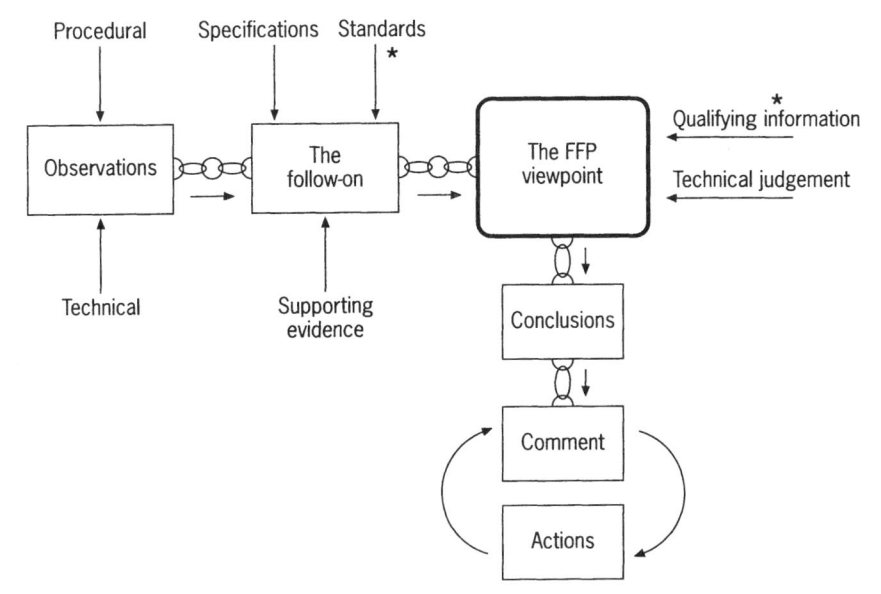

Fig 15.4 Aim for a logical progression

others). Try and see through these to get to the real core structure of the statement. Do you see how there is no clear focus on FFP? Can you recognize the poor structure? Please try to eliminate points like these from your reports.

Be accurate and consistent

An inspection report is the place to demonstrate accuracy and precision, not to impress others with your particular specialism (if you have one). You can improve your reports by paying attention to the following points:

- *Technical accuracy*. Refer to components by their correct names. It is not difficult to use typical engineering terms (such as webs, fillets, lugs, projections, flanges, and walls) loosely, as their meanings can be similar. Try to use the accurate one. Refer to the equipment general arrangement drawing or an engineering handbook if you are in any doubt.
- *Consistency*. Stick to the same terminology throughout a report. If you are describing, for instance, a crane hoist, make sure you don't suddenly refer to it as a winch or a drum. This doesn't mean that you can't develop your technical arguments throughout your report. It is

'Thought stages'	Abstract A	Abstract B
Your procedural observations	A visual examination was carried out.	Note how abstract B has not used the logical thought stages.
Your technical observations	Attention to areas of ruling section. Defect found on inside at the 1 o'clock position.	and
Mention supporting evidence	Type of defect.	Where is the FFP focus?
Mention specification and standards(*)	Spec. section XXX and standard MSS-SP-55.	
Introduce FFP	The defect affects mechanical integrity.	
Reference qualifying information(*)	ASME VIII Stress calculations	
Use technical judgement	The defect is in a high stress area	
State your conclusions	This defect is unacceptable. NCR issued	
Expand (only if necessary)	The defect is above critical crack size. Circumferential orientation	
Recommend action	Submit repair procedure: – Excavate – Repair/initial NDT – Re heat-treat/final NDT	

Fig. 15.5 Organizing your thoughts

acceptable for a casting *defect* to be examined, compared with an engineering standard and be diagnosed as a *crack* – this is good logical progression as I have previously described. Once you have described it as a crack, though, don't change it back to a defect. Reread your reports and aim for consistency.

Abbreviations

Random abbreviations spoil the look of your inspection report and cloud the meaning of what you want to say. This is particularly the case if your client speaks a different 'mother tongue' to you, which is commonplace in the inspection business. My best advice is that you follow the Système International (SI) system of abbreviations for

Abstract A

A visual examination was carried out on the h.p turbine casing No. 31248/X. No defects were found in areas of ruling section. A significant defect was found on the inside surface at a point adjoining the third guide blade carrier locating flange (at the 1 o'clock position looking in the direction of steam flow). This was a linear defect approximately 20 mm long, orientated circumferentially around the casing surface (see sketch).

Section XXXX (clause YYYYY) of the contract specification prohibits linear defects above 2.0 mm in length. The standard MSS–SP–55 also classifies such defects as unacceptable. The size and position of the defect will affect the mechanical integrity of the turbine casing. (ASME VIII is used as a normative reference for calculating stresses in turbine casings).

This defect is unacceptable. A NCR was issued. The NCR requires the manufacturer to submit (within 7 days) a repair procedure covering excavation of the defect, weld repair, heat treatment, and appropriate NDT procedures.

Abstract B

An examination was made of the turbine casing (No. Off 1) under a good light and with Mr Jones of the Excellent Steam Turbine Manufacturing Company Limited (and Inc.) present at all times. The casing was stated to be of good commercial quality – surfaces were smooth with only a few grinding marks. The undersigned questioned a defect on the inside of the casing near the G.B.C.F. Mr Jones confirmed that defect acceptance levels were under discussion as no specific mention had been made of this in the contract specification. It was also stated that the casing had been hydrostatically tested to twice design pressure and therefore must be of sound construction. This will be resolved between design engineers.

Following the successful inspection, a very detailed review of the documentation dossiers was performed. These had been prepared to ISO 9001. It was noted that on page 67, line 14.3 there was only a black stamp and a blue stamp, the little red one was missing. Also Mr Jones had neglected to sign page 21. A little improvement is required.

All results were satisfactory. *Oh, really?*

Fig. 15.6 Content, structure, and style (good and not-so-good)

engineering units as rigidly as you can. It is acceptable to use any common abbreviations which are well understood in your client's industry, but don't make up your own. Also, don't be too keen to use, blindly, manufacturers' abbreviations in your report – they may be very specialized and not well understood outside the manufacturing environment. For instance:

- I would certainly be happy to say that a turbine rated speed was 3000 *rpm* and that its steam inlet condition was 60 bar g/550 °C. I am not quite so convinced that 'T/B spd of 3000 revs' would be clear to all clients.
- HSS is an accepted abbreviation for high speed steel, used in machine cutting tools. 316L will be widely understood to represent stainless steel, but 'SS' will not, nor will 'st. stl'.
- I have difficulty believing that even the most imaginative client would know that GBCF referred to a turbine guide blade carrier flange (remember Fig. 15.6?).

Get to the point

The most logical technical report structure, the most polished and clipped technical text, is largely wasted if you have difficulty getting to the point. I am afraid that you *have* to be conclusive in what you say. The ability to write delicately qualified and contingent statements is no doubt a fine gift, it's just that it is better suited to other businesses. Verifying fitness for purpose in your own mind is one thing, but all your client sees is the report. The physical act of writing down a firm technical decision will help you consider if what you are saying really gets to the core of FFP.

If you genuinely have difficulty in feeling confident about getting to the point, there are a few mechanisms you can use to lead yourself in gently. Try these:

- Keep the sentence containing the point as simple as possible. If you look again at Fig. 15.6 you will see the statement I have made: 'The size and position of the defect will affect the mechanical integrity of the turbine casing'. Note how there is no qualification in this sentence – I am not inviting queries. Long sentences containing multiple points *will* produce more valid queries than a simple statement.
- Use a negative. You have to do this very sparingly, otherwise you won't be getting to the point. Consider the following negative

statement in relation to my abstract A in Fig. 15.6. I could have said: 'There was no evidence provided by the manufacturer that the standard MSS-SP-55 considers such defects acceptable'. I am still getting to the point but introducing a defensive barrier for my decision. Use your own judgement on this. Don't go so far as to be evasive.
- If you have a shadow of doubt in your mind as to the validity of your decision (and which of use have perfect knowledge?) then be subtle. Be ready to use a *logic-shift*. Take two of the logic steps shown in Fig. 15.4 and consciously transpose them in the sense of the way in which you write the text. This will disrupt the logic thought pattern of potential critics of your statement (about 40 percent of them will initially misinterpret what you are really saying) but do little to soften the impact on those that wish to agree with it. You should only transpose two steps, no more. I transposed the two steps shown (*) in Fig. 15.5 when I wrote the abstract A of Fig. 15.6. I will leave it to you to find the result of this in the text.

Style

Even in the engineering world, style is not so much what you do, as the way that you do it. This also applies to inspection reports.

This does not pretend to be a book on the mechanics of writing and I cannot recommend to you a particular style that you should follow. The points I have made earlier in this chapter on description, logical progression, and careful use of abbreviations should do much to generate a good style. In this sense, 'good' means no more than that which allows clear communication with your client. I can demonstrate some things which you *shouldn't* do – whatever your personal writing style you should check how you are addressing the following points:

- Do not use long sentences. Too many inspection reports contain 20 to 30 word sentences to try and explain things. Keep it short and sharp. It should be taut. Like this. It is easier to understand.
- In general, use the third person and be direct. Say 'the turbine was tested' not 'we tested the turbine'. When you are getting to the point, say 'this defect is unacceptable' rather than 'on balance the defect can be considered as unacceptable'.
- Keep away from obsolete or legal phraseology such as:
    — 'We, the undersigned'

- 'On behalf of our principals'
- 'Further or in the alternative'

You are not writing a binding contract.
- Beware the meaningless or woolly statement (if you look you will find these in lots of inspection reports), such as:
  - 'The test was carried out in an expeditious manner'
  - 'This defect can be rectified in the future (when?)'
  - 'Clearly'
  - 'It should be noted that ....'
- Say what you did but *describe* what you found. Not the other way round.

A good reporting style will not in itself make for an effective inspection report. It will, however, polish an already good one. There is absolutely no doubt that if you get it right it will add value to what you have done and help your competitive position.

The report itself

Remembering all that has gone before about logical progression of technical argument and the importance of style, we can progress to the inspection report itself. To structure a report effectively it is best to see it as having three main sections – a beginning, a technical middle, and a conclusive end. These are accompanied by a well-defined set of smaller parts such as the executive summary, preamble, contents list, attachments, and in some cases your non-conformance/corrective action report sheets.

I have shown this basic structure in outline form in Fig. 15.7. Use a single sheet, divided roughly as shown to plan the report structure. If you keep the layout approximately the same size, this will help keep the sections in balance. Use this planning sheet to make notes – which you will then write into the appropriate section of the report. It should be clear, from what I am saying, that I do not accept it is possible to write a really effective report *without* first organizing the major points on a planning sheet. To draft a report in sequence (i.e. from beginning to end) is to invite poor technical descriptions. Technical accuracy will suffer and the logical progression towards the conclusions may be flawed. Your report will have weaknesses.

You can see the actual planning sheet I used for the specimen report at the end of this chapter.

The work of most inspection engineers involves the writing of a lot of relatively short inspection reports rather than the very detailed and analytical documents found in other mechanical engineering disciplines. A positive aspect of this repetition is that once working to a clear reporting structure you can quickly develop your reporting efficiency, better explaining FFP issues and learning from the comments of your readers.

The basic outline of what goes where in the report structure (refer to Fig. 15.7) is as follows.

Executive summary

There is a strong argument for an executive summary to be included in all but the very shortest of inspection reports. An average length inspection report of, say, 600–800 words should have an executive summary of perhaps two or three paragraphs. The summary should be an absolute maximum of one page long, even for the most detailed inspection report. The executive summary is an exercise in contracted English. Its purpose is to be a concise summary of conclusive points of the rest of the report, not to introduce new conclusions or decisions. Make sure that your executive summary follows the principles of logical thought progression. Arguably, the logic needs to be even more taut in the executive summary than in the body of the report – any errors will stand out clearly in the very contracted text. You may find it helpful to concentrate on the paragraph content of the executive summary to help with the structure and logic. Most summaries will fit into three paragraphs.

The first paragraph

Start with a brief but accurate reference to the equipment that has been tested. The objective is to enable your reader to identify the equipment, not to describe it. Then:

- Say what tests were done. Summarize them or use a collective term if there are a lot of them, but don't lose the sense of what *types* of tests were involved.
- Mention specifications, standards, or acceptance levels that you used to help interpret the results.
- Make an *allusion* to the main FFP criteria. Keep this short and as near the end of the first paragraph as you can. Its purpose is to form a link to the second paragraph – to make the text 'flow'.

(1) Executive summary	
(2) Preamble and attendees	**(3) Contents sheet**
(4) The input material	**(5) Test procedures and results**
Previous activities and reports Specification clauses Contract amendments Standards ITPs content Working document and drawings	The test procedure Say what you did but describe what you found. Report by description then by exception (as we have discussed)
(6) Conclusions Start with FFP issues Put your findings in the context of FFP Draw conclusions State a single course of action	NCRs CAs Itemize the points Itemize the points • • • • • • • • (see Chapter 2 for guidance on these)
(7) Drawings and attachments	
Drawings help explain test results Photographs are always useful Enclose test result sheets (keep to a minimum, do a results summary sheet if necessary)	

Fig 15.7   How to structure report content – the planning sheet

The second paragraph

State the results, and *only* the results, in here. As part of mentioning the results, you need to make an identifiable but slightly shielded reference to fitness for purpose. There are good reasons for this. Try also to stick to a single result and FFP 'implication' per sentence. Use two or three separate sentences if necessary.

The last paragraph

If you are reporting completely satisfactory results, keep this to a maximum of two sentences. Simply relate your conclusions. Don't feel that you have to elaborate. If you have identified non-conformances then the last paragraph needs to be slightly longer. The best technique here is to provide a concise reference to the non-conformance but only hint at the corrective action. Leave your detailed statement on corrective actions to the body of the inspection report as a single explanation of the consensus on what is to be done.

I have explained in some detail a recommended approach to the structured executive summary, which is a small part of the final report. Treat this part with respect though, in a controversial report it will be more widely read and dissected than the body of technical detail it accompanies.

Write the executive summary last, after the rest of your report is written, and corrected. You can see how I have done this in the specimen report at the end of this chapter.

The input material

This is information and data which forms the background to the works inspection activities. It is differentiated by the fact that it is *given* information so it does not need to be derived, or need any judgement. It is normal to state the input material used near the beginning of an inspection report. Typical material will be:

- Reference to previous inspections and reports relating to the equipment (it is good to achieve continuity with earlier activities).
- Specification sections, and specific clauses, that are relevant to the equipment under test. Note the document hierarchy (see Chapter 3) defining how documents relate to each other and which are the most important.
- Technical standards.
- Inspection and test plans.
- Working documents such as manufacturers' equipment test procedures.
- Drawings.

There is some freedom in the report headings under which the input material can be referenced. It is equally acceptable to use separate headings for 'specifications' and 'technical standards', or to reference everything together under a general heading of 'reference documents'.

Test procedure and results

This is the middle ground of the report. Try to describe simply what tests were performed, placing emphasis on the engineering aspects of the tests rather than getting too involved in procedure. Don't make the awful mistake of describing a test and then saying that it was the incorrect one. You should have already resolved such issues using the provisions of the contract specification and its explicit document hierarchy.

A simple rule to follow in this section is to *say* what you did but *describe* what you found. It is not a disaster if you concentrate a little too much on the results – they are, after all, the important point. In Fig. 15.3 we looked at the balance of reporting by description and reporting by exception. The core of this exercise is, of course, your appreciation of the FFP criteria of the equipment being inspected. If you struggle with identifying the correct FFP criteria this will shine through clearly in this part of your report. Inevitably, your report will be read by engineers with good specialist knowledge and experience of the particular equipment – they will have an intuitive feel for the real FFP issues. Make a real effort to address the *correct* FFP issues in this section of an inspection report.

Voluminous tables of results should be relegated to the 'attachments section'. If results can be quickly summarized and presented in tabular or graphical form then it is acceptable to include them here, but aim to make the 'test procedure and results' section as self-supporting as you can. Do not give up too easily on this – practice using diagrams instead of lots of text to say what you mean.

Finally, check that you have compared actual test results or readings with acceptance limits or levels. Don't leave the reader in any doubt about whether the results are acceptable or not.

Conclusions

So much depends on whether or not you have issued a non-conformance report. If not, then a simple statement saying that all tests were satisfactory is all that is required. It is difficult to see why there should be any 'grey' areas that you need to explain. If there are non-conforming items then be prepared for some careful reporting technique. First two general points:

- Have you checked the NCR/CA content using the suggestions I made in Chapter 2?
- Have the above points been reflected in the NCR/CA report that you have written? It is easy for them to be subject of discussions but not be written down.

The absolutely key issue when writing the 'conclusions' section of your report is to not contradict *either* the technical statements made in the NCR, or the logic you have used to connect these to your decisions about which corrective actions are required. For this reason it is wise to keep the conclusions text as short as possible. Stick to broad technical fact only, letting the NCR/CA report hold your opinions. You can see this type of approach in the specimen report.

Attachments

Detailed technical results, datalogger print-outs, and other such information should be put in the attachments or appendices, rather than allowing it to detract from the main thrust of the report. Think of this information as the unsifted supporting evidence for your report's conclusions rather than being part of them. You should aim to use consistently the same order for your attachments. Generally, the shortest attachments should come first. A typical order would be:

- NCR/CA report (if applicable)
- signed test certificates
- detailed results sheets
- ITP extracts
- drawings and photographs
- test procedure documents.

It is very important that you keep the number of sheets of paper comprising the attachments to a sensible level. I work to an absolute maximum of eight to ten sheets. If it is likely to extend to more than this you should summarize the information and relegate the raw data to your filing system to be accessed if and when required. Treat attachments as an essential part of your report, but maintain a healthy optimism about the number of people that actually read them.

Some logistical points

Content aside, there are some logistical or housekeeping points that can have a very real influence on the effectiveness of inspection reports. Let economics be the watchword here, these logistical points can have a significant impact on the overall *costs* of your works inspection activities, so they are worthy of attention. Take this one step further – reporting time is one of the few controllable aspects of your works inspection role, perhaps because reporting is not subject to so many outside influences and uncertainties as activities that involve interaction with the other parties involved. This means that you are responsible for the financial viability of your own reports. There are three areas to consider.

Standard proforma reporting formats

You may find yourself assisted or, conversely, constrained by this approach. They are used by those inspection companies that employ a collection of freelance inspection engineers of variable ability and commitment. I feel that they probably keep reporting costs down, but at the expense of encouraging a really incisive verification of fitness for purpose. They sometimes encourage inaccuracy. Most of the standard formats I have seen used tend to concentrate on the collection of a large number of drawing and equipment serial numbers. This is fine, as long as you can accept that this is only a small part of effective inspection. I am sure that there are better standard reporting formats being used in the works inspection business.

If working to a standard reporting format does not suit you, then make sure you control two things; report 'length' and 'balance'.

Overlong reports lose their edge and are so often a vehicle for disguised indecision. Very short reports are unlikely to provide adequate description of important FFP criteria – this is taking the principle of exception reporting too far, giving shallow perception (and low report value). Aim for 600–800 words for straightforward works inspections, up to a maximum level of 1000 words for more complex testing procedures or where there are NCR/CA reports relating to FFP issues. In any event do not go below 500 words, even for the simplest works inspection – you just won't have addressed the issues properly.

It is important to maintain balance between the sections of the report. Note the word distribution that I have used in the specimen report – this is suitable for an 'average' inspection report containing one or two NCR/CA reporting points. Feel free to vary the word distribution between the 'test procedure and results' and 'conclusions' sections to better suit the context but don't go so far that the results do not adequately support your conclusions, or vice-versa. Perhaps the most common example of poor reporting practice is to make the 'input material' section too long, containing a lot of 'given information', drawing numbers, and lists of standards. Perceptive clients will soon identify this as mostly padding, included to compensate for inherent weaknesses in the real business of FFP verification. Beware of this.

Production

Choose the most effective way to produce your inspection reports. It is also likely to be the cheapest way. Think carefully about the relative advantages of producing reports by hand drafting, or by typing directly into a computer. I do not know which is the best for you – all I can ask is that before you decide you consider these points, some of which we have already discussed in earlier parts of this chapter. Whichever method you use:

- You need a planning sheet to help you plan a logical technical progression.
- Every inspection report will be different, not least because every piece of equipment is different, even if in a small way.
- Do aim for a consistent corporate style to your reports. This means paying attention to page formats, margin sizes, typeface size and spacing, and heading/sub-heading layout.

Even works inspection reports benefit from just a *little* style in their presentation. Use good paper and careful binding.

Try to give your report that 'precision look'. Aim for consistency throughout – nothing looks more amateurish than a pile of inspection reports containing hand-written proforma sheets in different formats, or those using front sheets with most headings crossed out as 'not applicable'.

Distribution

Only send your works inspection report to your client. Never be persuaded to let the contractor, manufacturer, or third-party organization have a copy 'for information'. The non-conformance/corrective action report sheet that has been agreed is sufficient for their needs. Distribute this whilst you are in the works. Don't delay your report. Plan it, write it, correct it, and distribute it within *three working days*. Every time.

A Specimen Report

The Planning Sheet

Centrifugal pump performance test

Here is the actual *planning sheet* that I used for this report. Total word budget 870 words.

Preamble (100 words)	Executive summary (100 words)	Attendees (20 words)
First of 8 identical pumps Suction condition (installed) Full test on first, duty point only on others Document review	P1: q/H and NPSH to ISO 3555. Check stability P2: NPSH exceeded guarantee by 0.3 m (site margin) P3: NCR issued. No concession, likely impeller replacement	(Contractor) (Manufacturer) (Third party)

Input material (150 words)	Test procedures and results (300 words)

• Previous report balancing and hydrostatic tests – OK.
• Spec. Clause 1.1.1 requires test to ISO 3555 (class B)
• Vibration VDI 2056
• ITP ref: 1234/xyz step 10.0
• Manufacturers' test procedures ref: CP/2
• GA drg no. D3814/J
	Test rig to ISO 3555: accuracy OK
	q/H – 100% point OK
	    – 110% : H drop OK (slope)
	    – Shut-off OK. (rising)
	    – Vibration 1.0 mm/sec (OK)
	NPSH test – throttled suction method
	– stabilization time
	– units conversion
	– required 4.0m, test 4.3 m
	– quick H-drop
	Stripdown
	Assembly OK, bearing OK.
	Seal face OK, impeller size check OK.

Conclusions (200 words)		
NCR NPSH 0.3 m too high, test accuracy OK.	CA Mnfr to recheck impeller profile and report (replace) Design shows < 1m NPSH margin available. No concessions	Text NPSH does not comply – test accuracy OK. – not marginal

Drgs and Attachments	Sheets
NCR/CA report q/H curve only NPSH: curve only Stripdown report	1 1 2 1 Total 5 sheets

Inspection reports 451

The Front Sheet

Inspection report No. Specimen / 01

Project title:	Effective works inspection	Contractor: Contracting Co. Ltd
Date of visit:	13 July	Manufacturer: Slightly Lacklustre
Attendees:	The Inspector The Manufacturer	Manufacturing Co. Ltd
	The Contractor The Third Party	Sub-manufacturer: N/A

Equipment inspected

1 off (of 8) seawater pump rated 500 m^3/hr at 38m head. Serial No. KKS/123

Standard Nos.	Drawing Nos.	Specification Nos.
ISO 3555 Class B (test) VDI 2056 (vibration) ISO 5167 (flow measurement)	D 3814 /J	Spec Clause 1.1.1. *et al.*

Executive summary and conclusions

Sea-water pump KKS/123 was subject to a full performance and NPSH test. Particular attention was paid to its ability to meet suction conditions. q/H, vibration and noise performance was acceptable.

The NPSH exceeded the maximum acceptable value (at 100 percent q duty point) of 4.0 m by 0.3 m. This will result in an unacceptably small margin between the pump NPSH 'required' and the system NPSH 'available'.

An NCR was issued. A manufacturer's concession application is unlikely to be acceptable. The manufacturer will submit a written report on the pump performance. It is likely that impeller redesign will be required.

Non-conformance report issued : YES

Attachments

1: Non conformance/corrective action report (1 sheet)
2: q/H curve (1 sheet) } see typical Figs 11.1, 11.3, and 11.4 in Chapter 11 of this book
3: NPSH curve (2 sheets)
4: Stripdown report (1 sheet)

| **Inspector** | **Date report issued:** |
| The *effective* inspection engineer | 14th July |

The Report Itself

1. Preamble

The inspector attended the works of The Slightly Lacklustre Pump Manufacturing Company Limited to witness a full performance test on the first (of eight) condenser cooling sea-water pumps. These are single stage pumps rated 500 m^3/hr at 38 m total head (100 percent duty guarantee point). NPSH available is stated as 5.0 m on design drawing DD/01. Subsequent pumps will be tested only at the 100 percent duty point. A full documentation review will be carried out at the final inspection stage.

2. Previous reports

Previous inspection reports relating to this pump are:

Report No.	*Date of visit*	*Subject*	*Result*
01	02 Jan	Pump rotor balance	Satisfactory
02	03 March	Casing hydrostatic test	Satisfactory
03	18 June	Casing MPI and radiograph review	Satisfactory

3. Relevant documents

3.1 Specification clause 1.1.1 specifies pump performance testing to ISO 3555 (Class B).
3.2 Housing vibration to be assessed to VDI 2056 group T
3.3 ITP No. 1234/xyz, step 10.0 (Rev A)
3.4 Pump general arrangement drawing no. D3814/J (Rev B)

4. Test procedure and results

The pump was installed in a closed loop test rig. Measurement methods and general layout were checked for compliance with ISO 3555.

Flowmeter layout was seen to be compliant with ISO 5167. No excessive measurement errors were observed.

4.1 q/H performance test

A q/H test was performed using seven points from 120 percent q to $q = 0$ (shut off). ISO 3555 (Class B) tolerances were applied (± 2 percent q, ± 1.5 percent H). The q/H curve was seen to have an adequate 'rising' gradient throughout the flow range to shut-off. Results were as follows:

Guarantee requirement	Test reading	Result
Duty point: $q = 500$ m^3/hr	520 m^3/hr	Satisfactory
$H = 38$ m	40 m	Satisfactory
Minimum shut-off head = 44 m	45 m	Satisfactory
Continuously rising curve to shut-off	Continuously rising characteristic	Satisfactory

Vibration at 100 percent q was measured at 1.0 mm/sec rms. This is well below the acceptance limits of VDI 2056. Noise was measured at 85 dB(A), which is within the maximum acceptable level of 90 dB(A).

There is no efficiency or power consumption guarantee for this pump.

4.2 NPSH test

The pump was subject to an NPSH test using the 'throttled suction' method. Flow breakdown was assumed to correspond to the 3 percent total head drop point, as defined by ISO 3555. Flow stability up to the point of breakdown was good. Suction and discharge pressure gauge fluctuations were within the tolerances allowed by ISO 3555.

The 3 percent head drop was well defined and occurred at a suction gauge reading corresponding to an NPSH of 4.3 m (H$_2$O). This exceeds the maximum acceptable guarantee level of 4.0 m (H$_2$O) at 100 percent q. The test was repeated three times to ensure reproducibility of the NPSH reading. Prior to the third test, the suction and discharge gauges were removed for a calibration check. Errors were recorded at < 0.2 percent of full scale deflection. These are second-order quantities having a negligible effect on the results. The test rig was checked for factors which would cause an external influence on the pump suction flow regime. None were found – it was agreed that the NPSH test was giving representative results within the provisions of ISO 3555.

4.3 Stripdown examination

The pump upper casing, rotor and bearing/seal components were removed and all parts were subject to close visual inspection. Assembly accuracy and casing joint face condition were good. No problems were evident with bearing or shaft alignment. Mechanical seal faces were in good 'running' condition. Impeller size was checked and found to be as per datasheet (two sizes below the maximum). Impeller surface finish was generally good, with only superficial marks left from hand grinding.

5. Conclusions

The pump performance test was carried out in full accordance with ISO 3555 (Class B). The pump q/H performance was compliant with the guarantee requirements. Noise and vibration readings were well within acceptance levels. The pump NPSH (R) reading was 4.3 m. This exceeds the maximum acceptable guarantee level of 4.0 m. An NCR was issued.

Specimen Report: Attachment 1

Non-conformance report

Sea-water pump (no. 1 of 8) No KKS/123...... rated 500 m³/hr at 38 m head.

1. Pump NPSH measured at 4.3 m (H_2O) at 100 percent q duty point. This exceeds the 4.0 m (+ 0.15 m tolerance) guarantee requirement at this duty. Test results were assessed for accuracy and reproducibility in accordance with the provisions of ISO 3555 (Class B).

No mitigating factors were evident in the test installation or test conditions.

Corrective actions

1. A concession application is not acceptable due to the low system NPSH margin available.
2. Manufacturer to submit report on impeller design and performance within 14 days and propose timescales for new impeller design. Distribution to inspector, contractor, and third-party organization.
3. Retest with new impeller to be witnessed by all parties. Full retest required (not just NPSH).

Specimen Report: Attachment 4

Pump stripdown report

Sea-water pump no. KKS/123.

Stripdown

- Jacking screws: 2 fitted
- Casing lifting lugs: 1 fitted
- Bolt hole alignment: no misalignment
- Seal canister fit: sliding
- Casing face condition: upper OK/lower OK
- Casing surface defects: no visible defects > 3 mm deep
- Casing wear-ring fixing: shrink fit, grubscrew fix

Rotor

- Free rotation: yes, by hand
- Impeller size fitted/maximum: 210 mm/250 mm
- Impeller fixing: keyless
- Impeller wear ring fixing: shrink fit, grubscrew fix
- Wear ring contact: no
- Impeller finish: hand finished, good
- Shaft finish: < 0.4 μm R_a: OK
- Seal face condition: < 0.2 μm R_a: OK
- Seal type checked: mechanical sleeve type
- Flushing arrangements: not required

Key document review: quick check

- Casing hydro-test: yes
- Component balance: ISO 1940 G2.5
- Rotor balance: ISO 1940 G6.3
- Material certificates: reviewed

Comments: Pump stripdown result is acceptable.

KEY POINT SUMMARY: INSPECTION REPORTS

1. Your inspection report is your *product*. Make a professional job of it.
2. There are three main objectives of a good report:
    - to communicate
    - to explain FFP
    - to define one (and only one) course of action.
3. *All* your writing should have a logical progression. I have shown you what I consider good and bad examples of this.
4. *Get to the point*, even if you sometimes find it difficult.
5. Inspection reports benefit from a bit of style, but don't overdo it.
6. Loose abbreviations look awful and confuse your readers. Use them sparingly, and rigidly.
7. A good framework is:
    - Front sheet (with executive summary).
    - Preamble (but no 'padding').
    - Input material (possibly under several sub-headings).
    - Test procedures and results.
    - Conclusions.
    - Attachments (NCR/CA report first).
8. Pay particular attention to the Executive Summary. Use a three paragraph format and choose the content carefully.
9. If your reporting time costs become too high you will start to lose business. Keep them under control.

Appendix

Technical standards mentioned in this book are available from the following sources:

British Standards Institution (BSI)
British Standards House
389 Chiswick High Road
London. W4 4AL.
Tel: 0181 996 9000

Note that BSI is one of the few organizations in the UK permitted to purchase all international and most national standards direct from source, including ISO, IEC, DIN (English translations), ASME and API.

American Society of Mechanical Engineers (ASME), New York, USA. ASME publications are available from Mechanical Engineering Publications Limited, Northgate Avenue, Bury St Edmunds, UK, who are the exclusive European stock holding agents for ASME books, conference and symposia publications. The more popular ASME codes and standards are also available.

American Petroleum Institute (API) standards are available from ILI, Index House, Ascot, Berkshire, UK.

Index

Acceptance criteria 58, 85, 125, 137, 178
  for defects 136
  for hardness levels 145
Acceptance guarantee 56, 274, 312, 366, 389, 338
  couplings 302
  gas turbines 206
  tests 242, 315, 329
  gas turbine 208
Accessory base functional test (GT) 235
Active wear mechanism 278
Adhesion tests 398
AGMA range of standards 288
Agreement:
  between manufacturer and purchaser 139
  contractual 19
  of technical matters 40
  verbal 19
Air-drying paints 417
Air receivers 149
  to BS 5169 174
Alignment procedure 270
Alkyd resins 417
Allowable pressure rise 186
All-weld tensile test 142
Amount of inspection done 13
Angle probes 126
Angles, weld preparation 105
Applicability of standards 98
Application standard 125
Applications of pressure vessel codes 161
Approval testing of welders 99
Approximation to tensile strength 75
A-Scan 117
  presentation 124
  pulse-echo method 118
Asking and listening 17, 18, 25
ASME 154
ASME VIII 156, 159, 162, 163, 165, 185

ASME VIII Div 1 171
ASME VIII intent 172
Assembly of moving parts 12
Assessing turbine casing radiographs 247
ASTM standards 58
Austenitic castings 68, 120
Auxiliary power consumption 206
A-weighted sound pressure level 229, 231
Axial natural frequency (ANF) 304

Background noise corrections 231
Balance limits, concessions on 255
Balance:
  quality grade 251
  weights 214
Balancing:
  dynamic 211
  grade 302
  machine accuracy check 214
  of diesel engines 263
  of rotors 209
Bearing:
  housing vibration 209, 225, 242, 251, 297
  metal temperatures 233
  surface wear 281
  temperatures 276, 296
Bedding-in wear 278
Bedplate alignment, diesels 267
Bend tests, materials 146
Blade clearance checks, turbines 216, 252
Body grades, steel 116
Boiler plate 54
Boilers 149
Bottom-end strip, diesels 277
Brake :
  constant (diesel test) 275
  operation (cranes) 373

Index

Bridge rail misalignment, cranes 383
Brinell hardness test 69
Brittle fracture 73
Broad-band noise 228
Broken teeth, gears 301
BS 2790, shell boilers 178
BS 5169, air receivers 174
BS 5500 151, 157, 156, 162, 164, 169, 195
BS 5500 Form X 153, 197
BS 709 95
Bulging of vessels 189
Business pressures 10
Butt welds 128, 129
Buyer confidence 9, 25

Calculating test pressure 182
Carbon steel tube 55
Carbon–Manganese plate 54
Casing casting defects 246
Casing vibration 287, 315
Castings:
  austenitic steel 68, 120
  critical areas 120
  Cr–Ni steel 53
  defects 255
  examination of 117
  ferritic steel 238
  grades of 118
  steam turbine 239
  ultrasonic examination of 118
Catenary tables 268
Cause of weld defects 86
Caveat emptor 9
Centrifugal pumps 4, 311, 315
Certificate:
  chasing 197
  of conformity 62
  requirements 66
Chain of responsibility 29, 46, 48
Chain of traceability 62
Chain questioning 18, 22
Charpy V-notch test 72
Checklists 106, 277, 379, 378
Chemical analyses 199
  variation in 239
Chlorinated rubbers 418

Choice of material specimens 67
Chrome plating 404, 408
Circuit checks, pumps 317, 322
Circumferential (mis)alignment, vessels 102
Cladding, loose 400
Classification Society 265
Clearance checks, gas turbine 218
Coarse vacuum 184
Coating uniformity test 404
Coating weight test 404
Cobalt 60 133
Code:
  compliance 151, 178
  intent 176, 351
  requirements 200
Coded welders 95
Codes for engine power 265
Commerce:
  chain 10
  rules of 9, 13, 25
Commercial quality 248
Commercial risks 33
Common technical objectives 10
Commonality of inspection activities 260
Communicate your decisions 19, 22
Comparative assessment, surfaces 409
Comparator gauges 411
Comparisons between hardness scales 74
Competent organization 154
Competent person 154
Compressor testing 327, 330, 331
Concessions to vessel codes 198
Condensers 350, 353, 354
Confidence, buyer 25
Conformity 9
Consensus on FFP 137
Consultants 9
Contact checks, gears 292
Contact markings, gears 294
Contamination of lubricating oil 300
Contract:
  administrators 42
  specification 27, 28, 38, 43, 48, 50

Index

Contractors	33
Contractual agreements and protocol	19
Control valve sizing	332
Corporate style, reports	449
Corrective actions (CAs)	7, 21, 23, 455
objective of	21
Corrosion	12, 56, 190
intercrystalline	145
Cosmetic painting defects	430
Cost implications	11
Coupling design standards	303
Crack propagation	200
Cracked teeth, gears	301
Crack-like flaws	112
Crane:	
design classification	359
ITP	369
load test	375
mechanisms	363, 366
rail alignment	377
structure	366
Cranking test, GT	220
Crankshaft:	
construction, diesels	260
deflections, diesels	268
Credibility	16
Creep	57
Critical crack size	109
Criticism	24
Crosshead clearances, diesels	270
Dampers	337, 356
flap	338
guillotine	338
louvre	338, 346
Data sheets	35
Dead zone	124
Decibel scale	229, 231
Decisions	19
Defects	399
acceptance criteria	85, 97, 136, 157, 256
classification	136, 138, 243
common visible	106
evaluating	126
levels, steam turbines	239
longitudinal	126
map, steam turbines	245
photographs of	112
sub-surface	111
volumetric	128
Defined yield points	70
Deflection measurements	376
Degree of rusting	421
Densitometer	135
Describing defects	112
Description, reporting by	8
Design:	
appraisal	153, 178
check	304, 305, 365, 383, 359
features	390, 391, 413
of overhead cranes	361
review	340, 385
Design-based codes	161
Destructive testing of welds	137
Detailed plant specification	27
Detection limitations, surface cracks	106
Diesel engines	259
FFP criteria for	259
technical standards for	263
Dimensional check, vessels	188, 190, 191, 342
DIN 8570	191
Directional properties, materials	70
Discontinuities:	
in traceability	65
non-planar and planar	120
sizing	124
Discontinuity stresses, vessels	198
Dissimilar materials, welding	87
Distance gain size techniques	124
Distribution of reports	449
Document hierarchy	29, 33, 34, 46, 115
Documentation	61
incomplete	382
review	181
welding	92
Double helical gears	286
Double wall technique, radiography	130
Downgrading of vessels	200
Droop, governor	277
Dry film thickness (dft)	422, 428

Index

Ductile fracture 73
Duplication of inspectors 18, 150
Duty of care 151
Duty point, pumps 312, 314
Dye penetrant (DP) 99, 110
Dynamic balance of a gear rotor 291
Dynamic balancing tests 211, 289, 290

Ebonite 388
Edge grades, steel 116
Edge laminations, steel 110
Effective amount of inspection 14
Effective inspection 19, 28, 39, 41, 65
Electrical runout, rotors 295
Emission levels, diesels 262
End of manufacturing report 41
Enforcement 10
Engine:
  adjustment factors, diesels 262, 275
  guarantee schedules, diesels 261
  performance, diesels 259
  power, codes for 265
Engineering materials 50
Enhanced visual assessment 112
Environment of business 10
Epoxy paints 418
Equipment inventory checks 223
Equipment:
  release 43
  supply, responsibility of 13
  works tests 36
Essential variables 96, 199
Etching 143
Exception, reporting by 8
Excessive bridge deflection, cranes 383
Excessive crankshaft web deflections 281
Excessive vibration 234, 281, 301, 326
Executive summary 443, 451
Exhaust gas temperatures 233, 276
Extended guarantees 33
Exterior welding 189

Fabrication standards 263
Failed mechanical tests 255
Fall-back clauses 97

Fatigue 57
FATT 56, 73, 78
Feedback, lack of 27
Ferritic stainless steel 87
Ferritic steel castings 238
FFP, see fitness for purpose
Field tester, MPI 112
Film:
  density, radiographic 135
  location, radiographic 133
  thickness, radiographic 300
Final NDT 245
Finding a solution 21
Fine finishes 409, 411
Fine vacuum 184
First ITP drafts 42
Fitness for purpose:
  assessment criteria 5, 15, 49, 50, 285, 302, 431, 348, 370
  compromise of 196
  for diesel engines 259
  for painting 415
  for pumps 347
  for linings 387
  problems with 7
Fitted parts, turbines 241
Flame-cut edges, steel 105
Flange bolt holes 191
Flange faces, flatness of 250
Flowrate measurement 332
Fluid systems 347, 345
Flushing 272, 275
Focus 16, 42
'Following' standards 39
Forgings 51, 52
  ultrasonic examination of 122
Foundry practice 240, 250
Fracture surface appearance 142, 145
Fracture:
  brittle and ductile 73
Free air delivery q (FAD) 328
Frequency band sound pressure level 229
Frequency distributions 298
Front sheet, reports 451
Full penetration welds 87, 88, 90, 105
Full traceability 37, 62
Function 11

Index

Fusion:
    faces   88
    lack of   126

Galvanizing   401, 402, 413
Gamma Rays   133
Gas dampers   337
Gas turbines   205
    acceptance guarantees   207, 208
    clearance checks   218
    ITP   209
    procurement   208
    rotor overspeed test   216
Gear:
    design standards   287
    inspection   288
    rotor, dynamic balance of   291
    train contact checks   292
    types   286
Gearbox test: monitoring   297
General technical requirements   36
Geometric unsharpness   134
Getting to the point   440
Gland clearances, turbines   252
Good manufacturing practice   5
Good reporting style   442
Governing characteristics,
    diesels   207, 276
Grades of castings   118
Guarantee of integrity   153
Guarantees   261
Gudgeon pin, diesels   278

Hard rubbers   388
Hardness   78
    gradient   144
    scales, comparisons between   74
    tests   69, 73, 95, 144
Head Drop method, pumps   321
Head-to-shell alignment, vessels   190
Heat affected zone, welds   89
Heat:
    exchangers   350, 352
    rate, net specific   206
    treatment condition   59
High speed couplings   305
Hoisting height, cranes   373
Hold points   181

Hook approach test, cranes   373
Hot dip galvanizing   402, 406
Hot functional test, dampers   341
Housing vibration   225, 263
Hydrostatic tests   182, 183, 245, 248, 256, 354, 201, 249

Image quality indicator (IQI)   133
Impact strength   200
Impact tests   69, 72, 78
Implementing a corrective action   21
Incomplete:
    documentation   382
    material traceability   197
    root penetration   127
    statutory certification   197
Incorrect:
    dimensions   198
    documentation   181
    material properties   197
    root gaps   198
    speeds   382
    weld preparations   198
    WPS   199
Indention hardness   391
Independent design appraisal   150, 151
Information references   4
Input material, reports   445
Inspecting plated components   405
Inspection:
    amount done   13
    and survey   154
    and test plans (ITPs)   40, 176, 209, 265
    business   9
    effective   19
    effective amount of   14
    mandatory   178
    of shafts and couplings   303
    of rubber linings   388
    points   66
    procedure   5
    reports   433, 457
    responsibilities   14
    responsibilities, structure of   13
Insulation test, cranes   371
Integrity:
    of manufactured pressure vessels   150
    of welds   136

Index

Intercrystalline corrosion test	145	contamination	278, 300
International Rubber Hardness		Lusec	186
Degree (IRHD) scale	395		
Irregular tooth contact pattern,		Macro examination	146
gears	301	Magnetic particle inspection	
ISO standards	5	(MPI)	99, 110
ITPs	29, 35, 48, 61, 316,	Main bearing clearance, diesels	270
content	41, 65	Making decisions	18
gearbox	289	Mandatory inspection	178
heat exchanger	352	Manufacture, responsibility for	9
steps, missing	196	Manufactured pressure vessels,	
technical steps	178	integrity of	150
gas turbines	209	Manufacturer's correction	
cranes	368	curves	206, 233
steam turbines	243	Manufacturer's own material	57, 58
		Marginal differences in tensile	
Judgement	98	strength	200
		Marginal results	69
Key point summaries	8	Marking up of the ITPs	42
Killed steel	54	Martensite	146
		Matching of the PQRs and WPSs	96
Laminations	115	Material:	
Large-bore pipe butt weld	129	certificates	68
'Leading' standards	39	grade designations	59
Leak:		properties, incorrect	199
rate	186	selection	49, 59
tests	355	specimens, retests of	76
tubeplate	354	standards	51, 57
Leaking gas	234	tests	67, 349
Levels of examination, 1, 2, and 3	125	traceability	37, 49, 61, 347, 369,
Levels of traceability : EN 10 204	62	traceability, incomplete	197
Licensors, technology	209	traceability, lack of	255
Light run test, cranes	372	Maximum unbalance	306
Linear indications	112	Maximum vibration levels	296
Lining:		Mechanical:	
defects	391	integrity	259, 285, 308
failures	387, 413	properties	50, 199
metallic	400	runout, rotors	295
Listening and asking	25	tests, failed	255
Location of test bars	53	on paint films	422
Location of vibration sensors	297	Mechanism class, cranes	365
Logical progression	436, 437, 442, 457	Medium speed engines	260
Longitudinal datum, vessels	191	Metallic linings	400, 413
Longitudinal defects	126	Metallurgy	72
Loose cladding	400	Microphone positions, noise tests	231
Louvre dampers	338, 346	Microseizure	299
Lubricating oil	223	Microstructural weakness	142
consumption	262	Microstructure of weldments	144

Index

Millscale	419	Over-thickness defects	426
Minimum defect size	109	Overhead cranes	359
Misalignment, circumferential	102	technical features	362
Missing documents	196, 197	designs of	361
Missing ITP steps	196	Overspeed:	
Missing NDT	200	protection	265
Monitoring the painting process	416	settings	207
Most effective solution	10	test	216, 252, 298
Moving parts, assembly of	12	trips	224
Nameplate details	192, 194	Package contractors	9
National standard	5	Padding, reports	448
Natural rubbers	388	Paint:	
NDT	38, 289	application	419
results	201	colour	421
standards	99	failures, reasons for	417
tests	104	films, mechanical tests on	422
Near surface defect test	123	types	416
Near vicinity noise	231	Painting	415, 431
Need for certification	152	defects	427
Net positive suction head (NPSH)	312	process, monitoring	416
Net specific heat rate	206	Paintwork repairs	371, 429
Noise	234, 308	Paragraph content	443
levels	207, 209, 226, 262, 287, 326	Partial code compliance	166
measurement	226, 320	Partial traceability	166
tests	231, 289, 298	Payment stages	43
background	231	Peak cylinder pressures	281
surface originated	229	Penalty clauses	33
No-load running test	208, 210, 222, 232, 289, 308	Penetrameter	133
Non-conformance	7, 45	Phantom indications	112
reports (NCRs)	19, 455	Photographs of defects	112
Non-conforming mechanical properties	199	Pinholing	300
Non-metallic inclusions	112	Pipe-to-flange joints	130
Non-relevant indications	112	Piping	347, 348
Non-sinusoidal vibrations	209	Piston speeds, diesels	260
Non-synchronous vibration	226	Pitch errors, gears	299
NO_x emission level	206	Planar discontinuities	119
NPSH test	313, 314, 321, 323, 453	Planning sheet, reports	444, 450
		Plate:	
Objectives	7, 433	condition	189
Octave band	231	courses	189
One-third octave band	231	edges	105
Operator errors	86	Plated components, inspecting	405
Organizing inspection activities	40	Plating classification	405
Organizing your thoughts	438	Pneumatic testing	184
Orientation of tensile specimens	70	Poor adhesion, linings	399
		Poor fusion, welds	141
		Poor impact strength	200

Poor logical progression	436	fitness-for-purpose criteria for		347
Poor manufacturing practices	87	Purchase specification, GT		206
Positive identification, materials	51, 61	PWHT		92
Post-assembly checks, turbines	243			
Post-weld heat treatment	89	q (FAD) performance		336
PQRs	84, 94, 137, 198	q/H:		
Preece test	404, 406	characteristic		312, 323
Prelimary volumetric NDT	245	curve		453
Preliminary governor tests	273	guarantee		319
Preliminary scan, ultrasonic	119	performance test		453
Pre-running tests	220	Qualified vessel certificates		197
Pressure:		Qualified weld procedure		98
drop test	187	Qualifying information		15, 261
parts	52	Quality		12, 25
testing	181	Questioning, chain		22
vessel codes	157, 156, 203			
applications of	161	R_a		410
defect acceptance criteria	163	Radial clearances, turbines		252
material requirements	162	Radiographs		136
NDT requirements	163	Radiographic testing		247
dimensional tolerances for	191	Radiography		99, 113, 128
integrity of manufactured	150	Radius, effective		213
practice	128	Rail distortion, cranes		376
typical ITP content	180	Range of approval, welding		99
works inspections	150	Rapping test, rubbers		398
Pre-weld heating	89	Rated speed run		295
Primers	419	Reasons for paint failures		417
Principal stresses, static	11	Recalculation exercises		200
Principles of works testing	286	Receiver sizing, compressor tests		332
Probe, 0 degree	124	Reciprocating parts alignment,		
Problems with FFP	7	diesels		270
Procurement standards	242	Reference radiographs		136
Product liability	84	Registration schemes		154
Production control test plate	137	Repair:		
Proforma reporting formats	448	procedures		245
Properties of rubbers	389	welding		89
Protocol, contractual	19	Repeated running-in tests, diesels		273
Published material standards	59	Reporting		7
Pull off test, paint	422	by description		8
Pulse echo method, ultrasonic	117	by exception		8
Pump:		of defects		422
acceptance guarantees	311, 314	Reproducibility in weld defects		87
efficiency	312	Residual unbalance		290, 306
noise	321, 326	Responsibilities, inspection		14
performance testing	314	Retests		76, 200
stripdown report	455	of material specimens		76
test circuit	317	Retrospective solutions		195
centrifugal	4, 311, 315	Retrospective testing		62

Rimmed steel	54	Single course of action	434
Risk	85, 96	Single helical gears	286
Roof clearance, cranes	370	Single wall technique, radiography	130
Root:		Sinusoidal vibration	225
concavity	127	Site acceptance, material	44
incorrect gap	88, 198	Size of test specimens	143
land	130	Sizing discontinuities	124
penetration, incomplete	127	Slow roll runout, rotors	301
Rope grooves, crane hoists	363	Slow speed diesels	260
Rotors:		Small-bore full penetration pipe	
balancing of	209, 211	butt weld	129
parts	240	Solutions to non-conformances	195
runout measurement	210	Soundness of weldment and HAZ	142
tests	251	Source certification, materials	61
visual inspection	216	Spark testing, rubbers	391, 397
unbalanced	290	Special design features, rubbers	389
Rounded indications	112	Special engineering requirements	36
Rubber:		Specific fuel consumption	
hardness check	395	(SFC)	259, 262
lined vessels	392	Specific works test requirements	35
linings	387, 388	Specification:	
Rule of thumb	4	clash	37
Rules of commerce	9, 13, 25	clauses	34
Ruling section, castings	240	detailed and turnkey	27
Run-down vibration analysis	226	Specified tests	29
Running integrity	205, 235, 237	Spur gears	286
Rusting, degree of	421	Stabilization, rotors	213
R_z	410	Staining, galvanised coatings	402
		Stainless steel, ferritic	87
Safe working load (SWL), cranes	362	Standards	4, 27, 34, 348
Saying 'no'	23	State-of-loading factor, cranes	365
Scratch test	422	Static principal stresses	11
Scuffing of gear teeth	299	Stator parts	240
Sealing efficiency, dampers	338	Statutory	28, 29
Seat leakage	350	bodies	9
test, valves	349	certification	149, 150, 188
Sensitivity, radiographic	133	compliance	361
Service lifetime	12	equipment	92, 359
Severity level	238	inspector	150
Shaft vibration	209, 225, 242, 251, 287, 288, 296, 298	pressure vessels	99
		Steam turbine:	
Sheared edges, plates	105	acceptance guarantees	241
Shell boiler to BS 2790	178	casting, problems with	239
Sheradizing	402	component design	238
Shore D scale	395	corrective actions	255
Shotblasted surface finish	422	defect levels in	239
Shut-off condition, pumps	319	ITPs for	243
Side bend test, welds	143	non-conformances	255

parts, FFP of	241	Technology licensors	209
procurement	242	Tensile properties	70, 78
technology	241	Tensile strength, extra tolerance of	142
Steel:		Tensile strength, marginal	
carbon–manganese	52	differences in	200
killed	54	Tension, axial, couplings	304
rimmed	54	Terminology	5, 21
Strategy	12, 25	Test:	
String test	303	acceptance criteria	140
Stripdown tests	234, 259, 378	bars, location of	53
Structural welds, ultrasonic		circuit, pumps	316, 332
techniques	115	codes, performance	208
Structure of inspection		pieces, materials	98
responsibilities	13	plans for shafts and couplings	303
Style, reports	441	plate arrangement	139
Sub-orders	35	pressure, vessels	184, 248
Sub-surface defects	111	procedure and results	446
Suction control loop, pump tests	316	procedures	5
Superheater headers	176	reporting	145
Surface:		specimen, size of	143
crack detection limitations	106	Thermal efficiency, GT	205
examination	246	Thick welds	126
finish	189, 407	Third party certification	
finish, shotblasted	422	body	6, 95, 137
finishes, classification of	409	Third party inspector	66
indications	109	Third-party bodies	32
preparation	419, 431	Thought stages	436
roughness	410	Three percent head drop, pumps	453
SWL performance test	373	T-joints	126
Synthetic rubber linings	388	Torsional vibration	263, 265
Systems function	205, 235	Total head (H), pumps	312
		Total Indicated Runout (TIR)	210, 306
Tactics	14, 25	Total pressure (p)	328
Technical:		Traceability	153
differences	260	of materials	150
accuracy	437	chain of	62
argument	436	discontinuities in	65
contradictions	36, 59	Transition temperature	73
description	435	Transverse:	
documentation	241	bend test	142
information	3, 4	datum, vessels	191
objectives	10	defects	126
presentation	436	tensile test	140
risk	13, 51, 67, 87, 241	TRD boiler code	160, 176
standards	4, 29, 34, 38, 48, 241, 367, 422	Tubeplate leaks	354
		Tubes	51, 55
standards for diesel engines	263	Turbine casing radiographs	247
Technique information	130	Turbine:	

Index

manufacture	240
mechanical integrity of	222
Turbines, gas	205
Turbines, steam	237
Turbocharger surging, diesels	276
Turnkey specifications	27
Two-pack paints	418
Types of welds	87
Ultrasonic examination:	
of castings	118
of forgings	122
of welds	115, 124
of plate	115
standard	245
Unbalanced rotors	290
Unloaded length, couplings	304
Unrecorded engine adjustments	275
Unrecorded repairs, vessels	201
Upper line strip, diesels	277
Using standards	38
Utility	10
Utilization factor, cranes	365
Vacuum leak test: guidelines	187
Value for money	8
Valves	347, 348, 350
Variation:	
in chemical analyses	239
in mechanical properties	239
Verbal agreements	19
Vessel:	
bow measurements	192
code, concessions to	195, 198
codes as an intent	161
design	157
integrity	151
markings	192
rejection of	192
Vessel's documentation package, content of	157
Vibration	207, 242, 261, 308,
intensity	296
measurement philosophy	232
sensors, location of	297
spectrum analysis	226
bearing housing	209, 242, 251

casing	287, 315
non-synchronous	225
shaft	209, 251, 287
sinusoidal	225
torsional	263, 265
Vibrations, non-sinusoidal	209
Visible appearance of welds	106
Visible leaks	202
Visible surface defects	251
Volume flowrate (q)	311, 328
Volumetric defects	128
Volumetric NDT	113
Vulcanization, rubber	394
Wasting time	102
Wear	12
Weld:	
assessment routine	125
defects, cause of	86
heat treatment	89, 93
integrity	84
preparations	102, 104
preparations, incorrect	198
procedure specification (WPS)	92
root	125
seams	190
set-ups	105
test, destructive	145
Welder:	
approvals	99, 103
qualification	95, 96
Welding	38, 83
documentation	92
methods	86
/NDT tests	104
standards	98
supervision	83
Weldment/HAZ boundary	145
Welds	106
full penetration	90
visual inspection of	106
Wet film thickness	429
Wheel-space temperatures, GTs	233
Why defects occur	86
Witness point	42, 62, 66
Witnessing:	
a pneumatic test	184
pressure tests	181

responsibilities	178	X-rays	130
Word distribution, reports	448	Yield points, defined	70
Working documents	35	Yielding	378, 379, 383
Working to pressure vessel codes	156		
Works inspection, objective of	27		
WPSs	84, 198	Zinc plating	402